INTEGRATED CROP PROTECTION IN CEREALS

The Experts' Group Meeting on 'Integrated Crop Protection in Cereals' was organized by:

Commission of the European Communities
Directorate-General Agriculture

in collaboration with

Glasshouse Crops Research Institute
Department of Entomology & Insect Pathology

Proceedings of a Meeting of the EC Experts' Group / Littlehampton 25-27 November 1986

INTEGRATED CROP PROTECTION IN CEREALS

Edited by

R.CAVALLORO
Commission of the European Communities, Joint Research Centre of Ispra

K.D.SUNDERLAND
GCRI, Department of Entomology & Insect Pathology of Littlehampton

Published for the Commission of the European Communities by

A.A.BALKEMA / ROTTERDAM / BROOKFIELD / 1988

*The texts of the various papers in this volume were set individually
by typists under the supervision of each of the authors concerned.*

Publication arrangements: *P.P.Rotondó*, Commission of the European Communities,
Directorate-General Telecommunications, Information Industries and Innovation, Luxembourg

EUR 11525

Published by
A.A.Balkema, P.O.Box 1675, 3000 BR Rotterdam, Netherlands
A.A.Balkema Publishers, Old Post Road, Brookfield, VT 05036, USA

ISBN 90 6191 844 8

Foreword

In 1985 the twelve European Communities Member States grew 31.5 million ha of cereals (52% of the total crop area) and the total expenditure on pesticides to protect this area was 1,350 million ECU (ex manufacturer). In spite of this expenditure, farmers experienced recurring problems with pests (invertebrate pests, diseases and weeds). Tackling these problems with still greater inputs of pesticides runs the risk of selecting for pest organisms that are resistant to pesticides. In addition, public concern about food contamination and damage to the environment make an exclusively chemical solution to pest problems politically unacceptable.

With over-production of cereals in the Communities and intervention prices moving closer to world market prices, any solutions to pest problems that involve reductions in variable costs will also be welcomed by farmers. There is, therefore, considerable interest in making use of biological, biotechnical and cultural pest control methods, in combination with a limited and controlled use of pesticides. This is the integrated approach to crop protection, taking into consideration toxicological, economical and ecological aspects.

The aim of the meeting "Integrated Crop Protection in Cereals" was to bring together, from the EC-Countries, specialists in many aspects of integrated control, to exchange information, discuss and make recommendations for future research. Not all areas could be covered during the meeting, but the 33 papers presented included the most important recent developments in research into integrated control of invertebrate pests, diseases and weeds, together with economic appraisals of applying such research to model farming systems.

It is becoming clear from work on cereals aphids (Hemiptera: Aphididae) that forecasting schemes, natural enemies and partially-resistant cereal varieties can all play a valuable role in the reduction of direct damage and virus spread by these pests. Successful short-term forecasts of peak aphid abundance may soon be possible. Use of this information, together with economic thresholds, will enable farmers to spray in an efficient and timely way against a threatening aphid attack but will, at the same time, provide a sound basis for witholding such sprays on accasions when they are economically unnecessary. The resulting reduction of pesticide inputs will allow an increase in the numbers of the aphid's natural enemies. These include parasitic Hymenoptera, pathogenic fungi and many types of predator (more than 300 species of beetles, spiders, mites, etc.). There is a growing body of evidence that such natural enemies can maintain aphid populations below the economic threshold under a wide range of conditions. The frequency with which they are successful in doing this is likely to be increased if the crop has some degree of resistance to aphid reproduction (antibiosis) or increases aphid movement (antixenosis), thus enhancing the encounter rate of aphids with natural enemies. Work is underway to investigate the biochemical basis of crop resistance to aphids and also to assess the potential of resistance for reducing virus spread (especially barley yellow dwarf virus in maize). There are also some first indications that the natural enemies of aphids may play a part in reducing virus spread in some areas. Because of the perceived importance of natural enemies, attention is being given to the development of rigorous and repeatable methodologies for evaluating the side-effects of agrochemicals on beneficial organisms. Allied to this, research is being carried out on methods of mitigating the deleterious effects of any pesticides that must be applied; these methods include the use of low-dose pesticides and unsprayed headlands.

Similar approaches to those described above are being taken towards control of the corn stalk borer, Sesamia nonagrioides (Lef.) (Lepidoptera: Noctuidae), in maize. However, for this pest, there is also potential for control by mass-trapping individuals lured into fields by sex pheromones.

Some approaches to the integrated control of diseases are similar to those employed for invertebrate pests (e.g. the use of forecasting schemes and partially-resistant cereal varieties). Leaf spot (Septoria tritici Rob. ex Desm.) is spread by rain splashes which transmit spores from the basal leaves upwards into the canopy. A system that involves monitoring of inoculum levels before stem extension and the degree of rain splash (using simple apparatus) between stem extension and anthesis is showing potential for forecasting the need to apply fungicide sprays. It is hoped that such schemes will, in the future, provide a rational basis for reducing fungicide inputs without endangering the health of the crop. Research on some diseases of more recent economic importance, such as barley yellow mosaic virus, is concentrated, initially, on developing reliable bioassays. Several diseases are transmitted through the seed and there is still much work to be done (e.g. initiating certification schemes and seed treatment programmes) in some of the Member States to improve seed health. Seed treatments can reduce the requirement for foliar fungicides. This is advantageous in many ways; for example, it has recently been established that some fungicides are sufficiently insecticidal to disrupt biological control of pests in cereals. Many foliar diseases respond extremely rapidly to circumvent the effect of host resistance genes and fungicides. A powerful approach to reducing the severity of these problems is to increase the diversity of varieties and fungicides on various scales, including the within-field scale. It was found, for example, that, for plots of wheat inoculated with yellow rust (Puccinia striiformis Westend), a single fungicide spray given to plots of a varietal mixture (with different resistance factors) gave the same

yield advantage as three sprays given to a susceptible monoculture. Similarly, the yield of barley under attack by barley mildew (*Erisyphe graminis* DC. f. sp. *hordei* was the same whether the seeds of one or three components of a three variety mixture were treated with fungicide. Treating just one component, however, will reduce not only costs and environmental contamination but also selection pressure for fungicide insensitivity. The principle of diversification for reducing the incidence of barley mildew is also applicable on the European scale. Analysis of the progeny of spores sampled along transects from France to Denmark and from England to Austria, revealed distinct geographical patterns for virulence against barley resistance genes and for resistance
to the principal fingicide active ingredients. Spore populations of any given strain drift about 100 km per year, mainly from west to east on prevailing winds. Therefore, given sufficient trans-national coordination, a country about to be attacked by a particular strain, could be ready to deploy the optimal varieties and fungicides against that particular strain.

At present, chemical herbicides are very extensively used for weed control. Although the study of biological control of weeds in cereals is in its infancy, initial results are very promising. For example, the pathogenic fungus *Alternaria tenuissima* (Kunze *ex* Nees et Nees) Wiltshire gave biomass reductions of velvetleaf (*Abutilon theophrasti* Medicus), comparable to that of several types of commercial chemical herbicide.

In the complex cereal agroecosystem, many interactions are possible between the various measures taken to control invertebrate pests, diseases and weeds. In addition, cereals usually form part of a crop rotation, and pest control measures taken on one crop can influence events on the next. Acknowledging this dimension to pest control, long-term studies of farming systems have been undertaken in England, in Germany and in the Netherlands. These studies compare intensive farming systems, typified by high inputs of fertilizers and pesticides (often applied prophylactically), with supervised or integrated systems where pesticides are usually only applied if an economic threshold is exceeded. In such studies, it is not uncommon to find, in the intensive system, a progressive reduction from year to year in the numbers of beneficial organisms, accompanied by a greater severity of pest attack. Although yields are usually greater for the intensive systems, variable costs are also greater. Economic appraisals are showing that intensive systems are not always more profitable than integrated systems.

It is hoped that publication of this volume will give some impetus to the further development by researchers, and the adoption by farmers of integrated methods of pest control.

The Commission of the European Communities are to be congratulated for bringing together a group of specialists in this field and thereby stimulating greater efforts in this important area of crop protection.

R. Cavalloro
K.D. Sunderland

STRUCTURE OF THE MEETING

Scientific Committee

Cavalloro R.	- C.E.C. - Joint Research Centre, I-Ispra
Sunderland K.D.	- G.C.R.I. - Department of Entomology & Insect Pathology, UK-Littlehampton
Wratten S.D.	- Southampton University - Department of Biology, UK-Southampton

Sessions' Organization

Opening Session	Chairman: R. Cavalloro
Session 1 : Pests	Chairman: S.D. Wratten
Session 2 : Diseases and Weeds	Chairman: P. Lucas
Session 3 : Farming systems	Chairman: A. El Titi
Closing Session	Chairman: K.D. Sunderland

Local Secretariat

Miss Ruthanne Smith
Institute of Horticultural Research
Worthing Road
Littlehampton, West Sussex, BN17 6LP

Proceeding Desk

Mr. Pier Paolo Rotondò
Commission of the European Communities
Directorate-General Telecommunications, Information, Industries
& Innovation
2920 Luxembourg

Table of contents

Opening session

Session 1. *Pests*

Session 2. *Diseases and weeds*

Session 3. *Farming systems*

Closing session

Opening session

Chairman: R.Cavalloro

Towards the European development of an integrated and biological control in cereals

R.Cavalloro
CEC, Joint Research Centre, Ispra, Italy

The Commision of the European Communities, implementing decisions of the Council on plant protection researches at the Member Countries' level, set up a programme dealing with the defence of different crops, among which are cereals.

The aim was to study specific products and useful strategies of control respecting the environment, the agro-ecosystems in particular.

In order to act in full consciousness of the need for a more rational employment of pesticides, and the replacement of polluting chemical substances by means which respect the environment and which consume less energy, the recourse to the principles of integrated pest control management becomes very urgent and of paramount importance.

In this optic, taking into account toxicological aspects, ecological and economical requirements, and the possibility of rapid implementation of researches, the C.E.C. since 1979 is carrying out a joint programme in the Member States.

At first the actions considered mainly the development of prognosis and control strategies for aphid pests.

The activity developed at the EC-Member Countries' level is already a real success with the establishment of an 'EURAPHID' network of aphid sampling stations. By an unified trapping suction system, and with the same methodology, a common research programme throughout EC-Countries allows better biological and ecological knowledge of harmful aphids, forecasting in real time on the population dynamics, their behaviour, direct damage or viral diseases transmission as vectors, in order to adopt appropriate control methods.

A key of rapid identification of noxious species, and the inventory of existing biological possibilities of control by aphid pathogens, parasites or predators are also investigated, as well as pesticide effects on cereal fauna.

Subsequently, the principal wheat diseases were investigated too, with an attempt to protect by an inocolum of a hypoaggressive strain against attacks by an aggressive one.

In addition, the programme in progress envisages other insects,

Tab. I - EC-Institutes involved in common activities on integrated
crop protection in cereals

Belgium

. Rijkstation voor Nematologie en Entomologie - Merelbeke
. Laboratoire de Phytopathologie - Louvain-la-Neuve
. Station de Zoologie Appliquée de l'Etat - Gembloux

Denmark

. The National Research Centre for Plant Protection - Lyngby

Federal Republic of Germany

. Lehrstuhl fuer Pflanzenbau und Planzenzuechtung - Freising-Weihenstephan

France

. Association de Coordination Technico Agricole - Paris
. Centre National de la Recherche Agronomique - Versailles
. Institut Pasteur - Paris
. Laboratoire de Pathologie Végétale, INRA - Le Rheu
. Laboratoire de Recherches de la Chaire de Zoologie, ENSAR - Ren-
nes
. Laboratoire de Zoologie, INRA - Le Rheu
. Station de Zoologie, INRA - Versailles

Great Britain

. Glasshouse Crops Research Institute - Littlehampton
. Long Ashton Research Station - Bristol
. Rothamsted Experimental Station - Harpenden

Greece

. "Demokritos" Nuclear Research Centre - Aghia Paraskevi

Ireland

. Plant Pathology and Entomology Department - Carlow
. The Agricultural Institute - Dublin

Italy

. Dipartimento di Patologia Vegetale, Università - Bari
. Istituto Sperimentale per la Patologia Vegetale - Roma

Netherlands

. Experimental Farm 'Development of Farming Systems' - Nagele

----

as lepidoptera, and diseases, as mildew or rust, as well as the biological control of weeds cereals by plant and fungal pathogens and low doses of herbicides. That is infact, the development of an integrated crop production system for cereals.

This includes laboratory researches, pilot projects, field investigations, involving institutes of Belgium, Denmark, Federal Republic of Germany, France, Great Britain, Greece, Ireland, Italy and Netherlands. Nine among the twelve Member Countries are today directly involved in this programme, with a definite, common action that involve the principal institutes of researches (Tab I.). The two last Countries (Portugal and Spain), recent full Members of the European Communities, are now integrated with others in the same line towards an integrated and biological control in cereal cultivations.

The good formula of scientific cooperation adopted with success considers the realization of common activities on a contractual basis with chosen institutes, selected following a call for offer on the mainlines and the aim of the EC-programme. It is supported by financial contributions up to 50 per cent of the total cost of the researches, with the ramaining part at the charge of the contractant.

In parallel, an intense coordinated activity is considered and developed on the same topic. (Tab. II).

This consists chiefly of experts' meetings, for an exchange of information to verify the progress of the studies and the research-coordination, a more immediate transfer of knowledge and scientific results, and the possibility to put into practice in the field the laboratory or pilot test results.

Besides, training courses are carried out in order to prepare qualified research and agricultural personnel ready for the practical application of integrated pest control principles.

The CEC and IOBC collaborated in giving a training course on integrated crop protection in cereal systems. This was held in England at the Imperial College at Silwood Park and it had a graet attendance with much success.

The great effort on plant protection carried out by the Commission of the European Communities includes a very useful exchange of scientists between Member Countries (Fig. 1), contributing more and more to a better knowledge and cooperation, and in a greater mutual understanding at the community level for necessary convergence on the right solution of phytosanitary problems.

The diffusion of information by specific publications is considered as well.

Undoubtedly such a polyvalent action, will permit solution for healthy agricultural management in a short time by all the interested parties.

Tab. II - Coordinated activities dealing with integrated crop protection in cereals

---

- Meeting of experts

. INTEGRATED CONTROL IN CEREALS
  B - Gembloux, 15-16 May 1979

. BIOLOGICAL CONTROL OF PLANT PATHOGENS: PRESENT STATUS AND PERSPECTIVES
  B - Bruxelles, 24 September 1980

. FORECASTING PESTS AND DISEASES AS A MEANS OF AVOIDING UNNECESSARY APPLICATIONS OF PESTICIDES ON CEREALS
  GB - Harpenden, 27-28 November 1980

. THE EFFECTS AND USE OF SAPROPHYTIC MICROORGANISMS AND THEIR PRODUCTS FOR THE CONTROL OF CEREAL DISEASES
  B - Bruxelles, 25 June 1981

. ENVIRONMENTAL AND PESTICIDE INFLUENCES ON SOME PESTS OF CEREALS AND THEIR RELATED PREDATORS AND PARASITES
  IR - Dublin, 30 September-1 October 1981

. 'EURAPHID': AN APHID TRAPPING NETWORK
  B - Bruxelles-Gembloux, 30-31 March 1982

. APHID ANTAGONISTS
  I - Portici, 22-24 November 1982

. GENERAL MEETING OF THE CONTRACTANTS IN THE RESEARCH PROGRAMME 1979-1983
  B - Bruxelles, 14-16 November 1983

. APHID MIGRATION AND FORECASTING 'EURAPHID' SYSTEM IN EUROPEAN COUNTRIES
  F - Montpellier, 7-9 May 1985

. REGULATION OF WEED POPULATION IN MODERN PRODUCTION OF VEGETABLE CROPS
  D - Stuttgart, 28-31 October 1986

- Training course

. CEC/IOBC Training Course on Integrated Crop Protection in Cereal Systems
  Silwood Centre for Pest Management - Imperial College
  GB - Ascot, 1-13 July 1985

- Exchange of scientists

- Publications

6

Fig. 1 - Meetings (full points) and exchange of scientists (arrows) dealing with integrated crop protection in cereals

With this common purpose and fraternal collaboration our meeting takes place. Our objectives are not far from the action of the integrated control of cereal crops in view of reorientating the irrational practice of the systematic use of pesticides. A more efficacious strategy must be developed on the basis of the real evaluation of the needs of phytosanitary protection.

On behalf of the General-Directorate Agriculture of the Commission of the European Communities, it is an honour for me to thank the Glasshouse Crop Research Institute at Littlehampton, in particular the Crop Protection Division, for helping us to organize this meeting. And I would like to thank especially Dr. K.S. Sunderland, for his very collaborative effort in making fruitful our EC experts' group meeting.

The excellent organization, together with the cordiality shown by our English Collegues in welcoming us, are the prelude to a constructive, open, agreeable meeting. To all of them we express our feelings of gratitude and our keen thanks, with best wishes for useful work to all the participants present at Littlehampton, in this pretty corner of the very typical, dear, old Sussex county.

Welcoming address

K.D.Sunderland
GCRI, Department of Entomology & Insect Pathology, Littlehampton, UK

Firstly I would like to thank Prof. R. Cavalloro for giving us such a useful summary of previous meetings held in the framework of the "Energy in Agriculture" Programme of the Commission of the European Communities; this establishes clearly the context of the present meeting.

Prof. R. Cavalloro has also underlined the main aim of our meeting, which is to discuss the use of integrated methods with a view to the reduction of pesticide inputs and increases in efficiency of pest control.

Secondly, on behalf of the Director, Prof. T.R. Swinburne, the head of the Crop Protection Division, Dr. C.C. Payne, my colleagues and myself, I would like to welcome you most warmly to the Institute of Horticultural Research, Littlehampton.

The Institute of Horticultural Research is a major Institute within the UK Agricultural and Food Research Service, and is located at four sites: East Malling, Kent, Littlehampton, Sussex, the Hop Research Department at Wye College, Kent, and at Wellesbourne, Warwickshire. It employs about 240 research scientists engaged in research principally on vegetables, top and soft fruits, hardy ornamental stock, and protected crops and bulbs, but this expertise is also used to a limited extent to research problems on other crops, such as aphid control in cereals.

We are very fortunate in having the participation of 48 experts representing 10 EC-Countries. We will have an opportunity to hear and discuss 33 presentations covering a wide range of research activities and these will be divided into sessions on "Pests", "Diseases and Weeds" and "Farming Systems". This should provide us with a valuable overview of current research into integrated methods of crop protection in EC-cereals and enable us, as individual researchers, to see the relevance of our work in a wider context.

I hope we will have a very enjoyable meeting and a useful exchange of information and views, both during and outside the formal sessions, leading in our closing session to the formulation of recommendations for future common research objectives in the framework of the European agricultural policy.

Finally, I would like to thank both the Commission of the European Communities, Directorate-General Agriculture, and the Agricultural and Food Research Council of the UK, for making this meeting possible.

Session 1
Pests

Chairman: S.D.Wratten

Present status of cereal pests in Spain with special reference to cereal aphids

P.Castañera

Instituto Nacional de Investigaciones Agrarias, Madrid, Spain

Summary

Most of the winter cereal pests in Spain have not been considered economically important, probably because of the low-yielding/low-input cereal system traditionally used. Nevertheless, cereal yields have notably increased during the last two decades due to improvement in agricultural practices, so that farmers' attitudes to pest problems have also changed.

In this second context, sunn pest, Aelia spp. and Eurygaster spp., and cereal aphids will be considered as primary pests. Sitobion avenae, is the key aphid species, followed by Rhopalosiphum padi and Metopolophium dirhodum, though the relative abundance of the last two species varies by year and by location. The ecology and control of these species will be analysed.

Other primary sporadic pests like the Hessian fly, Mayetiola destructor, and the cereal leaf beetle, Oulema melanopus, and wireworms, Agriotes spp., will be examined. Secondary pests will also be described.

1. INTRODUCTION

The cereal growing area in Spain, which is around 7.5×10^6 ha, has remained practically unchanged during the last 30 years, but there has been a shift from winter wheat to barley. This area (Figure 1) represents about 40% of the arable land and 65% of the herbaceous crops (5). Hence, winter cereal crops are of great economic and social importance, especially if we consider that they are mainly restricted to very dry areas where the growing of more profitable crops is not feasible.

Most of the winter cereal insect pests in Spain have not been considered economically important, probably because of the low-yielding, low-input cereal system traditionally used, so that insurance treatments have not been a common strategy. Nevertheless, cereal yields have notably increased during the past two decades. In 1964 barley and wheat yields were, on average, 1.4 t/ha and 1.0 t/ha respectively, whereas in 1984 they reached 2.6 t/ha for barley and 2.5 t/ha for wheat (5). These increases in cereal yields can be greatly attributed to the introduction of new high-yielding varieties with better harvest index than the old ones, the increases in nitrogen fertiliser and changes in farming practices and mechanisation.

Therefore, the significance of insect pest problems in cereal systems may change, as well as farmers' attitudes to their control, by increasing the use of chemicals to reduce the higher potential losses.

In this new context, the present status of winter cereal pests and the prospects for integrated control will be analysed.

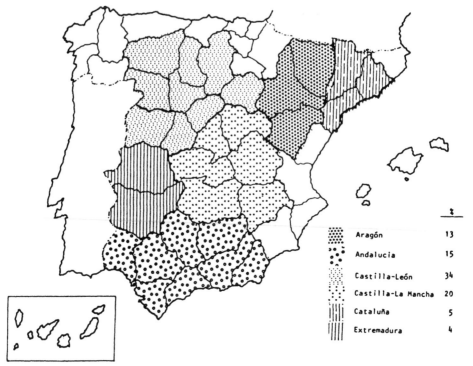

		%
	Aragón	13
	Andalucia	15
	Castilla-León	34
	Castilla-La Mancha	20
	Cataluña	5
	Extremadura	4

Fig. 1. Percentage over the total of the main cereal growing areas in Spain

2. PRESENT STATUS OF CEREAL PESTS OTHER THAN APHIDS

Little information on the quantitative importance of cereal pests is available, even for those traditionally considered most important. Because of that, the number of recordings of insect pests, their distribution and the major papers on them may be taken as an index of their relative importance throughout the time (Table 1).

According to this information, the key pest in this period seems to be Aelia rostrata (Boh.) and A. acuminata (L.) in most cereal growing areas, although they are particularly important in Castilla-León and western Andalucia in wheat fields, near uncultivated mountain areas (8). Studies on their biology, ecology and control were initiated at that time. There are also some references to another pentatomid, Eurygaster austriacus (Shrk.) of little importance and restricted to Toledo province in Castilla-La Mancha (8, 10).

Outbreaks of Aelia spp. have been observed periodically in different parts of the country (2, 7). At present, Aelia rostrata is the commonest and most important species of Aelia in Castilla-León and A. germari (Kust.) in Andalucia (26), where Eurygaster spp. has become the most serious sunn pest since 1970. An economic threshold for these pests has been developed (14). Recently, the biology and antagonists of A. rostrata have also been investigated in Castilla-León (20, 21), and a model of the population development is being prepared (19).

Another major pest of wheat, which used to be widely distributed, is Mayetiola destructor Say. There is little information on this insect in Spain, though its biology was studied and cultural methods of control were recommended (9, 3). Nowadays, this pest can be considered as occasional in

14

our country, and restricted to certain areas. Outbreaks of Mayetiola spp. in wheat have been recorded in Huesca in 1977 and in barley in Lérida and Huesca (Aragón) in 1983 (23).

Cereal leaf beetle, Oulema melanopus (L.), outbreaks and yield losses were reported periodically from 1938 to 1960 in various regions of Galicia, Castilla-León (León) and Castilla-La Mancha (Guadalajara) (18). In the last two decades economic losses have not been informed in Spain, and we have not detected any outbreak of this insect either, after five years of sampling, in one of the traditionally affected regions (Castilla-La Mancha), where small population densities have been observed every year.

The wireworm Agriotes lineatus (L.) used to be quite a damaging pest in Spanish cereal and horticultural crops. However, there is little information on damage assessment in cereal crops (16), where Agriotes spp. do not seem to be economically important in the last decade. Farming practices, like deeper ploughing than in the past, as well as an increase in the use of soil insecticides, could be the reason for their decrease.

The impact of other secondary insect pests, like Cephus pygmaeus (L.), Oscinella frit (L.) and thrips, has not been documented, though they have been recorded as pests of cereal crops (11, 6, 17).

Since the early 1970s some insect pests that were not economically important are becoming a problem in various parts of the country. That is the case for Cnephasia pumicana (Zell). This tortricid moth was first detected in Castilla-León in 1973 where yield losses between 20-80% were reported (22). Nowadays, it is widely distributed, though its economic impact is particularly serious in areas near strips of woodland, since they overwinter as first instar larvae under the bark of mature trees. The upsurge of this moth could be related to suppression of fallow land and the introduction of new cultivars in these areas. Another pest that is increasing in importance in the past decade is Zabrus tenebrioides, damage of this pest has been recently reported in the province of Lérida (Cataluña) (23).

3. CEREAL APHIDS

As shown in Table 1, cereal aphids were considered as a secondary pest, but in the past decade there is a major concern with direct and indirect damage produced by different species in Spain. It has been suggested that changes in agricultural practices may have contributed to the increased frequency and intensity of aphid outbreaks in recent years (25, 29), though one of the major reasons could be the increase in the nitrogen availability in cereal crops, which stimulates aphid reproduction and reduces development time and mortality (27).

Most of these agricultural developments have lately occurred in Spain and may have contributed to the increasing importance of cereal aphid populations. This new situation led us to undertake studies on the population dynamics of cereal aphids and their damage assessment in Madrid in 1980.

3.1 Study Area

All the studies related to the ecology and control of cereal aphids have been carried out at 'El Encin', Alcalá de Henares, Madrid, in a wheat field, of about 0.6 ha. The experimental design and the sampling procedure followed, have been described (12, 13). From 1985 onwards the autumn populations have been estimated with the Dietrick vacuum insect net (D-vac). Samples were taken from winter wheat volunteers, and a grass field near the study area to obtain information on the overwintering of cereal aphids.

15

TABLE 1. Number of Insect Pests Recorded in the Files of the Madrid Plant Protection Department from 1925 to 1974, and Number of Major Papers, as an Index of Their Relative Importance

Insect Pests	Recording year 1st	last	No. of recordings	Distribution (No. of counties)	Hosts	No. of major papers (years)
Aelia acuminata	-	-	7	7	mainly wheat	9 (1939; 47; 55 a; 72 a, b; 79 a, b; 81; 85)
A. rostrata	1925	1964	249	22	"	
Agriotes lineatus	1925	1967	88	18	cereals and other crops	1 (1949)
Agriotes sp.	1925	1969	36	15	cereals and other crops	
Anoecia vagans	1965	-	1	1	wheat and barley	
A. corni	1929	1949	5	4	wheat	
Calamobius filum	1934	1970	5	3	wheat	
Cantarinia tritici	1926	1934	3	-	wheat	
Cephus pygmaeus	1933	1973	14	8	wheat, barley and rye	2 (1947; 50)
Diuraphis noxia	1947	1948	3	3	wheat	1 (1947)
Eurygaster spp.	1935	1964	35	18	cereal crops	2 (1939, 41 b)
Forda sp.	1932	1966	3	3	barley	
Haplotrips tritici	1963	-	1	1	wheat	2 (1929; 47)
Limothrips cerealium	1944	1966	4	3	wheat	
Mayetiola destructor	1926	1972	130	28	mainly wheat, barley	3 (1941 a; 55 b; 85)
Oscinella frit	1948	1970	26	6	barley	1 (1949)
Oulema melanopus	1938	1965	53	10	oat, barley and wheat	2 (1940; 63)
Phorbia sp.	1968	-	2	2	wheat	
Rhopalosiphum padi	1948	1974	2	2	wheat and barley	
Schizaphis graminum	1949	-	1	1	wheat	
Sipha maydis	1928	-	1	1	oat	
Sitobion avenae	1927	1967	8	5	wheat and barley	
Zabrus tenebrioides	1926	1972	17	9	wheat and barley	2 (1947; 85)

TABLE 2. Cereal Aphid Species Found on Cereals and Grasses in Spain

Aphid species \ Host plants	Triticum aestivum	Triticum Turgidum	Hordeum vulgare	Avena sativa	Secale cereale	Triticale	Zea mays	Sorghum vulgare	Hordeum murinum	Avena spp.	Phalaris spp.	Bromus spp.	Lolium sp.	Hordeum bulbosum	Cyperus rotundus
Sitobion avenae	*	*	*	*	*	*			*	*	*		*		
Diuraphis noxia	*		*				*		*		*			*	
Schizaphis graminum	*						*	*	*		*				
Rhopalosiphum maidis	*		*				*	*							
Metopolophium festucae	*		*												
Sipha kurdjumovi	*														
Rhopalosiphum padi	*	*	*	*		*	*	*	*	*		*			*
Metopolophium dirhodum	*		*	*					*	*					
Sitobion fragariae	*		*						*	*	*	*	*		
Forda sp.			*												

(*) Aphid species collected on each host plant.

17

Forty samples, each of 0.1 m^2, were taken in a diagonal across the 1 ha field of grassland every 7-15 days. Besides that, occasional samples were taken in most of the main cereal growing areas to get information on the aphid species, their distribution, and the parasitoid complex on cereal crops and grasses.

3.2 Aphid Species

The main species observed in Central Spain between the end of April and the middle of June, in winter wheat, is Sitobion avenae (F.) followed by Rhopalosiphum padi (L.) and Metopolophium dirhodum (Wlk), though the relative abundance of the last two species varies, depending on the year. Similar results have also been obtained in the Lérida province in an irrigated field of winter wheat (1). Other species found, and the range of their host plants, are recorded in Table 2. The six former species shown are monoecious on Gramineae, whilst R. padi, M. dirhodum, S. fragarie are heteroecious between Rosaceae and Gramineae (28) and Forda sp. between Pistacia spp. and Gramineae (15). All the species are aerial feeders except Forda sp. which was found on the roots of spring barley.

3.3 Population Dynamics

The relative importance of the main aphid species found in Central Spain is shown in Table 3.

TABLE 3. Percentage Composition of the Main Cereal Aphid Species
Found on Ears and Leaves of Winter Wheat in 1980-85

	1980	1981	1982	1983	1984	1985
Sitobion avenae	92	83	88	66	54	90
Rhopalosiphum padi	2	16	1	2	2	-
Metopolophium dirhodum	6	1	11	9	44	10
Diuraphis noxia	-	-	-	23	-	-

S. avenae is by far the most important aphid species in winter wheat in Central Spain. The first alatae of S. avenae are usually detected about the middle of April, though the aphid density is very low until the beginning of May, increasing very rapidly from then on, to reach the maximum density about the end of May or the beginning of June (Figure 2) at plant growth stages 65/71 of Zadoks's scale (30). The collapse of the aphid population has always been very rapid, coinciding with high temperatures (T max. > 30°C.) and the ripening of the wheat.

After the breakdown of the aphid population, nearby maize fields were sampled regularly, but no aphids were found until September. This could be partially related to the high DIMBOA (2,4 dihydroxy - 7 methoxy - 2 H - 1,4 benzoxacin-3 (4 H) - one) content of some cultivated hybrids before tasselling (24). However, in most years, R. padi and R. maidis are very abundant in maize in September.

Other alternative plants, like grasses, are not suitable hosts during the summer in dry land, since they have similar phenologies to cereal crops.

The overwintering of cereal aphid populations is being studied. R. padi and S. avenae are the most abundant species collected so far, but,

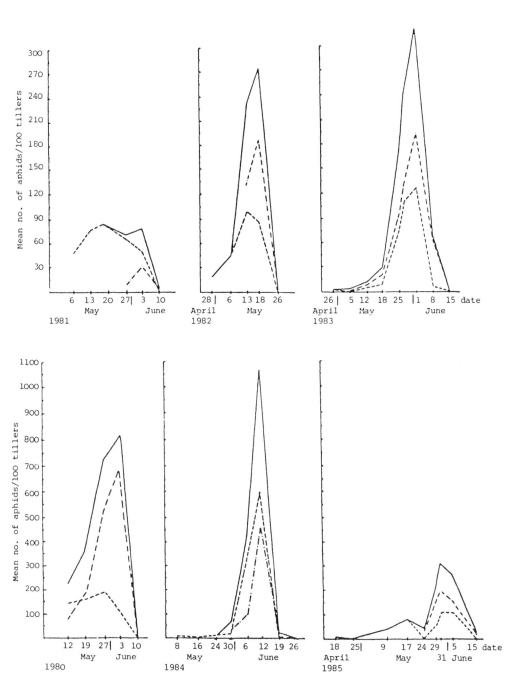

Fig. 2. Population development of the mean no. of S. avenae, M. dirhodum, R. padi and D. noxia on ears (-·-·-·-), on leaves (- - - -), and the total (____) in winter wheat in Central Spain in 1980-1985

19

in contrast to the wheat plots, the former species is much more abundant than the second one during autumn and winter months. Both species seem to be anholocyclic; more detailed information will be considered elsewhere.

3.4 Primary Parasitoids

Parasitoids were bred from samples of mummies and live aphids collected from the leaves and ears of wheat during the weekly sampling at 'El Encin', and from different cereal crops in other Spanish cereal areas.

Most of the parasitoids bred out, in all the areas, were mainly of Aphidus spp. Very few Praon volucre (Haliday) and Diaeretiella rapae (M'Intosh) have been collected. Ephedrus pagiator (Nees) has only been found in a barley field in Jaraiz de la Vera, Cáceres, on S. avenae.

The parasitoid composition obtained at 'El Encin' during 1980-83 is shown in Table 4.

TABLE 4. Percentage Composition of Primary Parasitoids Bred
from S. avenae, R. padi, M. dirhodum and D. noxia
Collected from Winter Wheat in 1980-83

	1980	1981	1982	1983
Aphidius ervi	46.9	60.4	23.5	43.7
A. uzbekistanicus-rhopalosiphi	49.6	39.6	76.5	53.1
Praon volucre	3.5	-	-	-
Diaeretiella rapae	-	-	-	3.2

Parasitism was assessed by counting mummified aphids on the plants and, when the aphid population was high enough, by rearing field-collected aphids in the laboratory. The number of aphids killed by parasitoids follows a similar pattern every year; the number of mummified aphids was very low during the first part of the population growth and reached maximum when the aphid population was declining. The values ranged from 2 to 18%, depending on the year.

The percentage parasitism obtained by aphid rearing in the laboratory ranged from 11.7 to 14% in 1980, 23.8% in 1982 and between 1.3-13.7% in 1985.

The percentage composition of parasitoids collected in other parts of the country is rather similar to the one mentioned. The two most common species collected were also A. ervi (Hal.) and A. uzbekistanicus-rhopalosiphi complex.

Activity of parasitoids has also been detected during the overwintering sampling. The species found, in order of abundance, were A. ervi, D. rapae and A. uzbekistanicus-rhopalosiphi.

Very few diseased aphids were found in all years, presumably due to the very dry weather conditions.

3.5 Predators

From the well known group of aphid-specific predators, the main species, collected at 'El Encin' during the weekly sampling, were Coccinella septempunctata (L.) and Adonia variegata (Goeze) (Coccinellidae), and the Syrphidae: Sphaerophoria scripta (L.), Metasyrphus corollae (F.), Episyrphus balteatus (Deg.) and Scaeva pyrastri (L.), though their relative abundance differed between years (12). Polyphagous predators' composition was also studied by using pitfall traps (14). Carabidae and spiders were the main predators collected, Staphylinidae were scarce in all years.

A total of 30 species of Carabidae was identified (14), the species number and their relative abundance differed between years, though Ditomus capito (Serv.), Harpalus distinguendus (Duft.), Harpalus anxius (Duft.), Poecillus crenulatus (Dej.), Bradytus apricarius (Payk), Harpalus rufipes (de Geer), Acinopus picipes (Oliv.) and Trechus quadristriatus (Schrank.) account for > 80% of the total number of captures in all years.

Of the 43 species of spiders collected, Thomisidae, Drassidae and Lycosidae represented more than 90% of the total in 1980 (14).

Further investigations on parasitoids and predators are needed to determine their role in regulating cereal aphid populations.

4. CONTROL STRATEGIES OF CEREAL PESTS AND PROSPECTS FOR INTEGRATED CONTROL

The substantial changes that have occurred in Spain in the past 25 years in cereal growing, namely increased inputs in mechanisation, fuel and fertiliser use, and the introduction of new high-yielding cultivars, have resulted in a great increase in cereal production. However, new insect pest problems may have been created. Therefore, the farmer's attitude is now more risk-adverse than it used to be, leading to a more intensive use of insecticides. Nevertheless, the main constraint for a high-yield strategy in Spain is rainfall, which is scarce and badly distributed in most of the cereal growing areas. Because of that, there would be little justification for the Spanish farmers to get involved in a maximum yield strategy.

The current practices for control of the main cereal pests are presented in Table 5. Chemical control is the only way still available to control sunn pests, as research has not yet produced the data that could permit reliance on alternative methods, but there is some information on the parasitoid complex, and fungal pathogens that may act as population regulators (21, 26). In addition, an economic threshold has been established (4), which allows a more rational use of the insecticides.

The control of the tortricid C. pumicana relies also on insecticides. However, treatments are restricted to very localised areas and based on a forecasting scheme. For all the other pests considered (Table 5), farmers do not usually apply chemical insecticides and, if they do, their use is very limited.

On the other hand, many modifications in crop production practices are designed to obtain cultural control of certain pests, as is shown in Table 5. Accurate timing of planting (e.g. Hessian fly and cereal aphids), ploughing to eliminate weeds or to make plant débris unsuitable for overwintering larvae (e.g. Hessian fly and wireworms) requires a deep knowledge of the pest's biology and ecology in order to couple techniques with agronomic practices. Certainly they are being used wisely and effectively with a concomitant reduction in reliance upon insecticide use.

Biological control in the form of man-manipulated parasitoids, predators and pathogens is not practised for any insect cereal pest in Spain, since data available on them are still very scarce.

In short, current practices of insect control in cereal systems in Spain are still more dependent on insecticides than on cultural or biological control, though some cultural practices have been very useful in the past, and even at present, providing the first line of defence against some pests (Table 5), and other suppressive measures (e.g. insecticides) have only been supplementary.

21

TABLE 5. Types of Control Used for the Main Insect Pests Attacking Cereal Crops in Spain and Provisional Damage Threshold

Pest/Pest complex	Chemical Control		Cultural Control		Biological control	Damage threshold
	Soil insecticides	Foliage insectides	Late planting	Straw destruction		
Aelia spp.		-			*	x
Eurygaster spp.		-				x
Mayetiola destructor		-	-	-		
Oulema melanopus		-		-		
Agriotes spp.	-					
Cereal aphids		-	-		*	
Cnephasia pumicana		-				x
Zabrus tenebrioides		-		-		

(-) Reduction in insect population or damage.

(*) Works on natural enemies.

(x) Damage threshold established in some areas.

Taking into account the Spanish limitation, already mentioned, for a high-yield system, a relatively low-input, low-yielding system should probably be adopted (compare with other European countries). This system will allow a major flexibility in pest control. In this context, the rational use of chemical insecticides will require the establishment of validated economic injury thresholds, and forecasting procedures for the key pests, as the primary tools for the development of integrated pest management in Spanish cereal systems; these measures will surely improve farmers' decision making.

REFERENCES

(1) ALBAJES, R. and PONS, X., 1985. Algunos datos sobre la biologia de los transmisores del enanismo amarillo de los cereales en Lérida. Actas II Jorn. Técn. Cereales de invierno, Pamplona, II, 59-68.

(2) ALFARO, A., 1955a. Notas sobre el garrapatillo del trigo Aelia rostrata en Aragón, Bol. Pat. veg. Ent. agric., XXI, 19-37.

(3) ALFARO, A., 1955b. Mavetiola destructor Say y M. mimeuri Mesnil en Zaragoza. Bol. Pat. veg. Ent. agric., XXI, 85-16.

(4) ALVARADO, M., 1983. Los chinches del trigo. las Jornadas Téc. Cereales de invierno. INIA. Madrid, II, 127-137.

(5) ANON, 1984. Ann. Est. Prod. Agr. para 1984, Ministerio de Agricultura, Pesca y Alimentación.

(6) BENLLOCH, M., 1949. Observaciones fitopatológicas en el año 1948. Bol. Pat. veg. Ent. agric., XVI, 203-242.

(7) CABALLERO, J.L., ALVARADO, M., ROMERO, J., DELGADO, J., DOMINGUEZ, A. and TOMAS, L., 1972. Biologia de la paulilla del trigo Aelia sp. en el año 1971 comparada con la de los años 1969 y 1970. Bol. Serv. Plagas, 96, 13-22.

(8) CAÑIZO, J. DEL., 1939. Pentatómidos perjudiciales al trigo. Bol. Pat. veg. Ent. agric., VIII, 15-26.

(9) CAÑIZO, J. DEL., 1941a. El mosquito del trigo (Mayetiola destructor Say) y la época de siembra. Bol. Pat. veg. agric., X, 256-263.

(10) CAÑIZO, J. DEL., 1941b. Notas sobre el 'sampedrito' del trigo (Eurygaster austriacus Seabrai Ch.). Bol. Pat. veg. Ent. agric., X, 264-274.

(11) CAÑIZO, J. DEL., 1947. Las plagas de insectos y su importancia en la economia agricola española. Bol. Pat. veg. Ent. agric., XV, 333-344.

(12) CASTAÑERA, P., 1983. The relative abundance of parasites and predators cereal aphids in Central Spain in 'Aphid Antagonists'. Edt. Cavalloro, 76-82.

(13) CASTAÑERA, P. and GUTIERREZ, C., 1985. Studies on the ecology and control of cereal aphid in winter wheat in Central Spain. IOBC/WPRS Bull., III/3, 45-47.

(14) CASTAÑERA, P. and DEL ESTAL, P., 1985. Studies of the soil fauna in winter wheat in Central Spain in 1980-83. IOBC/WPRS Bull., VIII/3, 140-141.

(15) DAVATCHI, G.A., 1958. Etude comparative sur la biologie et le polymorphisme des aphides gallicoles des Pistacia d'Asie Centrale, du Moyen-Orient, bu bassin méditerranéen et du Nord Africain. Rev. Pat. Veg. Ent. Agric. France, XXXVII, No.1, 85-146.

(16) DOMINGUEZ GARCIA-TEJERO, F., 1949. Los gusanos de alambre, elateridos de interés agricola. Bol. Pat. veg. Ent. agric., XVI, 119-156.

(17) DOMINGUEZ GARCIA-TEJERO, F., 1950. Tentredinidos perjudiciales a la agricultura. Bol. Pat. veg. Ent. agric., XVII, 163-208.

(18) DOMINGUEZ GARCIA-TEJERO, F., 1963. Crisomélidos de interés agricola., Bol. Pat. veg. Ent. agric., XXVI, 49-125.

(19) FERNADEZ-CANCIO, A., 1985. Approximación a la modelización del desarrollo de Aelia rostrata (Hemyptera, Pentatomidae). Bolm. Soc. port. Ent., I, 399-408.

(20) GALLEGO, C., 1978. Caracteristicas de los refugios de invernación del garrapatillo del trigo, Aelia rostrata Boheman (Hemiptera, Pentatomidae) en la Región Central. Ann. INIA, Ser. Prot. veg., 8, 33-44.

(21) GALLEGO, C. and SANCHEZ-BOCHERINI, J., 1979. Les punaises des céréales en Espagne. Caracteristiques des invasions dans la region Centrale. Etude du parasitisme naturel. Bull. SROP, II/2, 22-33.

(22) GARCIA-CALLEJA, A., 1976. Nuevas observaciones acerca de Cnephasia pumicana Zell (Lepidóptero, Tortricidae) en Valladolid. Bol. Serv. Plagas, 2, 205-223).

(23) GARCIA DE OTAZU, J., 1985. Mayetiola, Zabrus y Cnephasia en los cereales de invierno y su control. Actas II Jorn. Técn. cereales invierno, Pamplona, II, 71-82.

(24) GUTIERREZ, C. and CASTAÑERA, P., 1986. Efecto de los tejidos de maiz con alto y bajo contenido en DIMBOA sobre la biologia del taladro Sesamia nonagrioides Lef. (Lepidoptera, Noctuidae). Inv. Agrar., Prot. veg., I, 109-119.

(25) KOLBE, W., 1973. Studies on the occurrence of cereal aphids and the effect of feeding damage on yields in relation to infestation density levels and control. Pflanzenschutz-Nachr, Bayer, 26, 396-410.

(26) SANCHEZ-BOCHERINI, J. and GALLEGO, C., 1981. Biologie et lutte contre Aelia rostrata en Espagne. Bull. OEPP, 11(2), 43-46.

(27) VEREIJKEN, P.H., 1979. Feeding and multiplication of three cereal aphid species and their effect on yield of winter wheat. Agric. Res. Rep. No.888, Pudoc, Wageningen, 58pp.

(28) VICKERMAN, F.P. and WRATTEN, S.D., 1979. The biology and pest status of cereal aphids in Europe: a review. Bull. Ent. Res., 69, 1-23.

(29) WAY, M.J., 1979. Optimising cereal yield. The role of pest control. Proc. Brt. Crop. Prot. Conf., 3, 663-672.

(30) ZADOKS, J.C., CHANG, T.T. and KONZAC, C.F., 1974. A decimal code for the growth stages of cereals. Weed Research, 14, 415-21.

Towards the integrated pest control in cereals in Italy

S.Barbagallo
Istituto di Entomologia Agraria, University of Catania, Italy

L.Suss
Istituto di Entomologia Agraria, University of Milano, Italy

Summary

Several insects damage winter cereals in Italy (i.e. wheat, barley, oats and rye), but their infestation only occasionally reaches a high enough population density to be of any economic interest and, as a rule, it is confined to the major species. Among them, aphids undoubtedly represent the most constant and comprehensive group of cereal pests, followed by occasional outbreaks of some Pentatomidae bugs (i.e. Aelia rostrata) and of the Chrysomelidae beetle Oulema melanopa. At present, insecticide applications against such pests are not yet a routine practice - as for other intensive agroecosystems - but there is a general trend of farmers to increase their use. As to maize, severe damage is mainly linked with the European corn borer and Sesamia infestations. Such pests, as well as other harmful species (i.e. Elateridae and Noctuidae larvae, responsible for heavy attacks to the hypogeal plant organs), are usually controlled by applications of synthetic pesticides. All in all, it is advisable to limit the risk of a progressive utilisation of non-selective pesticides as the main method for pest control on cereals, to avoid the consequent deterioration of beneficial entomophagous effects. Present investigations aim, therefore, at promoting new methods of forecasting and control, in order to improve the standardisation of methodologies of supervised and integrated control.

1. INTRODUCTION

Cereal crops have traditionally been of great economic interest in Italy and have occupied a remarkable percentage of its growing surface. During the last few years there has been an increase of barley at the expense of wheat crops; an increase has also been registered for maize, while rice appears unmodified.

Some pests infest several cereal species indifferently, while others show a different degree of host preference and may be associated with a particular crop.

Modern changes in the growing techniques of cereals, have brought about variations in the ecosystem arthropod fauna, affecting the population dynamics and the interactions between pests and their entomophagous species.

2. THE MAIN CEREAL PESTS

Many insect species damage winter cereals (i.e. wheat, barley, rye, oats), and among them aphids are surely a group of primary economic interest, as they are responsible for direct damage and virus transmission (BYDV).

Sitobion avenae (F.) is generally the predominant and most harmful species in the northern part of the peninsula, while in the south and in the main islands (Sicily, Sardinia) *S. fragariae* (Walk.) is found, which becomes the commonest species there (1, 24). The two species overwinter as virginopara in most of our cereal areas and reach, as a rule, a high population density starting from early spring. *Rhopalosiphum padi* (L.) is another of the main cereal aphid species found everywhere in the Italian territory, while *Metopolophium dirhodum* (Walk.) is fairly common too, and develops with particular intensity on the leaves in southern areas. *Schizaphis graminum* (Rond.) and other secondary species appear to be less widespread. Crop fertilisation and parasitism play an important role in the development of aphid colonies. The application of nitrogen compounds has a positive correlation with the aphid population density and negatively affects the development of alatae (31) so that the expected increase of yields may be nullified or reduced by aphid damage if no convenient control of their population is applied. On the other hand, the increased applications of pesticides positively affect aphid abundance by reducing the number of their natural enemies. It is probably due to the combined effects of such factors that during recent years we have observed an increase in cereal aphid infestations, particularly in the northern districts of Italy. Together with the more intensive aphid population we have also followed the spread of barley yellow dwarf virus (BYDV) in our country.

The late-spring aphid infestations are essentially due to *S. avenae* or to *S. fragariae*, according to the territory areas; the aphids cause important damage after the appearance of the ears, where most of the specimens will concentrate. The effects of such infestation on wheat yield and quality was pointed out by Wratten (32). On the contrary, the most harmful infestations of *R. padi* are related to the autumn period, when the aphid moves from maize crops to wheat and barley, sown from the second half of September onwards in northern areas. It has been pointed out (17, 28) that early sowings are almost constantly subject to such infestations and consequently to BYDV infection, of which maize is a natural source.

The available data on the economic thresholds for cereal aphids in Italy relate only to wheat crops in northern districts; it has been found (24) that 4-5 aphids per tiller during flowering, together with an increasing aphid population, require control. It is recommended that selective insecticides be used if the aphid population exceeds the economic threshold, thus reducing any adverse environmental implications. Nevertheless, the beneficial effects of insecticide sprays may be reduced or nullified by wheeling damage during spraying (27).

The occasional pullulation of grain bugs (Pentatomids), such as some species of *Aelia* and *Eurygaster*, may be responsible for conspicuous economic losses. The bug infestation is of remarkable extent in the southern Mediterranean regions, affecting the meridional areas of our country as well. The chemical control of *A. rostrata* Boh., that heavily infested wheat in Sicily during the second half of the last decade (9) was carried out with malathion or other insecticide sprays.

The cereal leaf beetle, *Oulema melanopa* L., probably because of the high fertilisations supplied to wheat and barley, seems to become more harmful, year after year; the problem is particularly noticed in northern cereal areas.

Passing to maize pests, the European corn-borer (*Ostrinia nubilalis* Hb.) is undoubtedly the key species, at least for most of the Italian regions. Here the insect develops, as a rule, two generations a year, though three have sometimes been found and, recently, the presence of a univoltine lineage has been discovered (29). The pressure of its infestation appears to have increased recently, apparently linked with the improvement of some

26

growing techniques: high fertilisation, irrigation, higher sowing density, use of some herbicides and the technique of their application (11, 14, 22). Moreover, the increasing damage caused by corn-borer on plants other than cereals (Guinea pepper, French beans, fruit trees, vine shoots, etc.) should be borne in mind (4, 5).

The use of resistant maize hybrids for the control of such pests is not widespread and does not give fully satisfactory results, particularly as far as the second generation is concerned where damage has been found. Chemical control is therefore applied by farmers (with predominant use of organophosphorous compounds), but the need for more effective integrated control should be taken into consideration (30).

The corn stalk-borers, Sesamia cretica (Led.) and S. nonagrioides (Lef.), are responsible for damage comparable to that of O. nubilalis. The former species is predominant in the northern and central areas of the peninsula, while S. nonagrioides is widespread in the southern territories; it is particularly harmful in Sardinia and Sicily, where it develops up to four generations per year, heavily infesting maize, sorghum, and several other Gramineae (13, 20). The two species are usually controlled by insecticide treatments, similar to those used for the European corn-borer.

Infestations by cutworms (Noctuids) and wireworms (Elateridae) on maize are becoming common on large areas (25, 33). The damage mostly affects seedlings and young plants up to 3-4 weeks old. It is not easy to explain the reason for the increasing outbreaks of such hypogeal pests, but the change of growing practices is suspected to be once again, as for other pests, the main factor responsible for their pullulation. Localised applications of granulated insecticides have been tried to control them, but the effects have not always reached the desired results.

3. THE CEREAL AGROECOSYSTEM

From what we have briefly reported about the main cereal pests, we have noticed the interdependence of agricultural practices and the population dynamics of their pests. The lack of crop rotation through the years, or even worse, the succession of cereal crops in nearby areas of farms during the same year (i.e. winter cereals which follow the maize crop and so on) are sources of high pullulations of their own pests. At the same time, changes in growing techniques (use of new varieties, increment of fertilisation, growing density, irrigation, input of herbicide and other pesticide treatments), greatly affect the population density of pests and the interrelation with their entomophages; the implications of these may often lead to a higher percentage of yield losses due to the increased pest population.

Undoubtedly the key pests are aphids (particularly Sitobion and Rhopalosiphum) for wheat and barley crops, and the corn-borers (Ostrinia and Sesamia) for maize crops. Other insect pests produce only sporadic, though harmful, pullulations and may be responsible for damage which appears in different areas from year to year. Such fluctuations which may be partially ascribed to the activity of entomophages, climatic factors, and the effects of agricultural techniques, represent unforeseen events for the farmers who cannot avail themselves of timely control measures: for example, the cutworm and wireworm infestations on maize seedlings, as well as the bug outbreaks on wheat. Objective technical difficulties may be encountered also with large scale pesticide distribution, due to intrinsic crop features. Pesticide treatment, on the other hand, against late Ostrinia infestation on maize, or aphid pullulation on wheat after the ears have appeared, is not allowed in our country because insecticide application by aircraft is forbidden.

27

Nevertheless, considering the well known side-effects of the broad spectrum of insecticides, such limited use of pesticides helps to protect the beneficial arthropods which populate the same cereal agroecosystem (3, 7, 8, 15, 18, 23).

4. TOWARDS INTEGRATED PEST CONTROL

The risk of degradation in the biological equilibrium of the cereal agroecosystem, is mainly linked with the incorrect use of pesticides or to other growing practices, which have detrimental effects on the beneficial arthropods, and positive stimuli on the pest population.

Presently the cereal crops in Italy are indeed scarcely treated with pesticides, at least as far as winter cereals are concerned; the situation is not much different for maize crops, where insecticides are more frequently applied. On the other hand, supervised or integrated control are not widely used and are even unknown to most farmers.

Therefore, scientific institutions have promoted and encouraged research on alternative systems for pest control, in order to reduce the potential use and side-effects of broad spectrum insecticides.

Against aphid infestations on wheat and barley, the economic threshold has been estimated and the effects of chemical applications have been pointed out. The recent establishment of a monitoring system for aphid flights, by means of suction traps (Rothamsted type), may enable us to know the potential risks of aphid infestation on crops and the virus infection they may transmit to plants (2, 6, 16); in such a way, though on few occasions, their catching data have already been used to advise farmers about the suitable time for sowing autumn cereals, in order to minimise the risk of early BYDV infection by R. padi (17).

Several well-known predators and parasitoids play an important biological role against aphid development. Negative side-effects of different, commonly used pesticides, have already been investigated (20, 19, 26) and it has been shown that some aphicides supposed to be selective, have really different degrees of toxicity on endophagous microhymenoptera and other beneficial arthropods.

Among the predators, the beneficial effects of some beetles which belong to the soil fauna are worth mentioning. Several species of Carabidae and Staphylinidae are very efficient both on wheat (as well as on other winter cereals) and on maize, where they can even prey on colonies of R. padi protected inside the leaf sheath. Some Staphylinidae are also active predators of eggs and young larvae of the leaf beetle, thus contributing to a useful reduction of its early infestations on wheat. The use of soil insecticides (geodisinfectant) and herbicides negatively affects these predators, favouring an increase of phytophages.

All such knowledge appears, therefore, of great interest in order to make best use of pesticides and avoid the destruction of the biological equilibrium in cereal fields.

A few considerations should also be noted on the European corn-borer on maize; the employment of resistant varieties is still very scanty in our country and of little interest particularly against the damage produced by the second larval generation. The recent use of sex pheromone, together with a knowledge of the environmental effects (relative humidity, and particularly, temperature) on the insect stage development, allow more timely chemical interventions. The latter may be applied, at least alternatively, by selective microbiological products, such as Bacillus thuringiensis, whose use has recently been allowed in Italy. The possibility of integrating biological control, by the release of the oophagous Trichogramma maidis P. & V. is not negligible either (12). The opportune destruction of the infested plant remains, may efficiently contribute to a valuable integrated control against the same pest, as well as against the stalk-borer, S. nonagrioides, which in some areas becomes the main maize pest. The use of light traps, for the latter, was successfully experimented in Sardinia (21).

In conclusion, as the knowledge of the arthropod interrelations in cereal agroecosystems progressively increases, we can increase the application of integrated control methodologies against pests, supporting and improving the natural biological control that nowadays is often performed in our cereal fields. The present, rapid changes in farming techniques may cause degenerating variations in such a biological equilibrium, for which it is therefore necessary to encourage the correct use of chemical products (insecticides, fungicides, herbicides, etc.) and promote ecological methodologies (agronomical, biotechnical, biological methods) to achieve effective integrated pest control.

REFERENCES

(1)   BARBAGALLO, S. 1982. The proposed monitoring system for cereal aphids in Italy. In: Bernard J., Utilisation du piège à succion en vue de prévoir les invasions aphidiennes. Station Zool. Appl., Gembloux (B): 7-9.

(2)   BARBAGALLO, S. and PATTI, I., 1985. Progress towards the Euraphid network traps in Italy, with notes on the monitored aphid species during 1982-84 in Eastern Sicily. Proc. Experts' Meeting "Migrations aphidiennes et reseau 'Euraphid'", Montpellier 7-9 May, 1985 (in press).

(3)   BRYAN, K.M. and WRATTEN, S.D., 1984. The responses of polyphagous predators to prey spatial heterogeneity: aggregation by carabid and staphylinid beetles to their cereal aphid prey. Ecological Ent., 9: 251-259.

(4)   CIAMPOLINI, M., BARBIERI, M., CECI, D. and LAMBERTINI, F, 1981. La piralide del mais dannosa al fagiolino da industria. L'Informatore agrario, 37(9); 14349-14354.

(5)   CIAMPOLINI, M., ZANGRANDO, G.P. and SUSS, L., 1985. La piralide del mais (Ostrinia nubilalis Hb.) nociva alle piante madri dei vitigni portinnesti. L'Inf.tore Agrario, (41)31: 59-62.

(6)   COLOMBO, M. and LIMONTA, L., 1986. Cattura di afidi mediante trappola ad aspirazione di tipo Rothamsted nella Pianura padana. Atti Giornate Fitopatol. 1986. Coop. Libr. Un. Ed., Bologna, 1: 239-246.

(7)   EDWARDS, C.A. and GEORGE, K.S., 1981. Carabid beetles as predators of cereal aphids. Proc. Brit. Crop. Prot. Conf. 1981 (1): 191-199.

(8)   FEENEY, A.M., 1982. The occurrence and effect of pesticides on aphid predators in Ireland, 1979-1982. In: Cavalloro R., Aphid Antagonists, Proc. EC Experts' Meeting, Portici (I). Balkema, Rotterdam: 123-128.

(9)   GENDUSO, P. and DI MARTINO, A., 1974. Su una grave infestazione di Pentatomidi del frumento in Sicilia e sulla vegetazione rifugio. Boll. Ist. Ent. Agr. Oss. Fitopat. Palermo, 9: 81-100.

(10) HASSAN, A.S. et al., 1983. Results of the second joint pesticide testing programme by the IOBC/WPRS - Working Group "Pesticides and Beneficial Arthropods". Z. ang. Ent., 95: 151-158.

(11) LANDI, P. and MAINI, S., 1981. Infestazione da Ostrinia nubilalis Hb. (Lepidoptera, Pyralidae) su mais in relazione ad alcuni aspetti colturali e morfologici della pianta. Boll. Ist. Ent. Univ. Bologna, 36: 69-81.

(12) MAINI, S., CELLI, G., GATTAVECCHIA, C. and PAOLETTI, M.G., 1983. Presenza ed impiego nella lotta biologica del Trichogramma maidis

Pintureau e Voegele (Hymenoptera, Trichogrammatidae) parassita oofago di Ostrinia nubilalis Hb. (Lepidoptera, Pyralidae) in alcune zone dell'Italia settentrionale. Boll. Ist. Ent. Bologna, 37: 209-217.

(13) NUCIFORA, A., 1966. Appunti sulla biologia di "Sesamia nonagrioides" (Lef.) in Sicilia. Tecnica Agricola, 18: 395-419.

(14) OKA, N.I. and PIMENTEL, D., 1979. Ecological effects of 2,4 D herbicide increased corn pest problems. Contr. Centr. Res. Inst. Agric. Bogor, 49: 1-17.

(15) PAOLETTI, M.G., GIROLAMI, V., JOB M., ZECCHIN, F. and BIONDANI, E., 1981. I geodisinfestanti e la pedofauna nella coltura del mais. Atti Conv. "La difesa dei cereali nell'ambito dei Prog. Fin. C.N.R." - Ancona, 10-11 dic. 1981: 153-162.

(16) PATTI, I., 1985. Risultati di due anni di catture afidiche con trappole ad aspirazione (tipo Rothamsted) in Sicilia orientale. Atti XIV Congr. Naz. It. Ent., Palermo-Erice-Bagheria: 615-622.

(17) PERESSINI, S. and COCEANO, P.G., 1986. Incidenza delle infezioni di virus del nanismo giallo dell'orzo (BYDV) su orzo e frumento in rapporto a epoca di semina e località. Inf.tore Fitopatol. (1/86): 29-32.

(18) POWELL, W., DEAN, G.J., DEWAR, A. and WILDING, N., 1981. Towards integrated control of cereal aphids. Proc. Brit. Crop Prot. Conf. 1981 (1): 201-206.

(19) POWELL, W., DEAN, G.J. and BARDNER, R., 1985. Effects of pirimicarb, dimethoate and benomyl on natural enemies of cereal aphids in winter wheat. Ann. appl. Biol., 106: 235-242.

(20) PROTA, R., 1966. Osservazioni sull'etologia di Sesamia nonagrioides (Lefebvre) in Sardegna. Studi Sassaresi, 13: 336-360.

(21) PROTA, A., 1969. Nuove prospettive di lotta contro la Sesamia nonagrioides (Lef.) dannosa al granoturco in Sardegna. Atti Giornate Fitopatologiche 1969: 171-181.

(22) SHOWERS, W.B., BERRY, E.C. and VON KASTER, L., 1980. Management of 2nd generation European corn borer by controlling moths outside the cornfield. J. Econ. Entomol., 73: 88-91.

(23) SUNDERLAND, K.D. and CHAMBERS, R.J., 1982. Invertebrate polyphagous predators as pest control agents: Some criteria and methods. In: Cavalloro R., Aphid Antagonists. Proc. EC Experts' Meeting, Portici (I). Balkema, Rotterdam: 100-108.

(24) SUSS, L., 1980. Afidi del Frumento e soglie di tolleranza. Atti Giornate Fitopatol. 1980 (3): 231-240.

(25) SUSS, L., 1981. Considerazioni sull'impiego di formulazioni granulari contro gli Elateridi infestanti il Mais. Atti Convegno "La difesa dei cereali nell'ambito dei Progetti Finalizzati del C.N.R.", Ancona 10-11 dic. 1981. Parretti Grafiche (Firenze): 145-151.

(26) SUSS, L., 1982. Survival of pupal stage of Aphidius ervi Hal. in mummified Sitobion avanae F. to pesticide treatment. In: Cavalloro R., Aphid antagonists, Proc. EC Experts' Meeting, Portici (I). Balkema, Rotterdam: 129-134.

(27) SUSS, L. and COLOMBINI, F., 1981. Osservazioni sugli afidi del frumento e valutazione della soglia di danno. Atti Convegno "La difesa dei cereali nell'ambito dei Prog. Fin. C.N.R." - Ancona, 10-11 dic. 1981: 137-143.

(28) SUSS, L. and COLOMBO, M., 1982. Gli Afidi dei cereali. Inf.tore Fitopat. (6/82): 7-12.

(29) SUSS, L., ANTONELLI, F. and CIAMPOLINI, M., 1983. Presenza della piralide del mais (O. nubilalis Hb.) a ciclo monovoltino nell'Italia settentrionale. L'Inform.tore Agrario, 39(24): 26327-26333.

(30) TREMBLAY, E., 1986. Entomologia applicata, vol. 2^O -pt. 2^a. Liguori Ed., Napoli, 381 pp.

(31) VEREIJKEN, P.H., 1979. Feeding and multiplication of three cereal aphid species and their effect on yield of winter wheat. Centrum voor Landbouwpublikaties en Landbouwdocumentatie, Wageningen: 1-58.

(32) WRATTEN, S.D., 1978. Effects of feeding position of the aphids Sitobion avenae and Metopolophium dirhodum on wheat yield and quality. Ann. appl. Biol., 90: 11-20.

(33) ZANGHERI, S., CIAMPOLINI, M. and SUSS, L., 1985. I gravi danni causati al mais dall'Agrotis ipsilon. L'Informatore Agrario, 40(11): 71-78.

Forecasting of cereal aphid abundance

A.F.G.Dixon, J.C.Entwistle & M.T.Howard
University of East Anglia, School of Biological Sciences, Norwich, UK

Summary

An analysis of the 11 years of population estimates for the cereal
aphids, Sitobion avenae and Metopolophium dirhodum, on wheat in
south-eastern England, plus shorter runs of data for other areas, has
revealed easily measurable factors that are correlated with flight
activity and peak population size. These can be used to give a
relatively long-term forecast of the likelihood of outbreaks and a
more accurate short-term forecast of the peak numbers. Further
improvement is largely dependent on even longer runs of field data
and on the integration of data collected in different regions.

1.   Introduction

In the absence of an accurate forecasting system for aphids that
infest wheat farmers are likely to resort to the prophylactic application
of environmentally harmful aphicides, because the aphids are sometimes
very damaging (11,15). Several attempts have been made to forecast the
abundance of aphids on cereals in summer (1,6). However, a major factor
limiting success has been the absence of a long run of population data.
     The field data used in this paper has been collected mainly from
Norfolk and Suffolk by the Cereal Aphid Group at the University of East
Anglia since 1976, with the addition of data collected by N. Carter from
Holland for 1975 to 1979 and from outside Norfolk and Suffolk for 1979 to
1981. Simulation studies of cereal aphid population dynamics revealed a
good understanding of the system (2). However, these models are in-
complete, very complex and not suitable for forecasting. An alternative
approach is to identify easily measurable factors that are reliably related
to peak aphid population size. In this paper we outline methods for
forecasting the abundance of the grain aphid, Sitobion avenae (F.), and the
rose-grain aphid, Metopolophium dirhodum (Walk.), the commonest aphids
infesting cereals in south-eastern England.

2.   The grain aphid

2.1 Long-term forecast

The difference in the numbers of S. avenae caught by the Rothamsted
Insect Survey during the autumnal and spring migrations indicates that
recently this species has tended to increase in abundance over winter and
early spring throughout much of southern England but less so in the north

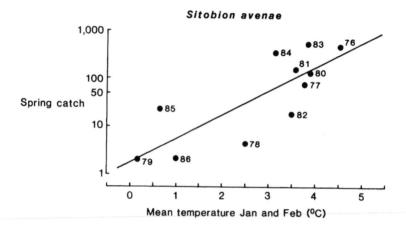

Fig. 1: The relationship between the number of S. avenae caught in the Broom's Barn suction trap up to the time of the end of flowering (G.S. 69) of the local wheat crop and the mean temperature in January and February for each year from 1976 to 1986

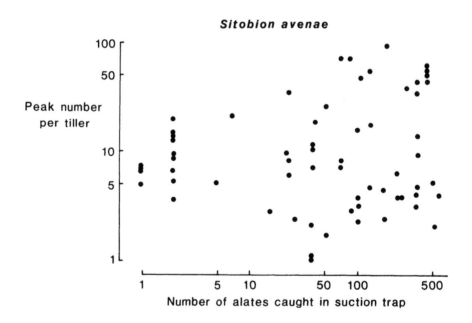

Fig. 2: The peak number of S. avenae per tiller in winter wheat crops in relation to the number of alates caught in the nearest suction trap up to the end of flowering (G.S. 69)

of the United Kingdom (22). This is associated with its mode of overwintering predominantly anholocyclic in southern England and holocyclic
in Scotland (17, 25). As the severity of winter varies and the survival
of anholocyclic clones is mainly dependent on temperature (29) one would
expect the size of the spring catch to be correlated with average winter
temperatures in southern England, and it is (Fig. 1). Thus by the end of
February it is possible to predict the size of the spring catch. Although
outbreaks are associated with large spring catches (Fig. 2),
it is a very weak association as has been previously pointed out (6). In
addition several of the cold winters of the 1950's and 1960's including the
very cold winter of 1963, were followed by outbreaks (5). Thus at
present it is not possible to make an accurate forecast early in the year
of the peak numbers of S. avenae on crops in summer. However, as an
outbreak is more likely to follow a high than a low spring catch, the
severity of the winter can be used to indicate the probability of an
outbreak, which is lower following a severe than a mild winter.

2.2 Short-term forecasting

The peak population density is positively correlated with the
population densities at the end of ear emergence (G.S. 59), mid-anthesis
(G.S. 65) and the end of anthesis (G.S. 69) (Fig. 3A). It is also
positively correlated with the observed rates of increase of the aphids
on the crop immediately before these stages (Fig. 3B). Both these
parameters have been incorporated into multiple regressions, which give
an accurate forecast of peak population density (Fig. 3C). This revealed that the surprisingly low peak densities in 1983, 1981 and 1980
outside Norfolk (6,26) were due to low rates of increase on crops during
anthesis rather than low initial populations.
The marked variability in the rate of increase (Fig. 3B) both between
and within years may be due to variations in weather (14,27), the
cultivars of cereals grown (16) or to the activity of natural enemies
(3,4,8,12). As there are considerable difficulties in assessing the
importance and abundance of the numerous species of natural enemies (6),
and in predicting the weather, the approach presented here, which
includes all these effects in the observed rate of increase, is currently
more viable (9)
This method of forecasting can be combined with a damage model
(damage relationship) similar to that of Watt et al., (28) to forecast
damage rather than peak aphid density. Although the two are related
(e.g. 24) the number of aphids present throughout the period of infestation must be taken into account in order to decide when and whether
to spray (28). This is especially important for S. avenae because
damage per aphid is very much higher during early anthesis than later
when peak densities usually occur (30).
The damage relationship and a control relationship, based on economic
considerations, have been combined to give a new forecasting system (10).
This is more reliable than the EPIPRE system (19,20,21), which tends to
overestimate the number of fields that require spraying, and the
economic threshold of rising from 5 per ear at the start of anthesis
(G.S.61) of George and Gair (11), which underestimates the number of
fields that require spraying.

35

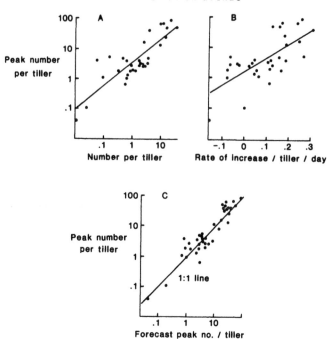

Sitobion avenae

Fig. 3: The peak number of S. avenae per tiller in relation to (A) the
number per tiller at the end of anthesis (G.S. 69), (B) the rate
of increase in mean numbers per tiller per day from mid anthesis
(G.S. 65) to G.S. 69 and (C) the forecast from the multiple
regression at the end of anthesis.

3.   The rose-grain aphid

     The only outbreak of this aphid in East Anglia in the last 15 years
occurred in 1979 following a cold winter, which delayed the development
of winter wheat and it is thought gave the aphid more time to reach outbreak levels (7).
     The Rothamsted Insect Survey suction trap catches of this aphid in
summer closely follow its changes in abundance on cereal crops. In most
years the trend in the catch each summer takes a definite form increasing
to a peak and then declining rapidly. The end of the season on the crop
is marked by most of the aphids developing into alatae, which as in
S. avenae, is determined by the maturation of the crop and population
density (13,23). For those years in which there is a definite trend in
the catch and a single peak there is an inverse relationship between the
time of the peak weekly suction trap catch and the number of day degrees
above 1°C that have accumulated up to the end of May. That is, the end of
the season on cereals for M. dirhodum comes earlier in years following
mild winters and springs in which the crop achieves an advanced state of
development before the end of May (13). The slopes of the relationships
between the suction trap catches and time for each season are also used

36

to obtain an indication of the maximum possible rate of increase.

The maximum rate of increase and the time to the end of the season can be used to predict the likelihood of an outbreak. If the weekly suction trap catch is equal to or greater than 100 (N_1) two weeks prior to the predicted end of the season then there is likely to be an outbreak (N_2) (Fig. 4). Two weeks prior to the end of the season is on average

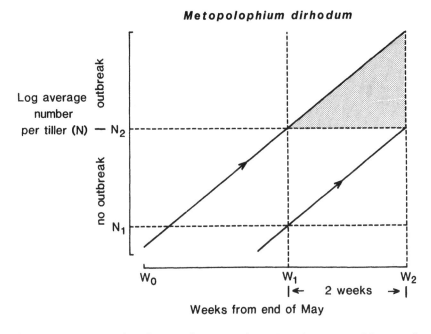

Metopolophium dirhodum

Log average number per tiller (N)

Weeks from end of May

Fig. 4: A schematic diagram illustrating that if the weekly suction trap catch of M. dirhodum alates is equal to or greater than 100 (N_1) two weeks prior (W_1) to the predicted end of the season. (W_2) then there is likely to be an outbreak (N_2) (W_1 is on average equivalent to G.S. 69)

equivalent to the end of anthesis (G.S. 69), but varies because the end of the season is not strongly associated with a particular growth stage. So in some years, for example 1980 and 1981, it can be determined before flowering that there will be no need to spray fields.

This system revealed that there was no danger of an outbreak in 7 out of the last 15 years. Of the remaining 8 potential outbreak years only one had an outbreak. In the other seven years population growth stopped before alate production curtailed further population increase. In certain cases (e.g. 1976, and some fields in 1985) the curtailment of population increase was associated with premature drying out of the leaves. However, in spite of its obvious shortcomings the system can be used to give an indication of the likelihood of an outbreak at least two weeks before damaging levels are reached.

Even weekly sampling of the buildup of the numbers of this aphid on cereal crops has not resulted in an early prediction of peak numbers, mainly because this aphid remains relatively rare on cereals until quite late in the season in most years. However, there is a high degree of association between the peak number per tiller and the number per tiller at early milky ripe (G.S. 73) (Fig. 5). If the numbers reach at least 10 per tiller by this particular growth stage the peak numbers will exceed 30 per tiller.

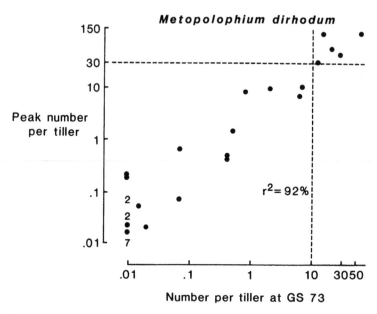

Fig. 5: The peak number of M. dirhodum per tiller in relation to the number per tiller at early milky ripe (G.S. 73)

4. Conclusions

An analysis of the 11 years of field data available for the Norwich area, plus shorter runs of data for other areas, has revealed easily measurable factors that are correlated with flight activity and peak population size. These can be used to give a relatively long-term forecast of the likelihood of outbreaks and a more accurate short-term forecast of the peak numbers in wheat fields. Although it is more profitable for farmers to use short-term forecasts than prophylactic applications of aphicides for control of S. avenae (10), further work is necessary to improve the accuracy of the short-term forecasts for both species and to test their reliability. The long-term forecasts give very approximate indications and so further work is required to reveal other patterns in data from the UK by means of analyses of the type performed by Pierre & Dedryver (18) for French data. The reliability of any long-term forecast must also be checked by continuously monitoring its accuracy. Therefore improvement in both short-term and long-term forecasting in the UK necessitates continuation of the longer runs of field data now available, and an integration of data collected in different regions. This would lead to the development of longer-term and more reliable forecasts, which would give even greater economic savings and reduce the use of aphicides. Preferably the sampling of aphid populations on cereals should be ongoing and an integral part of a research policy for cereals.

This raises an important question. Why is there no policy governing all research on cereals in the UK; the crop that accounts for 85 per cent of all arable land? The research on cereals needs clearly defined objectives that meet the requirements of the farming industry and the need to protect the environment. Such a policy would avoid the wasteful short-term expedient funding such as that which followed the 1968 cereal aphid outbreak. If there is to be a research policy and aphid population sampling is to be an integral part of that policy then further questions are raised - who in the United Kingdom should be responsible for collecting and collating the population data and what measurements should be taken? Unfortunately the large grain surpluses currently being produced in western-Europe have tended to obscure the fact that the cereal aphid problem still exists and is important. The first priority in determining how to spend money raised for research by the recently instituted levy should be to draw up and implement a research policy for cereals.

REFERENCES

1. CARTER, N. and DEWAR, A. (1981). The development of forecasting systems for cereal aphid outbreaks in Europe - pp. 170-173 (In) Kemmedahl, T. (Ed.). Proceedings of Symposia. IX International Congress of Plant Protection, Washington, D.C., U.S.A. 1979. Vol. 1. Plant Protection: fundamental aspects. pp. 1-411 Minneapolis, Burgess.
2. CARTER, N., DIXON, A.F.G. and RABBINGE, R. (1982). Cereal aphid populations: Biology, simulation and prediction. Simulation Monograph, Pudoc, Wageningen, The Netherlands. 91 pp.
3. CHAMBERS, R.J. and ADAMS, T.H.L. (1986). Quantification of the impact of hoverflies (Dipt. Syrphidae) on cereal aphids in winter wheat. II. Analysis of field populations. Journal of Applied Ecology 23:
4. CHAMBERS, R.J. and SUNDERLAND, K.D. (1983). The abundance and effectiveness of natural enemies of cereal aphids on two farms in Southern England (In) Aphid Antagonists - Proceedings. E.C. Experts' Group Portici, pp. 83-87. Ed. R. Cavalloro. Rotterdam; A.A. Balkema.
5. DEAN, G.J. (1974). The overwintering and abundance of cereal aphids. Annals of Applied Biology 76: 1-7.
6. DEWAR, A.M. and CARTER, N. (1984). Decision trees to assess the risk of cereal aphid (Hemiptera: Aphididae) outbreaks in summer in England. Bulletin of Entomological Research 74: 387-398.
7. DEWAR, A.M., WOIWOD, I. and de JANVRY, E.C. (1980). Aerial migrations of the rose-grain aphid Metopolophium dirhodum (Wlk.), over Europe in 1979. Plant Pathology 29: 101-109.
8. EDWARDS, C.A., SUNDERLAND, K.D. and GEORGE, K.S. (1979). Studies on polyphagous predators of cereal aphids. Journal of Applied Ecology 16: 811-823.
9. ENTWISTLE, J.C. and DIXON, A.F.G. (1986). Short-term forecasting of peak population density of the grain aphid Sitobion avenae on wheat. Annals of Applied Biology 109: 215-222.
10. ENTWISTLE, J.C. and DIXON, A.F.G. (1987). Short-term forecasting of wheat yield loss caused by the grain aphid Sitobion avenae in summer. Annals of Applied Biology.
11. GEORGE, K.S. and GAIR, R. (1979). Crop loss assessment on winter wheat attacked by the grain aphid, Sitobion avenae (F.) 1974-77. Plant Pathology 28: 143-149.
12. HOLMES, P.R. (1984). A field study of the predators of the grain

aphid Sitobion avenae (F.) (Hemiptera: Aphididae), in winter wheat in Britain. Bulletin of Entomological Research 74: 623-631.

13. HOWARD, M.T. (1987). The factors determining the pest status of Metopolophium dirhodum (Wlk.) the rose grain aphid. PhD Thesis, University of East Anglia.

14. JONES, M.G. (1979). Abundance of aphids on cereals from 1973 to 1977. Journal of Applied Ecology 16: 1-22.

15. KOLBE, W. (1969). Studies on the occurrence of different aphid species as the cause of cereal yield and quality losses. Bayer Pflanzenschutz Nachrichten 22: 171-204.

16. LOWE, H.J.B. and ANGUS, W.J. (1985). Grain aphid infestations of winter wheat variety trials, 1984. Annals of Applied Biology 106: 591-594.

17. NEWTON, C. (1986). Overwintering strategies of Sitobion avenae. PhD Thesis, University of East Anglia.

18. PIERRE, J.S. and DEDRYVER, C.A. (1984). Un modèle de régression multiple appliqué à la prévision des pullulations d'un puceron des céréales, Sitobion avenae F. sur blé d'hiver. Acta Oecologia/Oecologia Applicata 5(2): 153-172.

19. RABBINGE, R. and CARTER, N. (1984). Monitoring and forecasting of cereal aphids in the Netherlands: a subsystem of EPIPRE. (In) Pest and Pathogen Control: Strategic, Tactical and Policy Models. pp. 242-253. Ed. G.R. Conway. John Wiley and Sons.

20. REININK, K. (1985). Tarive Teeltbegeleidingssysteem: Bestrijding van Ziekten an Plagen Beschrjoing EPIPRE - Model. Internal Report. Proefstation voor de Akkerbouw en de Groenteelt in de Vollegrond, The Netherlands.

21. REININK, K. (1986). Experimental verification and development of EPIPRE, a supervised disease and pest management for wheat. Netherlands Journal of Plant Pathology 92: 3-14.

22. TAYLOR, L.R. (1986). Synoptic dynamics, migration and the Rothamsted Insect Survey. Journal of Animal Ecology 55: 1-38.

23. THORNBACK, N. (1983). The factors determining the abundance of Metopolophium dirhodum (Walk.) the rose grain aphid. Unpublished PhD Thesis, University of East Anglia. 233 pp.

24. VEREIJKEN, P.H. (1979). Feeding and multiplication of three cereal aphid species and their effect on yield of winter wheat. Agricultural Research Reports (Veralagen van Landbouwkundige Onderzoekingen) 888. Pudoc Wageningen, The Netherlands. 58 pp.

25. WALTERS, K.F.A. and DEWAR, A.M. (1986). Overwintering strategy and the timing of the spring migration of the cereal aphids Sitobion avenae and Sitobion fragariae. Journal of Applied Ecology 23:

26. WATSON, S.J. (1983). Effects of weather on the numbers of cereal aphids. Unpublished PhD thesis, University of East Anglia. 80 pp.

27. WATSON, S.J. and CARTER, N. (1983). Weather and modelling cereal aphid populations in Norfolk (U.K.). Bulletin of the European Plant Protection Organisation 13: 223-227.

28. WATT, A.D., VICKERMAN, G.P. and WRATTEN, S.D. (1984). The effect of the grain aphid Sitobion avenae (F.) on winter wheat in England: an analysis of the economics of control practice and forecasting systems. Crop Protection 3: 209-222.

29. WILLAMS, C.T. (1980). Low temperature mortality of cereal aphids. I.O.B.C./W.P.R.S. Bulletin 111/4: 63-66.

30. WRATTEN, S.D., LEE, G. and STEVENS, D.J. (1979). Duration of cereal aphid populations and the effects on wheat yield and quality. Proceedings 1979 British Crop Protection Conference - Pests and Disease 1: 1-8.

Potential interactions between varietal resistance and natural enemies in the control of cereal aphids

K.D.Sunderland, R.J.Chambers & O.C.R.Carter
Institute of Horticultural Research, Littlehampton, UK, and Coventry Lanchester Polytechnic, West Midlands, UK

Summary
A review of recent literature shows that partial resistance in cereals occurs frequently and can be due to any combination of four components (tolerance, developmental precocity, antixenosis and antibiosis). Known mechanisms of resistance relate to the gross structure, surface structure, fine structure and phloem sap chemistry of plants. Potential interactions are suggested (and evidence presented, where available) between these mechanisms and the role of predators, parasites and pathogens in suppressing aphid populations in the field. The relevance of these findings to plant breeding, variety screening and the development of resistance-inducing chemicals is discussed.

1. Introduction

A considerable body of information is now available concerning the incidence, components and mechanisms of partial resistance to cereal aphids. There is an equally large information base for the incidence, mode of attack and effectiveness of aphid antagonists. Little is known, however, of the way these factors interact in the field, despite the often-quoted speculation that such interactions could be of great value in integrated pest control. The aim of this paper is to review the literature on partial resistance and identify potential routes of interaction with natural enemies. It is hoped that this information will be useful during the design stage of future field experiments aimed at developing integrated control.

After brief reviews of partial resistance and natural enemies, attention is given to the possible interactions between these factors and the likely applications of this information.

2.1 Incidence of partial resistance

It is unlikely that agronomically useful varieties of cereal will be found that are completely resistant to aphids. For example, in a screen of 4238 varieties of winter wheat, none were completely uninfested by Rhopalosiphum padi (L.)(52). Partial resistance, however, is fairly common. We screened 14 varieties of winter wheat, in the glasshouse, for susceptibility to Sitobion avenae (F.); there was considerable variation in their ranking between trials, but Galahad was always the least infested and Rapier and Moulin were usually infested to a lesser extent than many of the other varieties (Table 1). There is no

Table 1. Mean number (log.(n + 1) and 95% CL) of Sitobion avenae (F.) per plant on a range of winter wheat varieties in four glasshouse trials at different aphid densities. Walkabout method (60), stem extension to booting (Zadoks d.s. 30-45).

Variety	Trial 1		Trial 2		Trial 3		Trial 4	
	Mean	95% CL	Mean	95% CL	Mean	95% Cl	Mean	95% Cl
GALAHAD	0.54	0.33-0.76	0.66	0.42-0.91	1.19	1.02-1.35	2.02	1.89-2.15
RAPIER	1.00	0.77-1.20	0.75	0.47-1.03	1.34	1.15-1.52	2.13	2.02-2.25
MOULIN	0.78	0.54-1.00	1.11	0.83-1.39	1.37	1.21-1.53	2.18	2.01-2.36
SLEJPNER	-		-		1.31	1.14-1.48	-	
BROCK	0.80	0.59-1.00	1.31	1.00-1.62	-		2.16	2.00-2.33
MISSION	0.71	0.52-0.89	1.44	1.28-1.60	-		2.22	2.05-2.39
NORMAN	0.87	0.66-1.08	1.16	0.84-1.47	-		2.31	2.23-2.39
FENMAN	1.02	0.80-1.20	1.02	0.67-1.37	-		2.33	2.27-2.40
HUNTSMAN	0.87	0.60-1.10	1.44	1.14-1.74	-		2.33	2.22-2.43
AVALON	0.89	0.64-1.15	-		-		2.38	2.22-2.53
MERCIA	-		-		1.56	1.41-1.72	-	
BRIMSTONE	0.98	0.80-1.10	1.23	0.92-1.54	1.50	1.36-1.64	2.45	2.38-2.52
AQUILA	1.10	0.91-1.28	1.41	1.26-1.57	1.77	1.67-1.87	2.42	2.31-2.52
LONGBOW	0.98	0.78-1.17	1.68	1.54-1.83	1.58	1.44-1.72	2.43	2.33-2.53

Table 2. Some recent examples of partial resistance in European cereals

TRIAL	AUTHORS	CEREAL	APHID	VARIETIES	
				Number tested	% partially resistant
G	68	O	SA	9	11
G	58	WW	SA	20	30
G	89	WW	SA	12	25
G	57	WW	SA	16	25
G	60	SW	SA	91	29
G	57	SW	SA	8	25
G	64	SW	SA	33	15
FP	12	O	RP	3	(33)
FP	13	WW	SA	22	32
FP	61	WW	SA	14	29
FP	65	WW	SA	9	22
FP	63	WW	SA	15	20
FP	24	WW	SA	10	10
FP	64	SW	SA	6	33
FP	1	SW	SA	8	13
F	12	O	RP	3	(33)
F	18	WW	SA	2	(50)
F	89	WW	SA	2	(50)
F	20	WW	SA,RP,MD	15	7
F	23	WW	SA	17	6

G = Glasshouse, FP = Field plots, F = Fields, O = Oats, WW = Winter Wheat, SW = Spring wheat, SA = Sitobion avenae (F.), RP = Rhopalosiphum padi (L.), MD = Metopolophium dirhodum (Wlk.)

objective definition of partial resistance. In our glasshouse trials
the difference in mean number of aphids per plant between Galahad
(partially resistant) and Longbow (susceptible), as a percentage of the
numbers on Longbow, varied from 59% to 92%. We also found 3 to 9 times
more S. avenae on Aquila than on Rapier in the field at the aphid peak
(89). Similarly useful levels of resistance have been recorded in the
literature during the last decade. Table 2 lists some examples of these
observations; in deciding which varieties to classify as partially
resistant we have been guided by the authors' comments (often guided, in
turn, by significance tests). In glasshouse trials and small-plot
trials in the field, about 25% of the varieties tested showed partial
resistance. Data are still too sparse where fields are the replicated
units.

2.2 Factors affecting the ranking of varieties

Differences between studies in the ranking of varieties for degree
of susceptibility to aphids can come about as the result of a large
number of factors, broadly classified into three categories; ie related
to the aphid, the plant and the environment (including cultural
practise). A variety that shows some resistance to one species of aphid
will not necessarily show the same amount of resistance to others (2,
44, 55, 57); there are also often (57, 62), but not always (68),
differences between clones in this respect. Sometimes, dramatic
differences in susceptibility are related to stable differences between
aphid clones (ie to biotypes), as has been described for the greenbug,
Schizaphis graminum (Rondani) (71, 96). Resistance rankings may vary
with aphid density (12), but not in every case (2). Rankings can be
affected by previous host for one aphid generation (44), but this is not
a problem in screens of longer duration (60). There is usually a better
differentiation between varieties if they have been vernalised (44, 56),
but this is not the case for some varieties (19). Before ear emergence,
there is a better differentiation (but no difference in ranking) on
older plants (59, 60); differences in ranking can occur, however, when
the comparison is made before and after ear emergence (2). The
fertilizer regime does not affect rankings in glasshouse screens (56,
60), which mainly use apterous aphids, but the relative attractiveness
of varieties to alatae in the field is affected by the level of nitrogen
(38). Plant growth regulators can reduce the susceptibility of some
varieties (28). Varieties that show some resistance to aphids are
sometimes susceptible to plant pathogens; causal links that have been
suggested are (i) fungal attack changes leaf pH (72) or increases rate
of leaf senescence (20) to the detriment of the aphid (ii) some species
of endophytic fungi can deter cereal aphids from feeding (42).

2.3 Components of resistance

The principal components of resistance are tolerance, pattern of
plant development, antixenosis and antibiosis.
Tolerant varieties are damaged less per unit aphid burden than
susceptible varieties. Tolerance of a range of cereal varieties to
greenbug attack has been screened in the glasshouse using a subjective
scale of damage rating (15, 83, 95, 96). There do not appear to have
been studies of the yield and quality aspects of tolerance in the field.
Early-earing varieties often have fewer aphids at the population

43

peak than varieties which develop less rapidly (11, 20). This form of
resistance is particularly relevant in areas where infestations are
normally initiated by a spring migration. In such areas early-maturing
varieties offer less time for aphid population increase at suitable host
developmental stages (3).

Antixenosis (= non-preference) refers to a variety having some
characteristic(s) resulting in reduced settling (or increased departure
rate) of alatae or increased movement of apterous aphids. It has been
measured in laboratory and glasshouse trials by recording the number of
alatae settling (49, 82) or taking off (21) from varieties in pots in a
flight chamber. Alternative methods include releasing aphids from the
centre of a ring of plants of different varieties and recording the
number settled on each (83, 95, 96) or counting the number of aphids
that settle on excised leaves held in agar in a perspex box (2, 59) or
on moist filter paper in petri dishes (42). Restlessness has been
assessed by regular recording of the position of apterous aphids on flag
leaves (2, 49); this is known to vary between clones (59). Antixenosis
towards immigrant alatae in the field could be estimated from the
abundance of alatae on plants during periods when none are being
produced in the field (ie alatiform fourth instars absent). Use of
alate abundance to measure antixenosis towards immigrant alatae at other
times (1, 24) rests on the assumption that immigrants remain on the
plants for much longer than alatae produced in the field; the evidence
for this appears to be slight.

The term antibiosis, as used in this paper, refers to any feature of
a variety which results in a reduced rate of growth, development or
fecundity of aphids that have settled and are engaged in active,
sustained feeding. This is apparently a common component of resistance
in cereals (but see also section 2.4). It has been detected in
laboratory and glasshouse trials by caging individual aphids on test
plants and recording adult weight or fecundity (2, 22, 59, 83, 94, 95,
96) or intrinsic rate of increase (1, 14, 82). It has been measured in
the field by recording adult weight and the number of large embryos in
dissected aphids (50, 86) or by caging aphids on plants and recording
fecundity and mortality (51).

Analyses of the results of some recent glasshouse studies show
that, although correlations are sometimes found between some of the
components of resistance, this is by no means always the case (Table 3).
This has important implications for the development of varieties with
stable resistance (see section 5)

2.4 Mechanisms of resistance

Resistance can be mediated through various aspects of the gross
structure, surface structure, fine structure and phloem sap chemistry of
plants affecting alatae and/or apterous aphids.

There is no unequivocal evidence that aphids are attracted by plant
odour in host-plant finding (25), and, although cereal aphid alatae
react to colours (5, 78) and may be attracted to yellowing cereals
(7) or deterred by the UV reflected from some varieties (67), this is
not yet a proven basis of resistance. Plant size can influence the
initial rate of landing of R. padi (5), but accumulation of aphids on
hosts, which is mainly due to rate of departure being less than rate of
arrival (25), is not known to be related to plant size. Varietal
variation in resistance can, however, be related to such aspects of ear

Table 3. Correlations between antibiosis (Ab), antixenosis (Ax) and
tolerance (T) in glasshouse trials.

AUTHORS	CEREALS	APHID	(BIOTYPES)	CORRELATION	df	r	P
83	R,W,T,B	SG	E	Ab,Ax	10	0.71	0.01
"	"	"	"	Ab,T	10	0.34	NS
"	"	"	"	Ax,T	10	0.36	NS
"	R,W,T	"	C,E	Ab,Ax	10	0.40	NS
"	"	"	"	Ab,T	10	0.80	0.01
"	"	"	"	Ax,T	10	0.23	NS
95	W,T	"	"	Ab,Ax	7	0.74	0.05
"	"	"	"	Ab,T	7	0.71	0.05
"	"	"	"	Ax,T	8	0.63	0.05
96	B	"	C	Ab,Ax	15	0.52	0.05
"	"	"	"	Ab,T	15	0.68	0.01
"	"	"	"	Ax,T	15	0.55	0.02
"	"	"	E	Ab,Ax	15	0.09	NS
"	"	"	"	Ab,T	15	0.58	0.02
"	"	"	"	Ax,T	15	0.37	NS
59	W	SA	–	Ab,Ax	5	0.14	NS

R = Rye, W = Wheat, T = Triticale, B = Barley, SG = Schizaphis graminum
(Rondani), SA = Sitobion avenae (F.), df = number of varieties –2

Table 4. Predator species: vertical distribution and aphid predation in
cereals. Number of species known to eat cereal aphids in the
field in parentheses.

GROUP	NUMBER OF SPECIES ON:		
	GROUND ONLY	GROUND & PLANT	PLANT ONLY
Araneae	61 (10)	38 (19)	10 (1)
Acari	0	2? (2)	0
Opiliones	7 (3)	1 (0)	0
Chilopoda	4 (2)	0	0
Neuroptera larvae	0	2 (1)	0
Formicidae	1? (1)	0	0
Dermaptera	0	1 (1)	0
Hemiptera	0	5 (4)	0
Diptera adults	0	5 (4)	0
Diptera larvae	0	14 (2+ ?)	0
Carabidae	51 (16)	6 (4)	0
Staphylinidae	25 (5)	12 (7)	0
Cantharidae	0	8 (2)	0
Coccinellidae	0	5 (5)	0
Coleoptera larvae	18 (7)	3 (2)	0
Total	167 (44)	102 (65)	10 (1)
% of Grand Total	59.9 (40.0)	36.5 (59.1)	3.6 (0.9)

Presence on plant from sweeping (4 years, 8 sites, 100 day and 28 night
sampling occasions). Presence on ground from quadrats (6 years, 9 sites
136 sample occasions) and pitfalls (6 years, 42 sites, 667 occasions)

structure as spikelet density (29) and presence of awns (1, 2, 94).
Aphid numbers are lower on awned than bare ears due to a combination of
antibiosis and greater dislodgement (4).

There is some evidence that non-glaucous varieties (lacking a waxy
bloom) are more resistant than glaucous varieties and this could operate
by deterring the initial landing of alatae (67, 85) or by repelling
aphids that have landed and tested the chemical composition of the wax
(47, 48). On non-graminaceous hosts some species of aphid prefer a
glaucous surface and this is related to their having a tarsal structure
that enables them to grip (45, 88); this aspect does not appear to have
been investigated for cereal aphids. Pubescence can be a barrier to
aphid movement (45) and feeding (25). R. padi preferred to settle on
glabrous wheat and had a higher fecundity on glabrous than on pubescent
wheat (79). Resistance of cereals to aphids in the field is known to
increase with increase in the density of silica-containing hairs
(39, 46), and increase in the amount of silica in epidermal cells (77)
and the pectin content of the intercellular matrix (28, 73). In
susceptible wheat varieties, cells adjacent to the stylet path are
damaged by salivary enzymes, but this does not always occur in resistant
varieties. It is possible that some salivary enzymes play a role in
preventing damaged phloem from forming the callose which stops sap
conduction and that resistant varieties inactivate these enzymes (6).

Constituents of the phloem sap that could influence resistance are
levels of soluble nitrogen and amino acids and the presence of
phagostimulants and repellents. Susceptibility of plants to aphids is
often positively correlated with the level of soluble nitrogen and amino
acids (25, 27). This was found to be the case in a few studies of
cereal aphids (29, 49, 73) but, in general, there is little information
for cereals. This may change with the recent development of electrical
recording of aphid feeding combined with stylet microcautery (49, 70)
which enables samples of phloem sap to be taken, analysed and related to
aphid performance. Various toxins in the leaves of cereals, such as
DIMBOA (8, 9, 14) and gramine (99, 100) are known to reduce survival,
feeding and increase rate of aphids, but the levels of these compounds
decline after the seedling stage. Phenols have similar effects on
cereal aphids (53, 73, 91) and their levels in cereal leaves increase in
response to attack by aphids and other herbivores (52). When the
feeding behaviour of aphids on cereals is observed and electronically
monitored it is often found that they spend much more time probing and
withdrawing stylets and less time feeding from the phloem on resistant
as compared to susceptible cultivars (16, 28, 71). They may move very
little between probes but nevertheless be semi-starved as a result of
failure to tap the phloem for long periods. Semi-starved aphids are
known to have reduced growth rates and fecundity (10). Thus much of the
resistance attributed to antibiosis could, more accurately, be
"micro-antixenosis" and this has implications for the interaction with
mortality factors in the field (see section 4).

3. Natural enemies

Cereal aphids are attacked by predators, parasites and pathogenic
fungi. The vertical distribution of these natural enemies is of
importance in relation to the antixenotic component of resistance (see
section 4). Aphids feeding on the crop are accessible to parasites and
pathogens, but not to all predators. Table 4 shows the vertical

distribution of predator species in cereals in a fifty square kilometre study area in West Sussex, UK. Two thirds of the total number of species were never recorded on the plant. In contrast, two thirds of those species known to eat cereal aphids in the field (and these include the majority of numerically abundant species) were found on the plant in addition to the ground. Nevertheless, it is clear that a considerable proportion of the predatory fauna will not encounter aphids, unless the aphids descend to the ground.

4. Plant-aphid-natural enemy interactions

Where varieties exhibit a high degree of resistance there is a strong selection pressure for the development of aphid biotypes which are no longer affected by this resistance. The appearance of new biotypes of greenbug, several times in recent years, has nullified years of research and resulted in the withdrawal of previously successful commercial "resistant" varieties (71, 83). It was suggested, over two decades ago, that partial resistance in combination with natural enemies could maintain aphid populations below the economic threshold whilst presenting less selection pressure for the development of resistance-breaking biotypes (33). This hypothesis received some support from glasshouse experiments with parasites (32, 49, 80, 84), but few field investigations have been made in cereals. Some evidence was found, in field studies, for an increased effectiveness of natural enemies on partially resistant varieties, relating to coccinellids on sugar beet (54), parasites and syrphids on rape (30), spiders on rice (43) and coccinelllids and syrphids on winter wheat (18). Interactions with fungal pathogens do not appear to have been studied, but it is possible that reduced infection rates will be found for rapidly-developing aphids on susceptible varieties because moulting can result in avoidance of infection, especially by strains of fungi that are slow to penetrate the aphid cuticle (41). When cereal plots are regularly sprayed with carbaryl, aphid populations increase more rapidly than on control plots, and this has been attributed to a greater mortality of natural enemies than aphids because the former are more active than the latter and make more contact with the insecticide (66). Carbaryl-enhancement of aphid populations was greater on semi-resistant than on susceptible spring wheat (1), suggesting that there is some synergism between varietal resistance and natural enemies.

There are many potential interactions between antixenosis and natural enemies. Antixenosis towards alatae, even if it does not result in emigration from the field, is likely to increase the amount of time alates spend flying amongst plants and hence their availability to predatory Diptera and web-making spiders. More alatae were caught on sticky traps in semi-resistant than susceptible spring wheat (4) and such traps provide a realistic estimate of aphid inputs into the horizontal hammock webs of linyphiid spiders (90). Intensive field observations have established that there is much aphid movement between ears and leaves (40); if the rate of this activity is increased by antixenosis, more aphids will be available to parasites, such as Aphidius rhopalosiphi (DeStefani-Perez), which concentrate their searching activity on the leaves (34). There is evidence that a variable, and sometimes very high, proportion of the aphid population move over the ground (36, 90, 93). This proportion is likely to increase in response to antixenosis and make aphids more available to

47

ground-based predators (Table 4) and perhaps more vulnerable to abiotic mortality agents, such as heavy rain. An increased proportion of aphids on the ground could result from (i) increased voluntary movement, in response, for example, to feeding deterrents such as DIMBOA, gramine and phenols or (ii) increased dislodgement; for example, "micro-antixenosis" (section 2.4) would result in aphids having their stylets withdrawn from the plant more frequently and hence render them more vulnerable to dislodgement in windy weather. More aphids are also dislodged from the ears of awned, as compared with awnless, wheat (4). Unfortunately, there are few field studies of these potentially important aspects of resistance. Although antixenosis in combination with natural enemies is likely to be of benefit to the farmer, antixenosis alone could be detrimental by increasing virus spread in autumn and spring.

Various aspects of plant structure associated with resistance could contribute to aphid control through interactions with natural enemies. Aphid colony destruction was greater on short-strawed than on long-strawed varieties of wheat during a period of windy weather and it was suggested that this may have been due to more predators being dislodged by wind from the long-strawed varieties (40). Semi-dwarf varieties would also be easier for natural enemies to search. In our glasshouse trials, where winter wheat varieties were grown up to booting under standardised conditions, the mean growth stage of the principal shoot was inversely related to the mean number of shoots per plant (8 varieties, $r = 0.84$, $P = 0.01$); ie the plant appears to put its resources either into rapid development of a few shoots or slower development of many shoots. Aphids distributed themselves amongst the available shoots. If this occurs in the field, natural enemies would have to search a much larger area to locate an aphid on some varieties than on others. Ears of two-row barley varieties were less infested with aphids than ears of six-row varieties and this could be because aphids on ears of the former were more accessible to large predators, such as coccinellids (69). The surface structure of plants (degree of hairiness or waxiness) is known to influence the searching ability of coccinellid larvae (81) and many other predators and parasites (75). A plant surface that allows easy movement of arthropods will be conducive not only to increased predation but also to increased disturbance of aphids (resulting, for example, from release of alarm pheromone at the approach of a coccinellid) and in this way the plant surface can have an indirect antixenotic effect. In fields split-sown with the glaucous winter wheat, Aquila, and the non- glaucous winter wheat, Rapier, we swept more than twice as many individuals of the aphidophagous coccinellid, Propylea 14-punctata (L.), from Rapier, even though the aphid density was 3 - 9 times greater on Aquila. Although more than one explanation can be offered for this result (for example, humidity, see below), the possibility that this coccinellid has difficulty gripping, and is easily dislodged from, the slippy surface of Aquila, cannot be dismissed. In contrast, the carabid, Demetrias atricapillus (L.) and staphylinids of the genus Tachyporus, are found very frequently on cereal plants (92) and possess adhesive climbing setae on their tarsi, enabling them to grip smooth surfaces (87). There is likely to be a considerable range of climbing ability amongst the many species of cereal aphid predator (Table 4). A study of the incidence of these predators on a range of cereal varieties, in relation to the surface structure of the plants and the modes of climbing and associated structures of the predators, would be rewarding. Pubescence, waxiness and the growth form of plants also affects the humidity below and around them (74), which will, in turn, affect the spread of Entomophthoraceous fungi (98) and probably also the activity of some predators, such as syrphid larvae, earwigs and Tachyporus.

Plant-herbivore-natural enemy interactions are under-researched in cereals. Several categories of interaction, additional to those already discussed, have been found to operate in other crops and could also occur in cereals. For example, various characteristics of plant structure and odour can directly attract or repel predators and parasites (32, 37, 75) and some predators require simultaneous reception of two volatiles, one from the plant (synomone) and one from the prey (kairomone) to stimulate search of a habitat (37). Although the interactions can undoubtedly be complex (for example, the effects of antibiosis in soybeans have been detected through four trophic levels (76)), research in this area is likely to be extremely cost-effective because the combination of natural enemies and partial resistance shows great promise as a powerful and environmentally acceptable method of pest suppression.

5. Applications

A knowledge of which plant resistance characteristics interact with natural enemies in the field to promote aphid control would be very useful at three levels of application; (i) long-term plant breeding programmes (ii) screening of existing varieties (iii) development of non-insecticidal sprays which either induce resistance in the plant or mimic the effects of resistance. Some of the useful characteristics mentioned in previous sections could feasibly be incorporated into mass screening programmes, either at the plant breeding or variety recommendation stages. Examples are; non-glaucousness associated with cuticular waxes deficient in diketones (67), assayable toxins, such as DIMBOA (14), developmental precocity (3) and presence of awns (4). It is clear from experience with resistance-breaking greenbug biotypes (83, 95) that it would be highly desirable to produce varieties with multiple resistance requiring the involvement of a wide range of genes. Fortunately, it appears that the various components of resistance can have separate genetic bases (section 2.3, Table 3)(59). Some guidance on the likely effects on aphid populations of various combinations of resistance components can be gained from population simulation models (17, 97) and these will become increasingly valuable when information can be incorporated on interactions with natural enemies in the field. There is evidence that application of some chemicals, such as silicates (39, 46, 77) and plant growth regulators (28) can affect plant structure and increase the resistance of susceptible varieties, whilst behaviour-modifying chemicals, such as (E)-β-farnesene (35) mimic antixenotic effects; full exploitation of these chemicals will require a knowledge of interactions with natural enemies. Whilst much remains to be learnt about the interactions between partial resistance and natural enemies in cereals, there are encouraging first indications from studies on other crops (26, 31) that these measures can complement low-dose applications of selective insecticides in the integrated control of aphid pests.

Acknowledgments

We are grateful to Mr Fenwick (National Institute of Agricultural Botany) for advice and the supply of seeds and to the farmers who cooperated with field trials. The work was financed by the Ministry of Agriculture, Fisheries & Food through the Agriculture & Food Research Council and by an EEC grant.

REFERENCES

1. ACREMAN, T.M. (1984). The contribution of resistance to cereal aphid control. Proceedings of BCPC 'Pests & Diseases' 1984 Conference, Brighton, UK, 1, 31–36.
2. ACREMAN, T.M. (1985). Resistance to cereal aphids in wheat. PhD thesis, University of East Anglia, UK.
3. ACREMAN, T.M. & DIXON, A.F.G. (1985). Developmental patterns in wheat and resistance to cereal aphids. Crop Protection, 4, 322–328.
4. ACREMAN, T.M. & DIXON, A.F.G. (1986). The role of awns in the resistance of cereals to the grain aphid, Sitobion avenae. Annals of Applied Biology, 109, 375–381.
5. ÅHMAN, I., WEIBULL, J. & PETTERSSON, J. (1985). The role of plant size and plant density for host finding in Rhopalosiphum padi (Hemiptera: Aphididae). Swedish Journal of Agricultural Research, 15, 19–24.
6. AL-MOUSAWI, A.H., RICHARDSON, P.E. & BURTON, R.L. (1983). Ultrastructural studies of greenbug (Hemiptera: Aphididae) feeding damage to susceptible and resistant wheat cultivars. Annals of the Entomological Society of America, 76, 964–971.
7. AJAYI, O. & DEWAR, A.M. (1983). The effect of barley yellow dwarf virus on field populations of the cereal aphids Sitobion avenae and Metopolophium dirhodum. Annals of Applied Biology, 103, 1–11.
8. ARGANDOÑA, V.H., CORCUERA, L.J., NIEMEYER, H.M. & CAMPBELL, B.C. (1983). Toxicity and feeding deterrency of hydroxamic acids from Gramineae in synthetic diets against the greenbug, Schizaphis graminum. Entomologia experimentalis et applicata, 34, 134–138.
9. ARGANDOÑA, V.H., LUZA, J.G., NIEMEYER, H.M. & CORCUERA, L.J. (1980). Role of hydroxamic acids in the resistance of cereals to aphids. Phytochemistry, 19, 1665–1668.
10. AUCLAIR, J.L. & CARTIER, J.J. (1960). Effets comparés de jeunes intermittents et de périodes équivalents de subsistance sur des variétés résistantes ou sensibles de pois, Pisum sativum L. sur la croissance, la reproduction et l'excrétion du puceron du pois, Acyrthosiphon pisum (Harr.) (Homoptères: Aphidides). Entomologia experimentalis et applicata, 3, 315–326.
11. BARABAS, I. & BENOVSKY, J. (1986). The effect of winter wheat cultivars on the occurrence of the English grain aphid (Sitobion avenae). Sb Uvtiz (Ustav Vedeckotech inf Zemed) Ochr. Rostl. 21, 195–199.
12. BASEDOW, T. (1985). Studies on the effect of partial host plant resistance on the population dynamics of the cereal aphids. Bulletin SROP/WPRS VIII/3, 120–122.
13. BASEDOW, T. (1985). Ergebnisse vierjähriger Freilanduntersuchungen zur Anfälligkeit verschiedener Winterweizensorten gegenüber der Grossen Getreideblattlaus, Macrosiphum avenae F. (Hom., Aphididae).

Gesunde Pflanzen, 37, 252–256.

14. BOHIDAR, K., WRATTEN, S.D. & NIEMEYER, H.M. (1986). Effects of hydroxamic acids on the resistance of wheat to the aphid Sitobion avenae. Annals of Applied Biology, 109, 193–198.

15. BURTON, R.L., SIMON, D.D., STARKS, K.J. & MORRISON, R.D. (1985). Seasonal damage by greenbugs (Schizaphis graminum) (Homoptera: Aphididae) to a resistant and a susceptible variety of wheat. Journal of Economic Entomology, 78, 395–401.

16. CAMPBELL B.C., McCLEAN, D.L., KINSEY, M.G., JONES, K.C. & DREYER, D.L. (1982). Probing behaviour of the greenbug (Schizaphis graminum, Biotype C) on resistant and susceptible varieties of sorghum. Entomologia experimentalis et applicata, 31, 140–146.

17. CARTER, N. & DIXON, A.F.G. (1981). The use of insect population simulation models in breeding for resistance. Bulletin SROP/WPRS IV/1, 21–24.

18. CHAMBERS, R.J., SUNDERLAND, K.D., STACEY, D.L. & WYATT, I.J. (1984). Aphid-specific predators and cereal aphids. Annual Report of the Glasshouse Crops Research Institute for 1983, 86–91.

19. CHEN, B.H., FOSTER, J.E. & OHM, H.W. (1984). Effect of wheat vernalisation on Rhopalosiphum padi survival. Crop Science, 23, 1125–1127.

20. DEDRYVER, C.A. & DIPIETRO, J.P. (1986). Biology of cereal aphids in the West of France: VI. Comparative study of the development of field populations of Sitobion avenae, Metopolophium dirhodum and Rhopalosiphum padi, on different winter wheat (Triticum aestivum) cultivars. Agronomie, 6, 75–84.

21. DENT, D.R. (1984). Components of host resistance to the grass aphid Metopolophium festucae (Theob.). Bulletin SROP/WPRS VII/4, 13–14.

22. DENT, D.R. & WRATTEN, S.D. (1986). The host-plant relationships of apterous virginoparae of the grass aphid Metopolophium festucae cerealium. Annals of Applied Biology, 108, 567–576.

23. DEWAR, A.M. (1984). Factors affecting cereal aphids in fields monitored by RISCAMS in 1983. Proceedings of the BCPC 'Pests and Diseases' 1984 Conference, Brighton, UK, 1, 25–30.

24. DEWAR, A.M., CARTER, N. & POWELL, W. (1985). Assessment of resistance of commercial varieties of winter wheat to cereal aphids. Tests of Agrochemicals and Cultivars No. 6 (Annals of Applied Biology 106, Supplement, 168–169.

25. DIXON, A.F.G. (1985). Aphid Ecology. Blackie, Glasgow & London, 157 pp.

26. DODD, G.D. (1973). Integrated control of the cabbage aphid (Brevicoryne brassicae (L.)). PhD thesis, University of Reading, UK.

27. DODD, G.D. & VAN EMDEN, H.F. (1979). Shifts in host plant resistance to the cabbage aphid (Brevicoryne brassicae) exhibited by Brussels sprouts plants. Annals of Applied Biology, 91, 251–262.

28. DREYER, D.L., CAMPBELL, B.C. & JONES, K.C. (1984). Effect of bioregulator-treated sorghum on greenbug (Schizaphis graminum) fecundity and feeding behaviour: implications for host-plant resistance. Phytochemistry, 23, 1593–1596.

29. DVORYANKINA, V.A. & DVORYANKIN, E.A. (1983). Resistance of wheat to Sitobion avenae. Selektsiya i Semenovodstvo USSR, 2, 21–22.

30. DUNN, J.A. & KEMPTON, D.P.H. (1969). Resistance of rape (Brassica napus) to attack by the cabbage aphid (Brevicoryne brassicae L.). Annals of Applied Biology, 64, 203–212.

31. van EMDEN, H.F. (1984). The anatomy of a pest management programme. In 'Statistical Mathematical Methods in Population Dynamics and Pest Control', Ed. by R. Cavalloro, Proceedings of E.C. Experts' Group, Parma, 1983, A.A. Balkema, 125-135.

32. van EMDEN, H.F. (1986). The interaction of plant resistance and natural enemies: Effects on populations of sucking insects. In 'Interactions of Plant Resistance and Parasitoids and Predators of Insects'. Ed. by D.J. Boethel and R.D. Eikenbary, Ellis Horwood, Chichester, UK, pp 138-150.

33. van EMDEN, H.F. & WEARING, C.H. (1965). The role of the aphid host plant in delaying economic damage levels in crops. Annals of Applied Biology, 56, 323-334.

34. GARDNER, S.M. & DIXON, A.F.G. (1985). Plant structure and the foraging success of Aphidius rhopalosiphi (Hymenoptera: Aphididae). Ecological Entomology, 10, 171-179.

35. GRIFFITHS, D. & PICKETT, J. (1986). Electrostatic sprayers for behaviour- controlling chemicals. In 'Science, Sprays and Sprayers'. Ed. by J. Hardcastle, Agricultural and Food Research Council, London, pp 18-19.

36. GRIFFITHS, E. (1983). The feeding ecology of the carabid beetle, Agonum dorsale, in cereal crops. PhD. thesis, University of Southampton, UK.

37. HAGEN, K.S. (1986). Ecosystem analysis: Plant cultivars (HPR), entomophagous species and food supplements. In 'Interactions of Plant Resistance and Parasitoids and Predators of Insects'. Ed. by D.J. Boethel and R.D. Eikenbary, Ellis Horwood, Chichester, UK, pp 151-197.

38. HANISCH, H.C. (1980). The influence of increasing amounts of nitrogen applied to wheat on the reproduction of cereal aphids. Zeitschrift für Pflanzenkrankheiten und Pflanzenschutz, 87, 546-556.

39. HANISCH, H.C. (1981). Zum Einfluss von Natronwasserglas auf die Populationsentwicklung von Blattläusen an Weizen mit unterschiedlich hoher Stickstoffdungung. Zeitschrift für angewandte Entomologie, 91, 138-149.

40. HOLMES, P.R. (1983). A field study of the ecology of the grain aphid, Sitobion avenae, and its predators. PhD thesis, Cranfield Institute of Technology, UK.

41. JACKSON, C.W., HEALE, J.B. & HALL, R.A. (1985). Traits associated with virulence to the aphid Macrosiphoniella sanborni in eighteen isolates of Verticillium lecanii. Annals of Applied Biology, 106, 39-48.

42. JOHNSON, M.C., DAHLMAN, D.L., SIEGEL, M.R., BUSH, L.P., LATCH, G.C.M., POTTER, D.A. & VARNEY, D.R. (1985). Insect feeding deterrents in endophyte-infected tall fescue (Festuca arundinacea). Applied and Environmental Microbiology, 49, 568-571.

43. KANEDA, C. (1986). Interaction between resistant rice cultivars and natural enemies in relation to the population growth of the brown planthopper. In 'Interactions of Plant Resistance and Parasitoids and Predators of Insects'. Ed. by D.J. Boethel and R.D. Eikenbary, Ellis Horwood, Chichester, UK, pp 117-124.

44. KAY, D.J., WRATTEN, S.D. & STOKES, S. (1981). Effects of vernalisation and aphid culture history on the relative susceptibilities of wheat cultivars to aphids. Annals of Applied Biology, 99, 71-75.

45. KENNEDY, C.E.J. (1986). Attachment may be a basis for specialization in oak aphids. Ecological Entomology, 11, 291–300.
46. KLEBER, U. (1983). Erhöhung der Pflanzenresistenz durch Blattapplikation kieselsäurehaltiger Präparate. Mitteilungen der Deutschen Gesellschaft für allgemeine und angewandte Entomologie, 4, 150–152.
47. KLINGAUF, F. (1972). Die Bedeutung von peripher vorliegenden Pflanzensubstanzen für die Wirtswahl von phloemsaugenden Blattläusen (Aphididae). Zeitschrift für Pflanzenkrankheiten und Pflanzenschutz, 79, 471–477.
48. KLINGAUF, F., NOCKER-WENZEL, K. & ROTTGER, U. (1978). Die Rolle peripherer Pflanzenwachse für den Befall durch phytophage Insekten. Zeitschrift für Pflanzenkrankheiten und Pflanzenschutz, 85, 228–237.
49. KUO, H.L. (1986). Resistance of oats to cereal aphids: Effects on parasitism by Aphelinus asychis (Walker). In 'Interactions of Plant Resistance and Parasitoids and Predators of Insects'. Ed. by D.J. Boethel and R.D. Eikenbary, Ellis Horwood, Chichester, UK, pp 125–137.
50. LEE, G. (1981). Host non-preference and antibiotic resistance to the aphid Metopolophium dirhodum in winter wheat cultivars. Tests of Agrochemicals and Cultivars No. 2. Annals of Applied Biology 97, Supplement, 68–68.
51. LEE, G. (1984). Assessment of the resistance of wheats to Sitobion avenae feeding on the ear. Tests of Agrochemicals and Cultivars No. 5. Annals of Applied Biology 104, Supplement, 100–101.
52. LESZCZYNSKI, B. (1985). Changes in phenols content and metabolism in leaves of susceptible and resistant winter wheat cultivars infested by Rhopalosiphum padi (Homoptera: Aphididae). Zeitschrift für angewandte Entomologie, 100, 343–348.
53. LESZCZYNSKI, B., WARCHOL, J. & NIRZ, S. (1985). The influence of phenolic compounds on the preference of winter wheat cultivars by cereal aphids. Insect Science and its Application, 6, 157–158.
54. LOWE, H.J.B. (1975). Crop resistance to pests as a component of integrated control systems. Proceedings of the 8th British Insecticide and Fungicide Conference, 87–92.
55. LOWE, H.J.B. (1980). Resistance to aphids in immature wheat and barley. Annals of Applied Biology, 95, 129–135.
56. LOWE, H.J.B. (1981). Resistance to Sitobion avenae in wheat. Bulletin SROP/WPRS IV/1, 47–50.
57. LOWE, H.J.B. (1981). Resistance and susceptibility to colour forms of the aphid Sitobion avenae in spring and winter wheats (Triticum aestivum). Annals of Applied Biology, 99, 87–98.
58. LOWE, H.J.B. (1982). Some observations on susceptibility and resistance of winter wheat to the aphid Sitobion avenae (F.) in Britain. Crop Protection, 1, 431–440.
59. LOWE, H.J.B. (1984). Characteristics of resistance to the grain aphid Sitobion avenae in winter wheat. Annals of Applied Biology, 105, 529–538.
60. LOWE, H.J.B. (1984). Development and practice of a glasshouse screening technique for resistance of wheat to the aphid Sitobion avenae. Annals of Applied Biology, 104, 297–305.
61. LOWE, H.J.B. (1984). Glasshouse and field assessments of resistance to grain aphid, Sitobion avenae, in winter wheat. Tests of Agrochemicals and Cultivars No. 5. Annals of Applied Biology 104,

Supplement, 108-109.

62. LOWE, H.J.B. (1984). A behavioural difference amongst clones of the grain aphid Sitobion avenae. Ecological Entomology, 9, 119-122.

63. LOWE, H.J.B. (1985). Observations on the occurrence and inheritance in wheat of resistance to the grain aphid Sitobion avenae. Crop Protection, 4, 313-321.

64. LOWE, H.J.B. & ACREMAN, T.M. (1984). Consistency among assessments of resistance to grain aphid in spring wheat. Tests of Agrochemicals and Cultivars No. 5. Annals of Applied Biology 104, Supplement, 106-107.

65. LOWE, H.J.B. & ANGUS, W.J. (1985). Grain aphid infestations of winter wheat trials, 1984. Annals of Applied Biology, 106, 591-594.

66. LOWE, H.J.B. & BENEVICIUS, L.A.D. (1981). Increase in numbers of cereal aphids by insecticide application as an aid to plant breeding. Journal of Agricultural Science, Cambridge, 96, 703-705.

67. LOWE, H.J.B., MURPHY, G.J.P. & PARKER, M.L. (1985). Non-glaucousness, a probable aphid-resistance character of wheat. Annals of Applied Biology, 106, 555-560.

68. MARKKULA, M. & ROUKKA, K. (1972). Resistance of cereals to the aphids Rhopalosiphum padi (L.) and Macrosiphum avenae (F.) and fecundity of these aphids on Graminae, Cyperaceae and Juncaceae. Annales Agriculturae Fenniae, 11, 417-423.

69. MARREWIJK, G.A.M., van & DIELEMAN, F.L. (1980). Resistance to aphids in barley and wheat. In 'Integrated Control of Pests in the Netherlands'. Ed. by P. Gruijs and A.K. Minks, Wageningen, Pudoc, 165-167.

70. MENTINK, P.J.M., KIMMINS, F.M., HARREWIJN, P., DIELEMAN, F.L., TJALLINGII, W.F., van RHEENEN, B. & EENINK, A.H. (1984). Electrical penetration graphs combined with stylet cutting in the study of host plant resistance to aphids. Entomologia experimentalis et applicata, 35, 210-213.

71. MONTILLOR, C.B., CAMPBELL, B.C. & MITTLER, T.E. (1983). Natural and induced differences in probing behaviour of two biotypes of the greenbug, Schizaphis graminum, in relation to resistance in sorghum. Entomologia experimentalis et applicata, 34, 99-106.

72. NIKOLENKO, M.P. & OMEL'CHENKO, L.I. (1985). Immunity characteristics of wheat with interaction of Sitobion avenae F., Rhopalosiphum padi L. and Puccinia triticina Erikss. Sel'skokhozyaistvennaya Biologiya, 2, 52-55.

73. NIRAZ, S., LESZCZYNSKI, B., CIEPIELA, A., URBANSKA, A. & WARCHOL, J. (1985). Biochemical aspects of winter wheat resistance to aphids. Insect Science and its Application, 6, 253-257.

74. NORRIS, D.M. & KOGAN, M. (1980). Biochemical and morphological bases of resistance. In 'Breeding Plants Resistant to Insects'. Ed. by F.G. Maxwell and P.R. Jennings, Wiley Interscience, New York, pp. 23-61.

75. OBRYCKI, J.J. (1986). The influence of foliar pubescence on entomophagous species. In 'Interactions of Plant Resistance and Parasitoids and Predators of Insects'. Ed. by D.J. Boethel and R.D. Eikenbary, Ellis Horwood, Chichester, UK, pp 61-83.

76. ORR, D.B. & BOETHEL, D.J. (1986). Influence of plant antibiosis through four trophic levels. Oecologia, 70, 242-249.

77. PIORR, H.P. (1986). Reducing fungicide applications by using sodium silicate and wettable sulphur in cereals. Mededelingen van de Fakulteit Landbouwwetenschappen Rijksuniversiteit Gent, 51,

719–729.

78. RAUTAPÄÄ, J. (1980). Light reactions of cereal aphids (Homoptera, Aphididae). Annales Entomologici Fennici, 46, 1–12.

79. ROBERTS, J.J. & FOSTER, J.E. (1983). Effect of leaf pubescence in wheat on the bird cherry oat aphid (Homoptera: Aphididae). Journal of Economic Entomology, 76, 1320–1322.

80. SALTO, C.E., EIKENBARY, R.D. & STARKS, K.J. (1983). Compatibility of Lysiphlebus testaceipes (Hymenoptera: Braconidae) with greenbug (Homoptera: Aphididae) biotypes "C" and "E" reared on susceptible and resistant oat varieties. Environmental Entomology, 12, 603–604.

81. SHAH, M.A. (1982). The influence of plant surfaces on the searching behaviour of coccinellid larvae. Entomologia experimentalis et applicata, 31, 377–380.

82. SOTHERTON, N.W. & VAN EMDEN, H.F. (1982). Laboratory assessments of resistance to the aphids Sitobion avenae and Metopolophium dirhodum in three Triticum species and two modern wheat cultivars. Annals of Applied Biology, 101, 99–107.

83. STARKS, K.J., BURTON, R.L. & MERKLE, O.G. (1983). Greenbugs (Homoptera: Aphididae) plant resistance in small grains and sorghum to Biotype E. Journal of Economic Entomology, 76, 877–880.

84. STARKS, K.J., MUNIAPPAN, R. & EIKENBARY, R.D. (1970). Interaction between plant resistance and parasitism against greenbug on barley and sorghum. Annals of the Entomological Society of America, 65, 650–655.

85. STARKS, K.J. & WIEBEL, D.E. (1981). Resistance in bloomless and sparse-bloom sorghum to greenbugs. Environmental Entomology, 10, 963–965.

86. STOKES, S., LEE, G. & WRATTEN, S.D. (1980). Resistance to cereal aphids in winter wheat cultivars 1978. Tests of Agrochemicals and Cultivars No. 1. Annals of Applied Biology 94, Supplement, 50–5.

87. STORK, N.E. (1980). A scanning electron microscope study of tarsal adhesive setae in the Coleoptera. Zoological Journal of the Linnean Society, 68, 173–306.

88. STORK, N.E. (1980). The role of waxblooms in preventing attachment to brassicas by the mustard beetle, Phaedon cochleariae. Entomologia experimentalis et applicata, 28, 100–107.

89. SUNDERLAND, K.D., CHAMBERS, R.J. & CARTER, O.C.R. (1986). Interactions between wheat varieties and grain aphid control. Annual Report of the Glasshouse Crops Research Institute for 1985, in press.

90. SUNDERLAND, K.D., FRASER, A.M. & DIXON, A.F.G. (1986). Field and laboratory studies on money spiders (Linyphiidae) as predators of cereal aphids. Journal of Applied Ecology, 23, 433–447.

91. TODD, G.W., GETAHUN, A. & CRESS, D.C. (1971). Resistance in barley to the greenbug, Schizaphis graminum. I. Toxicity of phenolic and flavonoid compounds and related substances. Annals of the Entomological Society of America, 64, 718–722.

92. VICKERMAN, G.P. & SUNDERLAND, K.D. (1975). Arthropods in cereal crops: nocturnal activity, vertical distribution and aphid predation. Journal of Applied Ecology, 12, 755–766.

93. WATSON, S.J. (1983). Effects of weather on the numbers of cereal aphids. PhD thesis, University of East Anglia, UK.

94. WATSON, S.J. & DIXON, A.F.G. (1984). Ear structure and the resistance of cereals to aphids. Crop Protection, 3, 67–76.

95. WEBSTER, J.A. & INAYATULLAH, C. (1984). Greenbug (Schizaphis

graminum) (Homoptera: Aphididae) resistance in triticale.
Environmental Entomology, 13, 444–447.
96. WEBSTER, J.A. & STARKS, K.J. (1984). Sources of resistance to two
biotypes of the greenbug Schizaphis graminum (Homoptera: Aphididae).
Protection Ecology, 6, 51–55.
97. WIKTELIUS, S. & PETTERSSON, J. (1985). Simulations of bird
cherry-oat aphid population dynamics: a tool for developing
strategies for breeding aphid-resistant plants. Agriculture,
Ecosystems and Environment, 14, 159–170.
98. WILDING, N. (1981). The effect of introducing aphid-pathogenic
Entomophthoraceae into field populations of Aphis fabae. Annals of
Applied Biology, 99, 11–23.
99. ZÚÑIGA, G.E. & CORCUERA, L.J. (1986). Effects of gramine in the
resistance of barley seedlings to the aphid Rhopalosiphum padi.
Entomologia experimentalis et applicata, 40, 259–262.
100. ZÚÑIGA, G.E., SALGADO, M.S. & CORCUERA, L.J. (1985). Role of an
indole alkaloid in the resistance of barley seedlings to aphids.
Phytochemistry, 24, 945–948.

Effects of hydroxamic acids on the resistance of wheat to the aphid *Sitobion avenae*

S.D.Wratten, D.Thackray & P.J.Edwards
Department of Biology, University of Southampton, UK
H.Niemeyer
Facultad de Ciencias, Universidad de Chile, Santiago

Summary

Resistance to the grain aphid, Sitobion avenae was assessed in relation to levels of hydroxamic acids, in particular DIMBOA (2,4-dihydroxy-7-methoxy-1,4-denzoxazin-3-one) in cultivars and species of Triticum and Aegilops. Antibiosis (measured as intrinsic rate of increase (r_m), and its components) was recorded. Strong correlations were demonstrated between r_m and acid levels, with correlation co-efficients up to 0.96. Genetic patterns of acid levels within the wheat genome have been detected, which increases the potential for a plant breeding programme aimed at improving levels of aphid resistance in modern wheats.

1.1 Introduction

Sitobion avenae F. is a sporadically damaging pest of wheat in temperate climates (5, 10, 3). Wheat extracts contain hydroxamic acids (Hx) (11) which have been shown to be important in resistance against insects in several Gramineae (6, 7, 8, 1). The most abundant of these acids in wheat is 2,4-dihydroxy-7-methoxy-1,4-benzoxazine-3-one (DIMBOA). This compound has also been shown to be involved in the resistance of several wheat cultivars to the aphid species Metopolophium dirhodum (Wlk.), Schizaphis graminum (Rond.) and Rhopalosiphum maidis (Fitch) (4), but S. avenae has not been investigated in this context.

The main objective of this investigation was to assess a range of Triticum species and cultivars in order to measure and rank them for anti-biotic resistance to S. avenae, and for hydroxamic acid levels at the seedling stage. Any correlations between aphid performance and acid levels would have implications for future screening of wheat material against this important European pest and would usefully reinforce the relationship demonstrated by (2).

1.2 Methods

Seed of Triticum material was germinated in a culture room based on the design of (9). The temperature was 20°C with a 2°C range. Light intensity was 75 $\mu Em^{-2}s^{-1}$. Plants were harvested at the two-leaf stage(G.S.12) (12). A portion 1.5 g, of plant tissue (10-15 plants)was then macerated with a mortar and pestel in water (6 ml total volume). filtered through cheesecloth and left for 15 min at room temperature. The extract was adjusted to pH 3 with 1 M HCl and centrifuged at 10 000 g for 10 min. The supernatant was extracted three times into equal volumes of ethyl ether and the organic

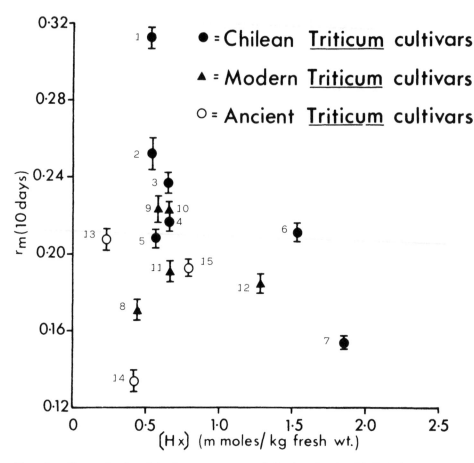

Fig. 1. The relationship between r_m and Hx levels in 13 Triticum taxa.

Chilean cultivars: 1) Huenufen; 2) Sonka; 3) Likay; 4) SNA2 (T. durum);
5) Naofen; 6) Quilofen; 7) SNA3 (T. durum)
Modern European Cultivars: 8) Jerico; 9) Armada; 10) Hobbit; 11) Musket;
12) Avalon.
Ancient Cultivars: 13) Triticum aestivum 15189: 14) Triticum monococcum
377666; 15) Triticum polonicum, 384345.

phases were evaporated to dryness. Hx forms a blue complex $(\lambda_{max}$ = 590 nm)
upon the addition of ferric chloride reagent (50 g $FeFl_3.6H_2O$, 500 ml 95%
ethanol and 5 ml 15 M HCl). The concentration of Hx in the tissues was
determined by comparing the absorbance of extracts with a standard curve made
with DIMBOA from maize (cv. Fola Union). Details of aphid culturing and
bioassay of the plant material are given in(2).

1.3 Results and Discussion

    Values for intrinsic rate of increase (with 95% confidence limits) and
hydroxamic acid levels are plotted in Fig. 1. The relationship between

intrinsic rate of increase and hydroxamic acid levels is much less clear
than that previously demonstrated (2). There are at least two possible
reasons for this. One is that the range of acid concentrations detected in
the current plant material is much smaller than that published in earlier
work (2). Secondly, the growth-form of the non-commercial Triticum species
used in this study differs markedly from the commercial Chilean and European
cultivars which are also included in Fig. 1. The closeness of the growth-
form of 'wild' Triticum species to that of non-agricultural grasses may mean
that there is a higher ratio of cell wall and fibrous material to that of
cell contents than in commercial cultivars. The crude acid assay, based on
fresh plant weight, may lead to difficulties of interpretation when old and
modern Triticum material is compared in the same data set, as in Fig. 1.

Differences between values calculated for the Hx concentrations in the
same cultivars grown in Chile and in Southampton may be attributed to
differences in assay techniques, and in plant ages. During Hx extraction
from different cultivars, ether evaporation in Chile was achieved rapidly
by rotary evaporation. This technique was not employed in Southampton, and
consequently the much slower evaporation may have allowed for considerable
Hx degradation. Although plants were harvested at the 2 leaf stage in both
laboratories, plant heights may not have been the same, and slight differences in age could lead to the detection of quite different Hx levels in
the same cultivars. Hx concentrations are known to decline rapidly in seedlings of most modern wheats.

Future work will expand the existing screen in four major ways:
1) increase the genetic range of the Triticum (and Aegilops) material
investigated; 2) investigate modified analyses of hydroxamic acid levels
which will permit more realistic comparisons between modern and ancient
cultivars and species; 3) screen the plants for acid levels and aphid
resistance at early and late growth stages; 4) re-evaluate the plant material
for resistance to dipterous and molluscan pests of seedling cereals.

1.4 Acknowledgements

We thank Dr. C. Law of the Plant Breeding Institute, Cambridge, Great
Britain for advice and for the provision of Triticum seed.

REFERENCES

1. BECK, D.L., DUNN, G.M., ROUTLEY, D.G. and BOWMAN, J.S. (1983). Bio-
    chemical basis of resistance in corn to the corn leaf aphid. Crop
    Science 23, 995-998.
2. BOHIDAR, K., WRATTEN, S.D. and NIEMEYER, H.M. (1986). Effects of
    hydroxamic acids on the resitance of wheat to the aphid Sitobion
    avenae. Annals of Applied Biology 109, 193-198.
3. CARTER, N., McLEAN, I.F.G., WATT, A.D. and DIXON, A.F.G. (1980). Cereal
    aphids - a case study and review. Applied Biology, 5, 271-348.
4. CORCUERA, L.J., ARGANDONA, V.H. and NIEMEYER, H.M. (1982). Effect of
    cyclic hydroxamic acids from cereals on aphids. In Chemistry and
    Biology of Hydroxamic Acids, 111-118 Ed. H. Kehl, Karger, Basel.
5. GEORGE, K.S. and GAIR, R. (1979). Crop loss assessment on winter wheat
    attacked by the grain aphid, Sitobion avenae (F.), 1974-1977, Plant
    Pathology, 28, 143-149.
6. KLUN, J.A., TIPTON, C.L. and BRINDLEY, T.A. (1967). 2,4-dihydroxy-7-
    methoxy-1,4-benzoxazin -3- one (DIMBOA). an active agent in the resist-
    ance of maize to the European corn borer. Journal of Economic

Entomology, 60, 1529–1533.

7. KLUN, J.P., GUTHRIE, D.D., HALLAUER, A.R. and RUSSELL, W.A. (1970). Genetic nature of the concentration of 2,4-dihydroxy-7-methyl-1,4-benzoxazin-3(4H)-one and resistance to the European corn borer in a diallel set of eleven maize inbreds. Crop Science, 10, 87–90.

8. LONG, B.J., DUNN, G.M., BOWMAN, J.S. and ROUTLEY, D.G. (1977). Relationship of hydroxamic acid content in corn and resistance to the corn leaf aphid. Crop Science, 17, 55–58.

9. SCOPES, N.E.A., RANDALL, R.E. and BIGGERSTAFF, S.M. (1975). Constant temperature ventilated perspex cage for rearing phytophagous insects. Laboratory Practice, 33–34.

10. VICKERMAN, G.P. and WRATTEN, S.D. (1979). The biology and pest status of cereal aphids in Europe: a review. Bulletin of Entomological Research, 69, 1–32.

11. WILLARD, J.J. and PENNER, R. (1976). Benzoxazinones: cyclic hydroxamic acids found in plants. Residue Reviews, 64, 67–76.

12. ZADOKS, J.C., CHANG, T.T., and KONZAK, C.F. (1974). A decimal code for the growth stages of cereals. Weed Research, 14, 415–421.

Importance of Microhymenoptera for aphid population regulation in French cereal crops

A.Fougeroux, C.Bouchet, J.N.Reboulet & M.Tisseur
ACTA, Paris, France

SUMMARY

Studies have been conducted to determine the main species of Microhymenoptera involved as parasitoids of cereal aphids in some areas of France, such as Normandy, Rhone Valley and the Centre of the Bassin Parisien. These studies confirm the results obtained by other researchers. These studies are aimed at determining the percentage of parasitism which prevents the increase of aphids on wheat during the ear stage. Finally, experiments have been set up to measure the effects of Microhymenoptera as regulators of aphid populations on winter wheat. All these studies conducted by ACTA enable methods to be proposed for determining the long-term effects of pesticides on Microhymenoptera.

1. INTRODUCTION

The development of integrated crop protection in cereals requires, initially, the improvement of forecasting methods (thresholds, models) and crop management and protection techniques using, when necessary, selective insecticides. Actually, about 2 million hectares of winter wheat are sprayed with insecticide in the spring. More and more frequently, these insecticides are mixed with fungicides applied at the end of stem elongation or at the earing stage. These treatments are sprayed to control both foliar diseases and pests such as aphids, Cnephasia pumicana (Zeller) cereal leaf beetles, or thrips.

Aphid populations are naturally limited by numerous parasites and predators. In the northern France cereal growing area, coincidence between aphid predators such as coccinellids, syrphids, or chrysopids and aphid population peaks is not the usual situation and very often these predators are efficient after the earing stage, i.e. after the main period of aphid increase between the end of May and the middle of June.

The efficiency of this type of predator is greater in the southern part of France where they are active during the month of May. For the northern part of France, polyphagous predators and Microhymenoptera are present and active at the end of the winter and are able to limit aphid populations during April and May. Thus it is necessary to improve our knowledge of the effect of parasitism on Microhymenoptera and to evaluate the effects of cultural practices on this parasitism.

To this end, the crop protection Service of ACTA has conducted studies between 1981 and 1985, to increase our knowledge concerning the role of Microhymenoptera on aphid populations.

2. PARASITISM OF CEREAL APHIDS BY APHIDIIDS

Although Rhopalosiphum padi L. can be observed on cereals in spring, Sitobion avenae F. and Metopolophium dirhodum WLK are the most frequent and damaging species during this period. Rabasse and Dedryver (9) showed that in the western part of France, cereal aphids are mainly parasitised by four species of aphidiids:

- Aphidius uzbekistanicus Luz. on S. avenae, M. dirhodum and R. padi,
- Aphidius ervi Haliday on S. avenae and M. dirhodum,
- Aphidius picipes Nees on S. avenae and M. dirhodum,
- Aphidius matricariae Haliday on R. padi.

Other aphidiidae can be observed more rarely on cereal aphids:

- Praon volucre Haliday on S. avenae and M.dirhodum,
- Ephedrus plagiator Nees on S. avenae.

A. uzbekistanicus seems to be the most important parasitoid attacking aphid populations. The other species are more variable from year to year and in response to a range of climatic conditions. The effect of these parasites can be estimated by counting the number of mummies during the aphid population peak. Generally, the percentage of parasitised aphids at this time does not exceed 10% (9).

To augment the information collected by these authors, a survey has been carried out in different cereal areas. By this means, we intend to assess the importance of the different species and to determine the percentage of parasitised aphids between the beginning of stem elongation and the earing stage. The regions concerned are: Normandy (Caen area), Bassin Parisien (Beauce and Brie), region of Lyon and region of Valence (Rhone Valley). The identification of aphidiid species was achieved by sampling mummies and identifying emerging adults. Results are summarised in Table 1 (4).

TABLE 1. Average Percentage of Aphidiid Species for 8 Fields Between 5/5/1984 and 13/6/1984

- Aphidius uzbekistanicus	57%
- Aphidius ervi	31%
- Aphidius picipes	8%
- Ephedrus sp	4%

The observations made in other areas confirm this tendency, i.e. during May, A. uzbekistanicus is resonsible for a large percentage of parasitism. In June, A. ervi and A. picipes parasitised the aphid populations to an extent varying according to year, region and field (Table 2).

In this situation, A. picipes is the most important parasite of S. avenae during the month of June.

In these two fields, between 2 and 35% hyperparastitism was detected between mid-May and the beginning of June (Cynipidae and Asaphes spp. at the beginning of June and Dendrocerus sp. at the end of June).

In the Rhone Valley (Valence area), observations made on 13 winter wheat fields (8) also show that A. uzbekistanicus, A. ervi and A. picipes

62

TABLE 2. Percentage Parasitism by the Three Main Aphidiid Species in Two Winter Wheat Fields in Normandy (Lacourte, 1983).

	Alencon						Cagny			
	14/5/83 %	24/5/83 %	3/6/83 %	9/6/83 %	23/6/83 %	28/6/83 %	1/6/83 %	10/6/83 %	21/6/83 %	27/6/83 %
A. uzbekistanicus	46	65	30	31	12	12	65	0	5	2.5
A. ervi	0	5	2	2	10	10	0	11	3	3
A. picipes	0	10	42	55	65	37	0	50	60	62

TABLE 3. Percentage Parasitism in Four Winter Wheat Fields in the Bassin Parisien

Dates 1983	Léchelle		Néron		Mainbervilliers		Chilleurs-aux-Bois	
	X	Parasitised aphids %	X	Parasitised aphids %	X	Parasitised aphids %	X	Parasitised aphids %
24/5-27/5	0.12	0	0.01	20.0	0.31	0%	0.15	2.0
6/6-10/6	0.36	1.1	0.02	8.3	0.50	0%	0.02	35.0
13/6-17/6	2.73	10.3	0.41	3.4	8.6	11.3	0.4	9.3
20/6-24/6	7.7	10.5	1.2	4.2	71.3	2.3	1.4	18.7
27/6-1/7	18.8	7.1	1.1	44.2	55.8	2.7	9.7	5.6
4/7-8/7	45.1	9.3	2.8	13.8	94.5	2.8	15.6	1.7
11/7-15/7	16.8	17.1	3.4	15.0	0.38	100.0	4.8	3.3

X = Number of aphids (average number per tiller)

are the most important parasitoids but the proportion of the different species was not evaluated.

3. LEVEL OF PARASITISM

Does the level of parasitism during May and at the beginning of June influence the aphid increase at the earing stage? If so, what percentage parasitism is necessary to prevent aphid increase? Over what period must we make the estimation of this parasitism in order to forecast parasite efficiency?

Field observations in Beauce and Brie showed an inverse relationship between early parasitism and the maximum of aphids present. These observations were made on 1000 tillers on each date and for each field. The results are summarised in Table 3 (3).

In Néron a high level of early parasitism (20%) was found in the field where the aphid populations stayed at a low level. On the contrary, at Lechelle and Mainbervilliers early parasitism was neglible and the aphid populations at the earing stage greatly exceeded the threshold (respectively 45 and 94.5 aphids per tiller).

Similar results were obtained in Normandy in 11 fields in the Caen region. In these fields, the percentage of parasitism was measured on different dates and compared with the maximum aphid populations (1) (Table 4).

TABLE 4. Percentage Parasitism in 11 Fields in Normandy

Date Field	D1 24.4 %	D2 6.5 %	D3 24.5 28.5 %	D4 30.5 31.5 %	D5 3.6 5.6 %	D6 10.6 11.6 %	Maximum aphid population
1	12.5	10.3	21.5	17	12	11	495
2	8.3	9.1	12.0	8	19	10	446
3	12.8	10.3	28.3	24.4	13	9	204
4	0	0	17.9	0	*	*	38
5	8.3	3.7	9.3	10	10	9	571
6	0	6.3	7	6	7	9	598
7	0	0	23.5	0	*	*	116
8	0	3.3	9.3	13	3	9	554
9	0	0	24	10	9.1	*	37
10	3.3	0	9	10	6	5	185
11	0	0	3.5	*	*	3	109

* = no sample

This relationship between the percentage parasitism during the last week of May and the maximum aphid populations shows that, except for field no. 1, in fields with a percentage between 15 and 30% a great increase of aphids does not occur. On the contrary, in fields with a low percentage of parasitism (15%) the populations of aphids can exceed the threshold (field nos. 2, 5, 6, 8). Results for fields 10 and 11 can be explained by the fact

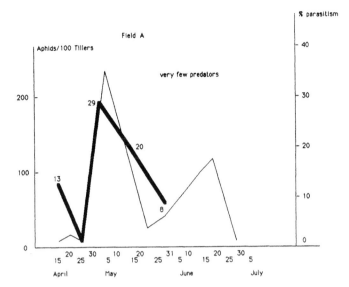

Fig. 1.

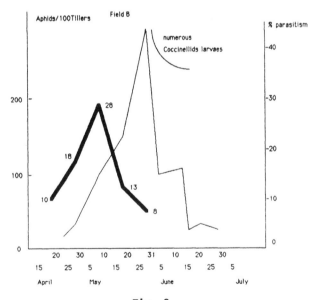

Fig. 2.

that other regulating factors are able to limit the aphid population growth.

A similar study has been conducted near Valence (8) in three fields. Results are summarised in Figure 1.

Parasitism by Microhymenoptera at the end of winter seems to influence the development of aphid populations. Fields A and B confirm this

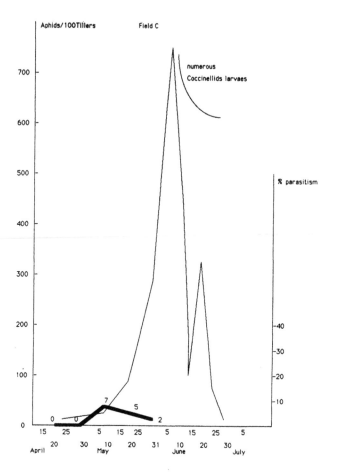

Fig. 3.

hypothesis. Field A, characterised by 13% parasitism and rather large populations of aphids (S. avenae mainly) at the end of winter, had no aphid infestation at the earing stage. Field C, with a low percentage of early parasitism and only few aphids at the end of winter, had a larger infestation of aphids at the earing stage (more than 7 aphids/ear).

Studies in these three regions confirm the importance of early parasitism during May in the northern part of France and April in the Rhone Valley. So, it is necessary, particularly during these periods, to avoid all cultural practices reducing the regulatory role of these Microhymenoptera.

4. EFFECTS OF CULTURAL PRACTICES ON PARASITISM BY APHIDIIDAE

In some studies (1, 4, 8) analyses have been conducted to determine the effects of cultural practices or environmental factors on parasitism.

Information concerning each field was collected to gather data on cultural practices (sowing date, fertilisers, varieties, pesticides...) and on environmental factors (landscape, surrounding crops, relief...).

Simultaneously, the main beneficials and the aphid population were observed. Correspondence Analysis (AFC) was used to analyse these data.

The aim of this study was to do a field typology with the beneficial fauna living in these fields on one hand and all the data concerning aphids, technical practices and environmental factors on the other hand. Only the results concerning Microhymenoptera are discussed in this paper.

Weeds surrounding the crops influence the level both of aphids and parasitoid populations. For example, Digitalis purpurea permits the development of Aulacorthum solani and its parasite Aphidius urticae (11).

More generally, adjacent crops of legumes or cereals have an important influence on the level of parasitism (12).

Rabasse and Dedryver (9) have shown the influence of hedges; they established that more parasites are found in parts of cereal fields bordered by hedges as compared with parts of fields without hedges.

The use of pesticides in intensive crop production causes decreases in parasitism. Delorme (2) showed that fungicides (benomyl - mancozebe) and acaricides (tetradifon) are slightly hazardous for Diaretiella rapae; insecticides such as pirimicarb, phosalone or lindane cause a reduction (60-70%) of aphidiid populations. A high toxicity is obtained with vamidothion and parathion (80-98% reduction of aphidiid populations).

All these results show clearly that aphid parasitism will be enhanced if all cropping techniques are suitable for aphidiids, in an integrated cropping system.

The results obtained with AFC (8, 4) suggest that early parasitism is linked with fodder crops (lucerne, meadows) and also hillsides. Hymenoptera are more effective in a low density of cereal vegetation. On the contrary the absence of fodder crops and high density of cereal vegetation seems to be unfavourable to aphidiids.

Parasitism is weakly linked to the different rotations in the region of Valence, except where the previous crop is winter wheat. This parasitism is linked with the presence of surrounding cereals, early sowing dates and a middle fertilisation.

In this study, no relationship is clearly shown between use of pesticides and parasitism.

5. CONCLUSIONS

All these results confirm the important role of aphidiids in limiting cereal aphid populations. Their effect varies according to field, region, year and environmental influences. Surrounding crops and cultural practices are very important. Amongst cultural practices, intensive methods of production, characterised by high crop densities, are unfavourable for parasitism. The main factors of intensive cereal production are sowing density, sowing date, fertilisation (especially nitrogen), fungicides, herbicides, insecticides and growth regulators.

It is important in the development of integrated crop protection to know the effects of chemical compounds on the main beneficials and especially on aphidiids. Some studies have been conducted in vitro and in semi-field tests to determine the short term effects of insecticides on Microhymenoptera (2, 6). No work has been done to test the effects of pesticides on the regulation by cereal aphids of aphidiids. To gather this information ACTA intend to develop a methodology to test the long-term effect on the regulatory role of beneficials (including Microhymenoptera) on aphids (10, 5).

Preliminary results show that in order to assess the effects of pesticides, this methodology needs to use a control (untreated plot) and a reference pesticide which is toxic to beneficials but non-toxic to aphids. The comparison of the evolution of the ratios

$$\frac{\text{number of beneficials}}{\text{number of aphids}} \times 100$$

obtained with each treatment will provide useful practical information on the effects of pesticides. Unfortunately it is impossible to test aphicides with this method because the total destruction of aphids precludes any conclusions on the effects of pesticides on beneficials. This methodology seems to be very useful for determining the long-term toxicity of herbicides, fungicides or non-aphicidal insecticides on beneficials and particularly on parasitism by aphidiids.

REFERENCES

(1)   DELION, D., 1985. Contribution à l'étude de l'influence des microhyménoptères sur le developpement des populations de pucerons du blé d'hiver. Memoire de fin d'études - ENSAIA-ACTA, 54 p.

(2)   DELORME, R., 1976. Evaluation en laboratoire de la toxicité pour Diaretiella rapae (Hym. aphidiidae) des pesticides utilisés en traitement des parties aériennes des plantes. Entomophaga, 21 (1), p. 19-29.

(3)   DUVERNET, L., 1983. Contribution à l'étude des parasites et prédateurs des pucerons en cultures de blé et de maïs. Mémoire de fin d'études - ENITA-ACTA, p. 51.

(4)   GAUTHERIN, J.C., 1984. Contribution à l'étude de l'influence du parasitisme par les hyménoptères sur les pullulations de pucerons en culture de blé d'hiver. Mémoire de fin d'études - ISARA-ACTA, 108 p.

(5)   GROS, V., 1985. Action secondaire à moyen terme des traitements sur ravageurs et auxiliaires. Mémoire de fin d'études - ENSAM-ACTA, 27 p.

(6)   HASSAN, S.A., 1985. Standard methods to test the side-effects of pesticides on natural enemies of insects and mites. Bulletin OEPP/EPPO, Bull. 15, p. 214-255.

(7)   LACOURTE, M., 1983. Contribution à l'étude des auxiliaires prédateurs et parasites des pucerons des céréales. Diplome d'Agronomie Approfondie. Mémoire de fin d'études INA-ACTA.

(8)   PONCELET, A., 1983. Ennemis naturels des pucerons des céréales: méthodes, évolution des populations, influence de l'agroecosystème. Memoire de fin d'études - ENSAM-ACTA, 79 p.

(9)   RABASSE, J.M. and DEDRYVER, C.A., 1983. Biologie des pucerons des céréales dans l'Ouest de la France. III - Action des hyménoptères parasites sur les populations de Sitobion avenae F., Metopolophium dirhodum WLK. et Rhopalosiphum padi L. Agronomie, 3 (8) p. 779-790.

(10)   REBOULET, J.N., 1986. Approche d'une méthodologie de plein champ pour définir les conséquences à moyen terme des pesticides sur la faune auxiliaire. Bulletin OILB/SROP 1986/XII 3, p. 46-54.

(11)   ROBERT, Y. and RABASSE, J.M., 1977. Rôle écologique de Digitalis purpurea dans la limitation naturelle des populations du puceron strié de la pomme de terre Aulacorthum solani par Aphidius urticae dans l'Ouest de la France. Entomophaga, 22 p. 373-382.

(12)   STARY, P., 1970. Biology of aphid parasites (Hymenoptera: aphidiidae) with respect to integrated control. Series entomologica 6. W. JUNK, La Haye 643 p.

The potential of *Tachyporus* spp. (Coleoptera: Staphylinidae) as predators of cereal aphids

G.P.Vickerman, W.Dowie & K.E.Playle
Biology Department, University of Southampton, UK

Summary

The mean numbers of I/II instar Metopolophium dirhodum consumed
by Tachyporus obtusus adults increased linearly with temperature
from 2.8 per day at 10°C to 8.9 at 25°C. At 20°C the mean numbers
of aphids consumed per day by T.hypnorum first-, second- and
third-instar larvae were 6.3, 9.6 and 13.2 respectively. The mean
total number of aphids eaten throughout larval life was 95.6. The
percentage of aphids which were totally consumed by the larvae
increased with age, from 8% on the day of hatching to over 97%
after the fourth day. Female T.hypnorum ate significantly more
aphids per day than males. Type III functional responses were
exhibited when T.hypnorum adults were confined with different
densities of aphids, either in petri dishes or on plants in the
laboratory. At high aphid densities up to 33 I/II instar aphids
were consumed in a day by an adult. The capture efficiency of
T.hypnorum and T.chrysomelinus adults was highest for I/II instar
aphids (60%) and lowest for adult alatae (c.35%). Handling times
for these two species varied from c.9 min with I/II instar aphids
to 20-30 min with adult aphids. T.hypnorum adults and larvae
preferred I/II instar aphids to III/IV instars and the adults also
preferred apterous to alate adult aphids.

1. Introduction

In recent years there has been increasing evidence that polyphagous
predators, such as members of the Staphylinidae, Carabidae, Araneae
and Dermaptera, may play a role in reducing the numbers of aphids, in
particular the grain aphid (Sitobion avenae (F.)), found in cereal
crops (4,5,6,12,16). Exclusion barriers have been used in some studies
(4,5,6,16) to reduce the numbers of polyphagous predators found in
areas of the crop; although in some cases this resulted in an increase
in the numbers of aphids, in others no such effect was observed. This is
not unexpected as the composition of the predatory complex may vary
markedly from site to site and there is the assumption that the density
of predators in the open crop is always sufficient to have an impact on
the aphid population; clearly, this may not always be the case. There
is, therefore, an urgent need to obtain more detailed information on the
ecology and predatory potential of the individual species in the
predatory complex. At present the lack of such information makes it
difficult to include data on polyphagous predators in simulation models
(2) and the ability to predict aphid population development is reduced.

In addition, it is difficult to predict the likely effects of changes in agricultural practices and pesticide use on the balance between cereal aphids and their predators.

The few studies that have been carried out on individual species have concentrated on the Carabidae (7,11). However, there are now indications (17) that the Araneae and Staphylinidae may have greater potential value as aphid predators and within the Staphylinidae attention has been given to beetles of the genus Tachyporus. In Britain the genus consists of fourteen species of which four, Tachyporus hypnorum (F.), T.chrysomelinus (L.), T.obtusus (L.) and T.nitidulus F., are usually found in cereal crops (12,14). These beetles are some of the most common Coleoptera found in cereals and both the adults and the larvae feed on fungi and on insects such as cereal aphids (15,18,19). Both the adults and the larvae climb cereal plants (8,9,19), in particular at night, and there are indications that they play an important role in reducing populations of the grain aphid (8,18). However, apart from some studies in Germany (10) and in Ireland (9), little is known about the ecology of these species. In the present study work was carried out in the laboratory to assess the potential of Tachyporus spp. as predators of cereal aphids. Details of the ecology of these species and of their predatory potential in the field will be published elsewhere.

2. Materials and Methods

Adult T.hypnorum, T.chrysomelinus and T.obtusus used in the laboratory experiments were collected by ground zone search in cereal fields. The three species were kept separately in ventilated plastic boxes containing a 2 cm layer of damp John Innes No.2 compost. The beetles were fed regularly with an excess of grain (S.avenae) or rose-grain (Metopolophium dirhodum (Walker)) aphids and maintained in an illuminated incubator at a constant temperature of 20°C and with a 16h photoperiod.

Tachyporus hypnorum larvae used in the experiments were bred in the laboratory. Batches of 5 male and 5 female beetles were confined on several pieces of damp filter paper in 9cm-diameter plastic petri dishes and fed cereal aphids. Under these conditions, T.hypnorum laid eggs readily in crevices in the filter paper around the edge of the dish. The eggs were transferred carefully to separate petri dishes. On hatching the larvae were fed with an excess of cereal aphids. Newly hatched or newly moulted larvae were used in the trials.

Aphids (S.avenae and M.dirhodum) used in the experiments were reared at 20°C on seedling wheat (cv Hobbit).

2.1 Aphid consumption

Tachyporus obtusus adults were confined individually in 9cm-diameter petri dishes on damp filter paper. The dishes were placed in an illuminated incubator (3230 lux; 16 h photoperiod) 72h prior to the start of the experiments and the beetles received no food during this period.

A section of wheat leaf with 15 M.dirhodum was then introduced to each dish. The trials were carried out using approximately equal numbers of first- (I) and second-(II) instar or third-(III) and fourth-(IV) instar aphids. Every 24h for the next 4 days the number of aphids eaten was recorded and fresh aphids were added to restore the original density. Any II instars that moulted to III instars or IV instars that moulted to adults were removed and replaced with fresh I/II or III/IV instar aphids as appropriate. As no aphids escaped from control (predators absent) dishes, aphids missing from the dishes containing predators were assumed to have been eaten.

The trials were carried out at constant temperatures of 10, 15, 20 and 25°C and in most cases there were 20 replicates for each aphid instar/ temperature combination.

Aphid consumption by T.hypnorum adults and larvae was assessed in a similar manner. Male or female beetles and newly hatched first-instar larvae were confined individually on damp filter paper in petri dishes in an incubator maintained at 20°C and a 16h photoperiod. A section of wheat leaf with 20 I/II instar M.dirhodum was introduced to each dish. Every 24h the number of aphids completely eaten (solid feeding) or with the body fluids removed (liquid feeding) was recorded and fresh aphids added to restore the original numbers. With the adults records were made over a period of 10 days. With the larvae records were made of the date of moulting and the trial continued until the pre-pupal stage was reached. In all trials there were 20 replicates.

2.2 Functional response

In an initial trial T.hypnorum adults were confined individually on damp filter paper in 9cm-diameter plastic petri dishes. The dishes were placed in an illuminated incubator (8240 lux; 16h photoperiod) at 20°C. I/II instar S.avenae were then added at densities of 1, 3, 5, 10, 20, 50, 100 and 200 per dish. After 24h records were made of the numbers of aphids consumed. The number of replicates varied from 55 at the lowest aphid density (1 per dish) to 7 at the highest aphid density (200 per dish).

In a second trial, a more complex experimental arena was used. T.hypnorum adults were confined individually, using plastic cylinders (31cm high; 9cm diameter), on 15 wheat (cv Hobbit) seedlings (2wk old) growing in plant pots in John Innes No. 2 compost. The trials were carried out at 20°C and under a 16h photoperiod (8240 lux). I/II instar S.avenae were introduced to the plants, using a fine paint brush, at densities of 1, 3, 5, 7, 10, 20 and 35 per cylinder. The number of aphids remaining on the plants was counted after 24h and each aphid density was replicated 20 times, with the exception of the highest density (35) which was replicated 10 times.

2.3 Capture efficiency and handling time

The experiments were carried out in a growth room with a 16h photo-period. The temperature in the light period ranged from 16 to 19°C. Prior to the experiment Tachyporus adults were starved for a period of 18h. They were then confined individually in petri dishes on damp filter paper. A section of wheat leaf with 5 aphids (S.avenae) was then introduced to each dish. Observations were then made of the number of attempts made by individual beetles to capture an aphid. Observations continued until 20 aphids had been captured. The trials were carried out using I/II instar and III/IV instar aphids and adult apterae and alatae. Tachyporus hypnorum and T.chrysomelinus adults were used in the trials.

Data on handling times were obtained using a similar procedure. The handling time was defined as the period between the successful capture of the aphid by the beetle and the time when the aphid was either completely consumed or the body fluids had been removed to leave only parts of the exoskeleton, such as the antennae and legs. Handling times of T.hypnorum and T.chrysomelinus were recorded for I/II instar and III/IV instar aphids, adult apterae and alatae and there were 20 replicates for each.

2.4 Feeding preferences

T.hypnorum adults or third-instar larvae were confined individually
on damp filter paper in petri dishes. The dishes were placed in an
incubator at 20°C and under a 16h photoperiod. With the adults,
preferences were tested for I/II instar aphids (S.avenae) compared with
III/IV instar aphids (no. replicates = 19) and for adult apterae compared
with adult alatae (n = 30). Larval preferences were tested for I/II
instar compared with III/IV instar aphids (n = 34). 20 aphids of each
morph were added to each dish and the numbers consumed recorded after 24h

3. Results

3.1 Aphid consumption

The mean numbers of I/II instar and III/IV instar M.dirhodum
consumed per day by Tachyporus obtusus adults are given in Table I. The
mean numbers consumed increased linearly with temperature, for example
from 2.8 I/II instars per day at 10°C to 8.9 at 25°C, and there was no
evidence of a decline in feeding rate at the higher temperatures.

Table I. Mean numbers ($\pm$ 95% C.L.) I/II instar and III/IV
instar M.dirhodum consumed per day by T.obtusus
adults at different constant temperatures

Temperature	Mean no. aphids eaten per day	
(°C)	I/II	III/IV
10	2.8 (0.44)	1.0 (0.28)
15	4.8 (0.60)	3.2 (0.44)
20	6.4 (0.71)	5.0 (0.63)
25	8.9 (0.82)	6.5 (0.70)

At all temperatures more I/II instar than III/IV instar aphids were
consumed. For both categories of aphid instar the differences between
feeding rates at different temperatures were highly significant (P<
0.001). The mean daily weight of aphids consumed as a percentage of the
predator's own body weight increased from 14% at 10°C to 46% at 25°C for
I/II instar aphids and from 8% at 10°C to 50% at 25°C for III/IV instar
aphids.

The mean numbers of I/II instar M.dirhodum eaten per day at 20°C by
T.hypnorum larvae and adults are given in Table II and shown in Fig. 1.
The mean daily consumption rate differed significantly (P<0.01) between
the different larval instars and was lowest for the first-instar larvae
(6.3) and highest for the third-instar larvae (13.2). The maximum number
of aphids consumed per day varied from 13 for first-instar larvae to 19
for second- and third-instar larvae. The mean total number of aphids
eaten throughout larval life was 95.6 (range 83 - 117) and increased from
about 16 (range 10 - 30) during the first instar to 58 (range 42 - 73) in
the third instar (Table II). Towards the end of the third larval instar
feeding rates dropped just prior to the pre-pupal stage (Fig. 1).

Table II. Extent of solid feeding and numbers of I/II instar
M.dirhodum consumed by T.hypnorum larvae and adults
at 20°C

	% solid feeding	No. aphids eaten / day		Total no. aphids eaten during instar	
		Mean	Max.	Mean	Range
First-instar larvae	31.1	6.3	13	16.4	10-30
Second-instar larvae	97.8	9.6	19	21.1	12-28
Third-instar larvae	99.8	13.2	19	58.1	42-73
Adults (♂♂)	96.4	6.8	14		
Adults (♀♀)	99.1	12.9	19		

The feeding rates of male and female T.hypnorum are given in Table
II. Female beetles consumed significantly more (P<0.001) aphids per day
than males and significantly more than either second- (P<0.05) or first-
instar (P<0.001) larvae. Male T.hypnorum ate significantly fewer (P<
0.001) aphids per day than the third-instar larvae (Table II).
On the first day after hatching only 8% of the aphids killed by T.
hypnorum larvae were totally consumed, but the percentage increased to
46% by the second day and 97% by the fourth day (Fig. 1). During the
first larval instar only 31% of aphids were totally consumed but during
the second and third larval instars at least 97% of the aphids were
completely eaten (Table II). Both male and female T.hypnorum completely
consumed at least 96% of the I/II instar aphids (Table II).

3.2 Functional response

The mean numbers of I/II instar Sitobion avenae consumed by T.
hypnorum adults when confined in petri dishes with different numbers of
aphids are shown in Fig. 2.
The mean numbers of aphids consumed per day increased from 0.56 at a
density of 1 aphid per dish to c.21 at densities of 100 and 200 aphids
per dish (Fig. 2). The response curve (Fig. 2) was sigmoid, indicating a
Type III functional response. The percentage of aphids consumed was
higher at a density of 1 aphid per dish (56%) than 3 aphids per dish
(41%) and reached a peak at 5 aphids per dish (72%). The percentage of
aphids consumed then decreased with increasing densities until only c.11%
were eaten at a density of 200 aphids per dish. At high aphid densities
up to 33 I/II instar S.avenae were consumed in a day by T.hypnorum
adults.
When T.hypnorum adults were confined with wheat plants infested
with S.avenae the number of aphids consumed per day increased from 0.25
at a density of 1 aphid per container to 9.0 at a density of 35 aphids
per container and the shape of the response curve was again sigmoid
(Fig. 3). The percentage of aphids consumed by T.hypnorum increased
to 41% at a density of 10 aphids per container and then declined with
increasing aphid density.

73

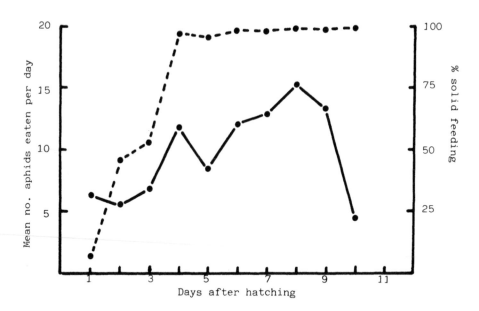

Fig. 1. Mean numbers (●—●) I/II instar <u>M.dirhodum</u> eaten by <u>T.hypnorum</u>
larvae on different days after hatching and percentage aphids
(●--●) completely consumed (solid feeding)

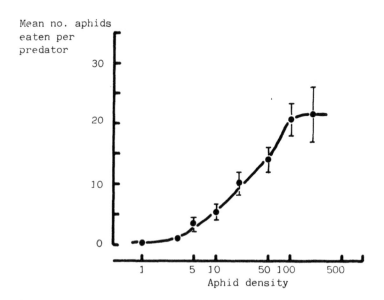

Fig. 2. Mean numbers I/II instar <u>S.avenae</u> ($\pm$ 95% C.L.) eaten per day by
individual <u>T.hypnorum</u> adults when confined in petri dishes with
different numbers of aphids

74

3.3 Capture efficiency

The capture efficiencies of Tachyporus hypnorum and T.chrysomelinus adults for I/II instar and III/IV instar aphids, adult apterae and alatae are illustrated in Fig. 4.

The capture efficiency of T.chrysomelinus was significantly higher for I/II instar aphids (69.0%) than for either III/IV instar aphids (48.8%) (X_i^2 = 5.20; P<0.05), adult apterae (43.5%) (X_i^2 = 4.64; P<0.05) or adult alatae (38.5%) (X_i^2 = 12.01; P<0.001). Capture efficiencies did not differ significantly for III/IV instar aphids, adult apterae and adult alatae.

Similarly, the capture efficiency of T.hypnorum adults was significantly higher for I/II instar aphids (63.6%) than for either III/IV instar aphids (44.4%) (X_i^2 = 3.90; P<0.05), adult apterae (45.4%) (X_i^2 = 3.87; P<0.05) or adult alatae (35.1%) (X_i^2 = 6.96; P<0.01), between which there were no significant differences.

Although the capture efficiencies of T.chrysomelinus for the different aphid morphs tended to be higher than those of T.hypnorum the differences were not significant.

3.4 Handling time

The mean handling times of T.chrysomelinus and T.hypnorum adults for different aphid morphs are given in Table III and shown in Fig. 4.

Handling times varied quite markedly, ranging from 4 to 15 min for I/II instar aphids to 16 to 44 min for adult aphids. There were highly significant differences (P<0.001) between the mean handling times of T.chrysomelinus for the different aphid morphs. Mean handling times were lowest for I/II instar aphids (9.4 min) and highest for adult alatae (26.0 min). Mean handling times were significantly longer (P<0.001) for adult alatae than for adult apterae (Table III).

With T.hypnorum there were also significant differences (P<0.001) between the handling times for the different aphid morphs. Mean handling times varied from 9.5 min for I/II instar aphids to 31.9 min for adult apterae (Table III). In contrast to the situation with T.chrysomelinus, the mean handling time with adult apterae was significantly (P<0.001) longer than with adult alatae.

There were no significant differences between the mean handling times of T.chrysomelinus and T.hypnorum for I/II instar, III/IV instar aphids and adult alatae. However, the mean handling time of adult apterae was significantly longer (P<0.001) for T.hypnorum than for T. chrysomelinus (Table III).

3.5 Feeding preferences

When given a direct choice between I/II instar and III/IV instar Sitobion avenae, adult T.hypnorum ate significantly more (t = 8.03; d.f. = 36; P<0.001) I/II instar ($\bar{x}$ = 11.1) than III/IV instar ($\bar{x}$ = 4.0) aphids.

Tachyporus hypnorum adults also consumed significantly more (t = 4.11; d.f. = 58; P<0.001) apterous adult ($\bar{x}$ = 5.4) than alate adult ($\bar{x}$ = 3.0) S.avenae when given a choice between the two morphs.

The third-instar larvae of T.hypnorum ate significantly more (t = 10.18; d.f. = 66; P<0.001) I/II instar ($\bar{\bar{x}}$ = 10.6) than III/IV instar ($\bar{x}$ = 5.1) S.avenae.

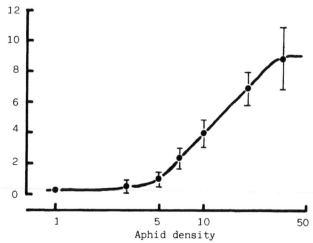

Fig. 3. Mean numbers I/II instar S.avenae ($\pm$ 95% C.L.) eaten per day
by individual T.hypnorum adults when confined on wheat plants
infested with different numbers of aphids

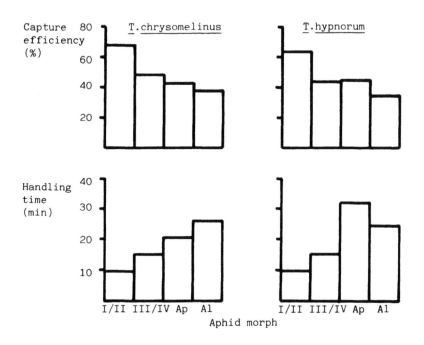

Fig. 4. Mean capture efficiency (%) and handling times (min) of
T.hypnorum and T.chrysomelinus adults for I/II and III/IV instar
S.avenae, adult apterae and adult alatae

Table III. Handling times of T.hypnorum and T.chrysomelinus
adults for I/II instar and III/IV instar aphids,
adult apterae and adult alatae

Tachyporus chrysomelinus

	Mean handling time (min)		95% confidence limits	Range of observed values (min)
I/II instar	9.4	a	7.98 – 10.82	4 – 14
III/IV instar	15.6	b	13.81 – 17.49	7 – 24
Adult apterae	20.6	c	17.49 – 22.71	16 – 34
Adult alatae	26.0	d	23.67 – 28.43	19 – 34

Tachyporus hypnorum

	Mean handling time (min)		95% confidence limits	Range of observed values (min)
I/II instar	9.5	a	8.08 – 10.88	4 – 15
III/IV instar	15.2	b	13.44 – 17.06	8 – 24
Adult apterae	31.9	c	27.86 – 35.94	19 – 44
Adult alatae	24.6	d	22.30 – 26.90	19 – 34

Note: For each species, means followed by the a different letter differ
significantly at $P < 0.001$.

4. Discussion

Beetles of the genus Tachyporus occur commonly in cereal crops. For example, over the period 1972 – 1979 mean peak numbers (per m^2) of first generation adults found in winter wheat fields in different years varied from 2 to 33 and peak numbers in individual fields varied from 9 to 52. Over the same period mean peak numbers of Tachyporus spp. larvae varied from 32 to 141 per m^2 and densities of up to 385 per m^2 were found in individual fields (20). Peak numbers of first generation adults were recorded usually in the last two weeks of May and of larvae in the last two weeks of June or in July , at a time when populations of both the grain and rose-grain aphid reach a peak. As the numbers of both Tachyporus adults and larvae found in cereal crops at night were from up to two to four times higher than during the day (19) this suggests that even these high densities may be considerable underestimates of the numbers actually present in cereal crops. Apart from spiders (Linyphiidae) and predatory flies (Empididae), that also feed on cereal aphids (17), these species are, therefore, some of the most common polyphagous predators found in cereals.

In July 1973 39% of T.chrysomelinus and 23% of T.hypnorum adults found in suction and sweep net samples taken from cereal fields had fed on aphids and in June 1974 14% of the Tachyporus spp. larvae found in sweep net samples (19). Averaged over the season as a whole, however, the percentage of T.chrysomelinus that had fed on aphids varied from 1% to 6% and of T.hypnorum from 1% to 16% in different years over the period 1972 - 1977 (18). However, when account was also taken of the density of the beetles they came very high on a ranking list of polyphagous predators that had potential for retarding aphid population growth (18). The results of the present investigation also indicate that the consumption rates of both the adults and the larvae are high, with the third-instar larvae and adults feeding on up to 19 and 33 aphids per day respectively. There was no evidence in the present study that feeding rates of the adults declined at higher temperatures as was found by Sopp & Wratten (13). Unlike the present study, when consumption rates at different temperatures were investigated simultaneously, the beetles used at different temperatures by Sopp & Wratten had been collected from the field at different times of the year. As pointed out by Sopp & Wratten (13), the decline in feeding rate at higher temperatures may have reflected changes in the physiological state of the beetles.

Sopp & Wratten considered that they had obtained maximum rates of aphid consumption for T.chrysomelinus and T.hypnorum but the feeding rates recorded in this study, with higher aphid densities, for T.hypnorum were up to two times higher. Sunderland & Vickerman (18) found that the percentage of Tachyporus adults feeding on aphids (a population response rather than a functional response) in the field did not increase with increasing aphid density when data from several years were combined. However, when confined with different densities of aphids in the laboratory there was evidence that T.hypnorum adults showed a Type III functional response; this was evident in both simple (petri dish) and more complex (wheat plants in cylinders) arenas. In the more complex arenas individual T.hypnorum adults found and consumed up to 14 aphids per day at the highest aphid density (35 I/II instar S.avenae per 15 wheat plants), suggesting that searching efficiency was relatively high. Evidence for such a functional response under field conditions must await the development of techniques to quantify the amount of aphid remains found in the guts of these predators. However, there is some evidence for a numerical response to aphid density in the field. In 1982 T. chrysomelinus and in 1983 this species together with T.hypnorum, T. obtusus and T.nitidulus were caught in significantly higher numbers in pitfall traps located in discrete patches of aphids in winter wheat than in traps in control areas (1). These predators may have the ability, therefore, to find localized aphid populations and to increase feeding rates in response to increasing aphid density.

Both T.hypnorum adults and third-instar larvae preferred I/II instar aphids to the larger morphs and capture efficiencies of both T.hypnorum and T.chrysomelinus adults for these small aphids were relatively high; as expected, handling times for the late instar and adult aphids were relatively long. Although capture efficiencies for the early instar aphids were less than those recorded for adult Linyphiidae those for the adults were considerably higher (35-45% c.f. 12%) (3).

The results of this and other studies suggest that both the adults and the larvae of Tachyporus spp. have considerable potential as aphid predators, and further studies on these species would be valuable.

5. Acknowledgements

The work of the senior author was financed by a research grant from the Ministry of Agriculture, Fisheries and Food.

REFERENCES

1. BRYAN, K.M. and WRATTEN, S.D. (1984). The responses of polyphagous predators to prey spatial heterogeneity: aggregation by carabid and staphylinid beetles to their cereal prey. Ecological Entomology 9, 251 - 259.
2. CARTER, N.,DIXON, A.F.G. and RABBINGE, R. (1982). Cereal aphid populations: biology, simulation and prediction. Pudoc, Wageningen.
3. CARTER, N., GARDNER, S., FRASER, A.M. and ADAMS, T.H.L. (1982). The role of natural enemies in cereal aphid population dynamics. Annals of Applied Biology 101, 190 - 195.
4. CHIVERTON, P. (1982). The effects of polyphagous predators on the establishment phase of bird-cherry oat aphid (Rhopalosiphum padi L.) infestations in spring barley. Växtskyddsrapporter, Jordbruk 20, 177 - 181.
5. DE CLERCQ, R. (1985). Study of the soil fauna in winter wheat fields and experiments on the influence of this fauna on the aphid population. Bulletin SROP / WPRS VIII / 3, 133 - 135.
6. EDWARDS, C.A., SUNDERLAND, K.D. and GEORGE, K.S. (1979). Studies on polyphagous predators of cereal aphids. Journal of Applied Ecology 16, 811 - 823.
7. GRIFFITHS, E. (1982). The carabid Agonum dorsale as a predator in cereals. Annals of Applied Biology 101, 152 - 154.
8. HOLMES, P.R. (1984). A field study of the predators of the grain aphid, Sitobion avenae (F.) (Hemiptera:Aphididae), in winter wheat in Britain. Bulletin of Entomological Research 74, 623 - 631.
9. KENNEDY, T.F., EVANS, G.O. and FEENEY, A.M. (1986). Studies on the biology of Tachyporus hypnorum F. (Col. Staphylinidae), associated with cereal fields in Ireland. Irish Journal of Agricultural Research 25, 81 - 95.
10. LIPKOW, E. (1966). Biologisch-ökologische Untersuchungen über Tachyporus-Arten und Tachinus rufipes (Col. Staphylinidae). Pedobiologia 6, 140 - 177.
11. LOUGHRIDGE, A.H. and LUFF, M.L. (1983). Aphid predation by Harpalus rufipes (Degeer) (Coleoptera:Carabidae) in the laboratory and field. Journal of Applied Ecology 20, 451 - 462.
12. POTTS, G.R. and VICKERMAN, G.P. (1974). Studies on the cereal ecosystem. Advances in Ecological Research 8, 107 - 197.
13. SOPP, P. and WRATTEN, S.D. (1986). Rates of consumption of cereal aphids by some polyphagous predators in the laboratory. Entomologia in press.
14. SPEIGHT, M.R. and LAWTON, J.H. (1976). The influence of weed cover on the mortality imposed on artificial prey by predatory ground beetles in cereal fields. Oecologia 23, 211 - 223.
15. SUNDERLAND, K.D. (1975). The diet of some predatory arthropods in cereal crops. Journal of Applied Ecology 12, 507 - 515.
16. SUNDERLAND, K.D., STACEY, D.L. and EDWARDS, C.A. (1980). The role of polyphagous predators in limiting the increase of cereal aphids in winter wheat. Bulletin SROP / WPRS III / 4, 85 - 91.
17. SUNDERLAND, K.D., CHAMBERS, R.J., STACEY, D.L. and CROOK, N.E. (1985) Invertebrate polyphagous predators and cereal aphids. Bulletin SROP / WPRS VIII / 3, 105 - 114.
18. SUNDERLAND, K.D. and VICKERMAN,G.P. (1980). Aphid feeding by some polyphagous predators in relation to aphid density in cereal fields. Journal of Applied Ecology 17, 389 - 396.

19. VICKERMAN, G.P. and SUNDERLAND, K.D. (1975). Arthropods in cereal crops; nocturnal activity, vertical distribution and aphid predation. Journal of Applied Ecology 12, 755 - 766.
20. VICKERMAN, G.P. (1986). The ecology of Tachyporus spp. (Coleoptera: Staphylinidae) in cereal and grass fields. Unpublished manuscript.

A contribution to a check-list of Staphylinidae (Coleoptera) of potential importance in the integrated protection of cereal and grass crops

J.A.Good & P.S.Giller
Department of Zoology, University College Cork, National University of Ireland

Summary

A checklist of common and abundant Staphylinidae in cereals and grassland in Northern Europe is presented. A summary of diet records for relevant genera is also included. This indicates that many adults are potential predators of various stages of cereal and grass pests, though less is known of larval diet. Evaluation of staphylinid predators and parasitoids in relation to chronotype categories and impact on pest populations is discussed.

1. INTRODUCTION

Staphylinid beetles are numerous and diverse in Northern European agroecosystems. However, the evaluation of their role in agroecosystems, and in particular their role as natural enemies of pests, is still very much in its infancy. This paper is a representative review of information currently available on cereal and grass crop Staphylinidae, contributing to the eventual preparation of a complete check-list of potentially important species in integrated pest management in these crops. Grassland has been included due to its importance both as a habitat for many cereal pests and their natural enemies (14, 36, 85, 114) and as a component of landscape heterogeneity affecting the abundance and diversity of cereal Staphylinidae (50, 117).

Predatory and parasitoid species within agroecosystems can be categorised in two ways. One is within a set of pestophagous guilds (i.e. the natural enemy complex associated with individual pest types, such as cereal aphids, shoot-boring flies, etc.). The other is within a set of agrohabitat assemblages (i.e. groups of species which have similar environmental requirements, such as a habitat free of a particular pesticide, a high humidity maintained by weed cover, etc.). Identification of staphylinid agrohabitat assemblages in relation to various agricultural management practices is currently in progress in cereal and grass systems of south west Ireland. However, a check-list of potentially pestophagous species is a necessary prerequisite to the assessment of the impact of these management practices on the role of staphylinids in the integrated protection of these crops.

To produce such a check-list, species abundance, diet, temporal and spatial overlap with pests, and the predators' efficiency in controlling prey, must be examined before an evaluation of the importance of the staphylinid predators can be made.

2. STAPHYLINIDAE COMMON AND ABUNDANT IN CEREAL AND GRASS CROPS

Values for total staphylinid biomass and density, and for the proportion of staphylinids in the polyphagous arthropod fauna of cereal crops and grassland, are given in Table 1. It is clear that they make up an important proportion of the secondary consumer fauna. Nearly 150 species of Staphylinidae have been recorded from these habitats in Northern Europe. However, many of these are vagrants from other habitats and represented by one or a few specimens only. Species occurring in less than 1-5% of the total staphylinid fauna (depending on sample size) are therefore excluded from the species lists presented in Tables 2-4.

TABLE 1. Selected Examples of Staphylinid Biomass, Density and Proportion of Polyphagous Arthropod Fauna in Cereals and Grassland

Field Type	A	B	C	Ref.
Arrhenatheretum meadow (unfertilised)	20.1	-	3.5	61
Arrhenatheretum meadow (fertilised)	39.9	-	20.9	61
Meadow (Staphylinid larvae)	13.1	7.31	-	77
Grazed pasture	-		c. 40	
Ungrazed pasture	-		c. 20	
Spring barley	-	-	10.8-29.9	12
Spring barley	-	46.1-55.3	-	118
Undersown spring barley	-	100.7-108.4	-	118
Cereals (mostly spring barley)	-		0.9-28.3	35
Winter wheat	-		8.6-41.2	82
Winter wheat (Tachyporus larve)	-	c.300-400		46,120

A: Biomass in mg. dry wt m^{-2};

B: No. individuals m^{-2};

C: % of No. or biomass of polyphagous arthropods (Carabidae, Staphylinidae, Aranae, Opiliones).

Figures represent adults unless otherwise stated.

A total of 45 species (two undetermined) are listed as common in cereals as adults, of which nine have been recorded as being abundant (greater than 20% of the total staphylinid fauna) in at least one survey (Table 2). Fewer species are listed from meadows and pastures (Table 3), as

TABLE 2. <u>Adult</u> <u>Staphylinidae</u> <u>Common</u> <u>or</u> <u>Abundant</u> <u>in</u> <u>Cereal</u> <u>Fields</u> <u>in</u> <u>Northern</u>
<u>Europe</u> [1]

Species included if more than 1% (n > 1000), 2.5% (500 < n < 1000), or 5%
(50 < n < 500) of total staphylinid fauna, or reported as being common or
abundant. Records for species which account for more than 20% of the total
fauna are marked with an asterisk, as are their references. Records
identified to genus but not to species are also included. Nomenclature
follows Kloet and Hincks (1977) Handbk. ident. Br. Insects 11,(3), 1-105.

Species	References
Omaliinae	
Anthobium unicolor (Marsham)	53
Lesteva longoelytrata (Goeze)	53, 108[2]
Omalium caesum (Gravenhorst)	108
O. rivulare (Paykull)	53, 108
Oxytelinae	
Coprophilus striatulus (Fabricius)	81
Anotylus complanatus (Erichson)	56
A. inustus (Gravenhorst)	81, 83, 85, 108
A. rugosus (Fabricius)	81, 83, 85, 108, 109
A. sculpturatus (Gravenhorst)*	74, 81*, 83, 85, 108
Oxytelus laqueatus (Marsham)	74
Steninae	
Stenus biguttatus (Linnaeus)	60, 81
Paederinae	
Lathrobium fulvipenne (Gravenhorst)	5, 60, 81, 108
L. longulum Gravenhorst	74
Staphylininae	
Gyrohypnus angustatus Stephens	108
Xantholinus linearis (Olivier)	60, 108
X. longiventris Heer	60
X. tricolor (Fabricius)	110
Philonthus cognatus Stephens*	7, 49, 100*, 103, 104, 108, 109,110
P. intermedius (Boisduval & Lacordaire)	58, 60
P. laminatus (Creutzer)*	58*, 60*, 108
P. politus (Linnaeus)	60
P. rotundicollis (Ménétriès)	58
P. varius (Paykull)	63, 108, 109
Tachyporinae	
Mycetoporus baudueri Mulsant & Rey	53
Tachyporus chrysomelinus (Linnaeus)	7, 67, 83, 85, 103, 108, 120
T. hypnorum (Fabricius)	5, 7, 56, 58, 60, 63, 64, 81*, 83, 84, 85, 89*, 100, 103, 104, 108, 120

TABLE 2. (continued)

T. obtusus (Linnaeus)	56, 60, 64, 108
T. solutus Erichson	81
Tachinus fimetarius Gravenhorst	53
T. signatus Gravenhorst*	5, 58, 60, 63*, 100, 103, 108, 109*

Aleocharinae

Aloconota insecta (Thompson)	63
A. gregaria (Erichson)*	84*
Amischa analis (Gravenhorst)*	67*
A. cavifrons (Sharp)*	67*
Dinaraea angustula (Gyllenhal)	84
Liogluta pagana (Erichson)	109
Atheta fungi (Gravenhorst)*	81*, 63*, 67*, 83, 84*, 85
Drusilla canaliculata (Fabricius)	108
Ilyobates subopacus Palm	108
Oxypoda exoleta Erichson	81, 84
O. lividipennis Mannerheim	53
Aleochara bipustulata (Linnaeus)	56, 108
A. brevipennis Gravenhorst[3]	108
A. curtula Goeze[3]	108

Micropeplinae[4]

Micropeplus[5]	64

Omaliinae

Omalium	64, 110

Oxytelinae

Oxytelus	35

Staphylininae

Xantholinus	64
Philonthus	35, 64
Quedius[5]	110

Tachyporinae

Tachyporus*	35, 110*
Tachinus	64

Aleocharinae[6]

Atheta*	53, 56, 58, 59, 74, 110*
Oxypoda	35, 110

1. Records from six countries, Belgium, Fed. Rep. Germany, Finland, Great Britain, Ireland, Sweden.
2. Tischler's list (108) includes species from other crop types as well as from cereals.
3. These two species may be associated with non-cereal crops (see note 2).
4. Now given family status by many authors.
5. Generic records only.
6. Many references do not identify Aleocharinae any further, but list it as the dominant group.

TABLE 3. Adult Staphylinidae Common or Abundant in Meadows and Pastures in Northern Europe

See Table 2 legend for explanation.

Species	References
Oxytelinae	
Carpelimus elongatulus (Erichson)	27
Platystethus arenarius (Fourcroy)	27
Anotylus rugosus (Fabricius)	24, 27
Steninae	
Stenus clavicornis (Scopoli)	24
S. impressus Germar /aceris Stephens	70
S. juno (Paykull)	24
S. subaeneus Erichson	70
Paederinae	
Sunius propinquus (Brisout)	91
Staphylininae	
Xantholinus lineris (Olivier)	111
X. longiventris Heer	24, 27, 70
Philonthus cognatus Stephens	24, 27, 33, 49
P. laminatus (Creutzer)*	24, 27*
P. marginatus (Strom)	111
P. splendens (Fabricius)	111
P. varius (Gyllenhal)*	24*, 27, 111
Staphylinus aeneocephalus DeGeer*	27, 111
S. olens Mueller, O.F.	45
Tachyporinae	
Tachyporus chrysomelinus (Linnaeus)	24, 27, 49, 70
T. hypnorum (Fabricius)	24, 27, 49
T. nitidulus (Fabricius)	27, 70, 91
T. obtusus (Linnaeus)	49
Tachinus signatus Gravenhorst	24, 27
Aleocharinae	
Amischa analis (Gravenhorst)*	32, 70
Atheta fungi (Gravenhorst)	70

less studies are available. Nevertheless, the importance of Stenus species in grasslands relative to cereals is apparent. Not many data are available on the occurrence of staphylinid larvae in agricultural soil (Table 4), and the neglect of larval taxonomy has not helped in this context.

TABLE 4. Staphylinid Larvae Common or Abundant in Cereal
Fields and Grassland in Northern Europe

As the complete larval staphylinid fauna has not been assessed in the literature available, percentage composition cannot be ascertained. Those listed as being common or abundant are included.

Genus or Species	References
Oxytelinae	
Oxytelus	85
Paederinae	
Lathrobium	53
Staphylininae	
Xantholinus	53
Philonthus	53, 104
Tachyporinae	
Tachyporus	85, 104, 120
Tachinus signatus Gravenhorst	24
Aleocharinae	
Atheta	85

3. POTENTIALLY PESTOPHAGOUS SPECIES

Problems of species identification and specialised feeding methods have made dietary studies difficult, therefore data on the food of staphylinid beetles and their larvae are scarce. A great many species are fluid-feeders (4, 18, 23, 39, 69, 78, 103), precluding the use of gut dissection, and even in species normally ingesting particulate matter, fluid-feeding may occur at times (64). This has led to the use of serological (9, 19, 20, 23, 28, 42, 44, 96, 104, 105, 120), electrophoretic (45) and radioisotope labelling (37) techniques, but some anomalies have developed. Sunderland et al. (104) recorded that more than 60% of individuals of several species gave negative results in ELISA tests immediately after feeding, and negative results were also obtained by electrophoresis under similar conditions (Good and Giller, unpublished). The widespread habit of extra-oral digestion among staphylinids (18, 38, 62, 78, 86) may account for these results, as well as for the rapid rates of digestion observed in these beetles (43, 97). The literature also contains many unfounded statements on their feeding preferences, in particular regarding the extent of saprophagy within the family (see review by Voris (121)).

A summary of diet records for adults of the genera listed in Tables 2-4 is given in Table 5. This is based on a more detailed list of staphylinid diet records to be published elsewhere (Good and Giller, in preparation). The assumption that different members of a polyphagous genus will have similar feeding habits is open to criticism, but given the paucity

TABLE 5. Summary of Diet Record for Staphylinid Genera (Adults) Common or Abundant in Cereal and Grassland

Genus	Recorded Food
Micropelus	Fungi
Anthobium	Collembola
Omalium	Oligochaeta
Anotylus	Diptera eggs and larvae
Oxytelus	Diptera and Coleoptera larvae, Nematoda
Stenus	Collembola, aphids, Elaterid and Curculionid larvae, Diptera
Lathrobium	Acari, Collembola
Gyrohypnus	Isopoda, Collembola, Lepidoptera pupae
Xantholinus	Formicid adults and larvae, immature Curculionids, Diptera larvae, Lepidoptera eggs
Philonthus	Diptera (all stages), aphids, Acari, Araneae, Collembola, Lepidoptera (all stages), Chrysomelid eggs and larvae, Isopoda, immature Elaterids and Curculionids, Staphylinid adults
Staphylinus	Nematoda, Diptera larvae, Formicid adults and larvae, Silphid larvae, Carabid adults, Dermaptera, Lepidoptera adults and larvae, Elaterid adults, Curculionid eggs, Heteroptera, Diplopoda
Quedius	Collembola, immature Muscids, Lepidoptera pupae and adults
Tachyporus	Collembola, aphids, Oscinellid eggs, Elaterids, Chrysomelid eggs and larvae, fungi
Tachinus	Collembola, Isopoda, Curculionid eggs, fungi
Dinaraea	Diptera larvae
Atheta	Green algae, Collembola, Diptera eggs, larvae and pupae, Enchytraeidae, Acari, Isopoda
Drusilla	Formicid larvae and adults
Ilyobates	Immature Curculionids
Oxypoda	Enchytraeidae, Nematoda, Lumbricidae, Diplopoda, Acari, Collembola, immature Curculionid, fungi, mouse carrion
Aleochara	Diptera eggs, larvae and pupae, Collembola, Acari, Lepidoptera eggs, bacon fat

(Based on full records to be published elsewhere (Good and Giller, in prep.))

87

of information currently available, a generic list is at present the only indication of the potential diet of a polyphagous species. On the basis of current information, three common genera must be omitted from the list of potentially pestophagous species. Micropeplus has been shown to have a potential role in the dispersal of certain fungi (54), and while beetle dispersal of cereal-associated fungi has been recorded (114), its role as a predator seems unlikely. Dennison and Hodkinson (23) recorded Anthobium as feeding solely on Collembola, and non-formicid records were not available for Drusilla canaliculata, so they are also excluded. No records have been found for the following genera: Carpelimus, Coprophilus, Platystethus, Sunius, Mycetoporus, Liogluta or Amischa. Until further data are available, they must also be excluded.

Relatively little is known of larval diet (Table 6); the ectoparasitoid larvae of Aleochara perhaps being an exception. Newton (75) noted that larval mouthparts appear to be more specialised than those of the adults (this may be due to the longevity of the adults in relation to specific food resources), and variation in mandible structure may be associated with feeding habits (17). Lack of mandibular brushes, presence of a nasale, an entire mandible without apical or preapical teeth, an articulated mala and a setose labium appear to be adaptations to extra-oral feeding and carnivory (62, 94). Such features are characteristic of the larvae of the Paederinae, Staphylininae and Steninae (62), all of which so far examined are predators. The mandibles of other larvae of unknown diet might be seen to possess predatory characteristics. Nevertheless, interpreting this as an indication of a predatory diet must be done with caution as much remains to be learnt about the functional morphology of staphylinid larval mouthparts. As can be seen from the diet records, the small size of the Aleocharinae does not exclude them from being potentially pestophagous. Their ability to feed on eggs and small stages of several pest species may be important.

TABLE 6. Summary of Diet Records for Staphylinid Genera (Larvae) Common or Abundant in Cereals and Grassland

Genus	Recorded Diet
Oxytelus	Insect eggs, Nematoda, Diptera larvae
Xantholinus	Diptera larvae
Philonthus	Acari, Diptera larvae, aphids, Coleoptera larvae
Staphylinus	Diplopoda, Lumbricids
Tachyporus	Aphids, Fungi, Collembola, Formicid larvae
Tachinus	Aphids, Collembola, Formicid and Diptera larvae
Atheta	Diptera larvae, Coleoptera eggs
Aleochara	Parasitoids of Diptera

(Based on full records to be published elsewhere (Good and Giller, in prep.)

4. OCCURRENCE IN THE SAME CHRONOTOPE AS THE PEST

To be an effective predator, a species must occur in the same place and at the same time as the pest (i.e. in the same chronotope). Therefore, a predator check-list should ideally be broken down into different chronotopes which will represent a particular vertical zone in the crop at a particular season. Little is known of the importance of Staphylinidae in the hypogeal habitat, but larvae are likely to be important. Several species have been associated with root aphids (30), and Philonthus decorus has been found to burrow into the soil to access hypogeal prey (43). Most species in Tables 2-4 have been recorded from the epigeal habitat (due to widespread use of pitfall traps), in particular the spring and summer epigeal habitat, and these are potentially important predators of dipteran larvae, dispersing aphids, etc. Winter epigeal species (Table 7) could only be important in relation to overwintering stages of certain pests; they will obviously not play a role in the dynamics of species like Oulema melanopa (L.) which overwinter away from the field (36). Plant-dwelling species will be most important in spring and summer (Table 7) as potential predators of leaf- and ear-feeding pests. The definition of plant chronotopes could be refined to include position of staphylinids on both crop and weed plants once more data are available on microhabitat preferences.

TABLE 7. Two Chronotope Lists for Cereal Staphylinidae

Winter Epidaphic (53)

Lesteva longoelytrata*	Tachinus fimetarius*	Lathrobium larvae
Omalium caesum*	Oxypoda lividipennis*	Philonthus larvae
Mycetoporus baudueri*	Atheta sp.	Xantholinus larvae

Summer plant (67, 112)

Anotylus complanatus*	Tachyporus hypnorum	Tachyporus larvae
A. sculpturatus	T. chrysomelinus	Amischa analis*
Philonthus varius	T. obtusus	A. cavifrons*
Atheta fungi	T. solutus	

Species marked with an asterisk recorded as common only in this chronotope.

5. EFFICIENCY OF PARASITISM

Five species of Aleochara are known to parasitise the wheat bulb fly and bean seed flies (Table 8) (see 19, 91, 101). Ryan (91) recorded 2-12.5% pupal parasitism of D. coarctata by Aleochara bipustulata and A. bilineata. Aleochara species have been found to be effective parasitoids in other hosts (41, 87, 107, 113). Few other detailed data are presently available.

6. EFFICIENCY OF PREDATION

There is an increasing awareness of the value of polyphagous predators in integrated pest management systems (71). Several studies based on exclusion experiments (11, 12, 21, 56, 98) and modelling (25) have shown polyphagous predators to be capable of reducing cereal aphid and other cereal insect populations. Direct observation (55) and the development of predation indices (106, 115) have contributed much to their evaluation. It might also be possible to evaluate the role of certain species by their selective removal from pitfall traps and suction samples, similar to the method of Clarke and Grant (13) for spiders. However, the ability of many staphylinids to colonise new areas rapidly by flight (in particular by Aleocharinae, Tachyporus and Philonthus (8, Good and Giller unpublished)

TABLE 8. Staphylinidae Recorded to Feed on Cereal Pests

Pest Species	Staphylinid Predators	References
Agriotes sputator (L.)	Philonthus spp.	42
	Staphylinus spp.	42
	Staphylinid larvae	42
Rhopalosiphum padi (L.)	Philonthus cognatus	96
	Tachyporus hypnorum	96
	T. chrysomelinus	96
Sitobion avenae (F.)	Philonthus cognatus	96, 104
	Tachyporus hypnorum	96, 110
	Tachyporus larvae	55, 104
Cereal aphids	Tachyporus obtusus	106
	T. chrysomelinus	106
	T. hypnorum	64, 103, 106
	Philonthus larvae	104
Oscinella frit (L.)	Tachyporus spp.	57
Delia coarctata (Fall.)	Aleochara verna larvae	65
	A. inconspicua larvae	29
	A. laevigata larvae	80
	Philonthus spp.	91
D. platura (Meig.)	A. bilineata larvae	80
	A. bipustulata larvae	80
D. florilega (Zett.)	A. bipustulata larvae	80
Oulema melanopa (L.)	Tachyporus spp.*	68

* Laboratory study

make exclusion experiments difficult to carry out. The possible value of flight interception barriers needs to be examined in this context.

A problem with generalising from the results of the above methods, and especially those based on laboratory consumption experiments for simulation models (97), is that they assume that predation rate on a given pest species is predictable at a given predator and prey density. For a polyphagous predator in the presence of alternative prey, this may not always be the case. For instance, the impact of Tachyporus larvae on cereal aphids is acknowledged (7, 55, 103, 106), but the role of fungi in their diet is not known. Despite Lipkow's (68) assertion that they are purely zoophagous (his fungal food tests were based entirely on baker's yeast), a very large proportion of larvae have been recorded with fungal material in their guts on some occasions (103). Of 20 cereal and grass sites surveyed in south west Ireland in 1985, the second highest plant-associated density of Tachyporus larvae was found in a field of winter wheat with very low aphid numbers. The gut contents of these larvae indicated that they had been feeding on ear and leaf fungi.

Staphylinids have been shown to be efficient predators in non-graminoid habitats (3, 4, 8, 9, 15, 16, 26, 30, 40, 41, 43, 51, 66, 73, 87, 90, 95, 107), but much more needs to be known of the nature of their natural food preferences. Several species have shown preferences in the laboratory (22, 34, 67, 73, 79), but the mechanisms governing diet choice and the possibility of switching (76) is not known. Functional responses by staphylinid predators have been demonstrated (66, 72), conforming to an asymptotic curve. Aggregative and reproductive numerical responses have also been demonstrated in this group (7, 66), and Kowalski (66) showed that a combined functional and reproductive numerical response by Philonthus resulted in density-dependent predation, thus the potential for pest-control is evident.

Three further points must be considered in the context of the role of staphylinids in integrated pest management programmes. The first is their trophic position. Large Philonthus and Staphylinus species may be important as tertiary consumers. Evans (39) recorded that Philonthus decorus would feed on Gyrohypnus and Stenus. Allen (1) found Philonthus tenuicornis feeding on Tachinus, and Staphylinus olens has been observed successfully attacking Carabus nemoralis (47) and the earwig, Forficula auricularia (48). Secondary predation, if common, would be a problem in interpreting results of biochemical diet analysis (105). The second point relates to virus spread (of importance if high predator populations are present with virus-carrying aphids). Roitberg and Myers (89) noted that predator-induced disturbance of aphids increased the spread of bean yellow mosaic virus, but the reverse was found in the case of beet western yellow virus when in the presence of predators (116). The effect of predators on the spread of barley yellow dwarf virus in Northern Europe needs to be clarified. The third and final point is concerned with interference competition at high predator densities. Such competition, while having a stabilising influence on the predator-prey interaction, lowers the numerical response (71). It is often manifested as cannibalism, which occurs in some Staphylinidae (10), even in the presence of surplus food (52).

7. EVALUATION

The evaluation of the importance of staphylinid predators will clearly depend on the major pests in a given region. In south west Ireland, cereal aphids (Sitobion avenae, Metopolophium dirhodum and Rhopalosiphum padi) and frit-fly (Oscinella frit) are the major pests of cereals and grasses. Simulation models of cereal aphid populations (25) and life table analyses for frit-fly (2, 99) both indicate an important role for polyphagous predators in their population dynamics. Both aphids and frit-fly almost always occur in the epigeal and plant zones, so hypogeal staphylinids are unlikely to be important. The importance of winter epigeal species in relation to aphid survival is not known, so they cannot be eliminated from a list of potential predators. Furthermore, both pest species occur in both cereal and grass crops. At this stage in our knowledge then, most of the species listed in Tables 2-4 could be included in the potential pestophagous guilds of both these pest species.

The importance of the total assemblage of polyphagous predators rather than any individual component species has recently been stressed (35, 88). Discussing spiders as polyphagous predators, Riechert and Lockley (88) state, "The spider buffering effect can only be achieved by the composite foraging activities of the assemblage of spider species in a given habitat: no given spider species, no matter how abundant, can hold a prey population in check, since its population does not track the density of the pest population. Thus community diversity must be maintained to maximise the number of predators that will encounter the pest species." Yet, management changes, even with the object of improving conditions for beneficial predators, are likely to affect certain species more than others. The evidence for preference given above indicates that a simple model of predation based on encounter rate is unlikely to be correct. If the impact of polyphagous staphylinid predators is to be predictable, the mechanisms governing preference will have to be understood for each species. Then, given the density of the pest species and of alternative prey, it may be possible to predict predation rate and the likely impact of the pestophagous staphylinid guild on the pest population.

8. ACKNOWLEDGEMENTS

We are most grateful to J. O'Halloran and K.G.M. Bond for comments on an earlier draft of this paper. We would also like to thank Ms. Irene O'Sullivan who typed this manuscript at short notice.

9. ERRATUM

Reference to *Omalium* and *Lesteva* is omitted from the discussion of staphylinid diet. Steel (102) records adults and larvae of both these genera as predators.

REFERENCES

(1) ALLEN, A.A., 1957. *Tachinus* species selected as prey by *Philonthus tenuicornis* Muls. and Rey (Col., Staphylinidae). Ent. mon. Mag. 93:94.

(2) ALLEN, W.A. and PIENKOWSKI, R.L., 1975. Life tables for the frit fly, *Oscinella frit*, in reed canary grass in Virginia. Ann. Ent. Soc. Amer. 68: 1001-1007.

(3) ANDERSEN, A., HANSEN, A.G., RYDLAND, N. AND ØYRE, G., 1983. Carabidae and Staphylinidae (Col.) as predators of eggs of the turnip root fly *Delia floralis* Fallen (Diptera, Anthomyiidae) in cage experiments. Z. ang. Entomol. 95: 499-506.

(4) BALDUF, W.V., 1935. The bionomics of entomophagous, Coleoptera. E.W. Classey, Hampton, Middlesex.

(5) BASEDOW, T., RZEHAK, H. and VOB, K., 1985. Studies on the effect of deltamethrin sprays on the numbers of epigeic predatory arthropods occurring in arable fields. Pestic. Sci. 16: 325-331.

(6) BOYD, J.M., 1960. Studies of the differences between the fauna of grazed and ungrazed grassland in Tiree, Argyll. Proc. Zool. Soc. Lond. 135: 33-54.

(7) BRYAN, K.M. and WRATTEN, S.D., 1984. The responses of polyphagous predators to prey spatial heterogeneity: aggregation by carabid and staphylinid beetles to their cereal aphid prey. Ecol. Entomol. 9: 251-259.

(8) BURN, A.J., 1982. The role of predator searching efficiency in carrot fly egg loss. Ann. appl. Biol. 101: 154-159.

(9) CALVER, M.C., MATTHIESSEN, J.N., HALL, G.P., BRADLEY, J.S. and LILLYWHITE, J.H., 1986. Immunological determination of predators of the bush fly, *Musca vetustissima* Walker (Diptera: Muscidae), in south western Australia. Bull. ent. Res. 76: 133-139.

(10) CAMPBELL, J.B. and HERMANUSSEN, J.F., 1974. *Philonthus theveneti*: life history and predatory habits against stable flies, house flies and face flies under laboratory conditions. Environ. Ent. 3: 356-358.

(11) CHIVERTON, P., 1982. Effekten av rovlevande skalbaggar och splindar på havrebladlusens (*Rhopalosiphum padi*) tidiga populationsutveckling i vårkorn. Växtskyddsrapporter, Jordbruk 20: 177-181.

(12) CHIVERTON, P., 1986. Predator density manipulation and its effects on populations of Rhopalosiphum padi (Hom., Aphididae) in spring barley. Ann. appl. Biol. 109: 49-60.

(13) CLARKE, R.D. and GRANT, P.R., 1968. An experimental study of the role of spiders as predators in a forest litter community. P.1. Ecology 49: 1152-1154.

(14) CLEMENTS, R.O. and HENDERSON, I.F., 1979. Insects as a cause of botanical change in swards. In: Changes in sward composition and productivity, Occasional Symposium No.10, British Grassland Society, York, 1978, 157-160.

(15) COAKER, T.H., 1965. Further experiments on the effect of beetle predators on the numbers of cabbage root fly, Erioischia brassicae (Bouché), attacking brassica crops. Ann. appl. Biol. 56: 7-20.

(16) COAKER, T.H. and WILLIAMS, D.A., 1963. The importance of some Carabidae and Staphylinidae as predators of the cabbage root fly, Erioischia brassicae (Bouché). Entomol. exp. appl. 6: 156-164.

(17) COFFAIT, H., 1972. Coléoptères Staphylinides de la Région Paléarctique Occidentale. 1. Généralités; sous-familles Xantholininae et Leptotyphlinae. Nouv. Rev. Ent. (Suppl.) 2,2,1-651.

(18) CROWSON, R.A., 1981. The biology of the Coleoptera. Academic Press, London.

(19) DANTHANARAYANA, W., 1969. Population dynamics of Sitona regensteinensis (Hbst.). J. anim. Ecol. 38: 1-19.

(20) DANTHANARAYANA, W., 1983. Population ecology of the light brown apple moth, Epiphyas postvittana (Lepidoptera: Tortricidae). J. anim. Ecol. 52: 1-33.

(21) DECLERCQ, R. and PIETRASZKO, R., 1983. Epigeal arthropods in relation to predation of cereal aphids. In: Cavalloro, R. (ed.) Aphid antagonists. A.A. Balkema, Rotterdam. pp.88-92.

(22) DELANY, M.J., 1960. The food and feeding habits of some heath-dwelling invertebrates. Proc. Zool. Soc. Lond. 135: 303-311.

(23) DENNISON, D.F. and HODKINSON, I.D., 1983. Structure of the predatory beetle community in a woodland soil ecosystem. 1. Prey selection. Pedobiologia 25: 109-115.

(24) DESENDER, K., MERTENS, J., D'HULSTER, M. and BERBIERS, P., 1984. Diel activity patterns of Carabidae (Coleoptera), Staphylinidae (Coleoptera) and Collembola in a heavily grazed pasture. Rev. Ecol. Biol. Sol 21: 347-361.

(25) DEWAR, A.M. and CARTER, N., 1984. Decision trees to assess the risk of cereal aphid (Hemiptera: Aphididae) outbreaks in summer in England. Bull. ent. Res. 74: 387-398.

(26) DICKER, G.H.L., 1944. Tachyporus (Coleoptera, Staphylinidae) larvae preying on aphids. Ent. mon. Mag. 80: 71.

(27) D'HULSTER, M. and DESENDER, K., 1982. Ecological and faunal studies on Coleoptera in agricultural land. III. Seasonal abundance and hibernation of Staphylinidae in the grassy edge of a pasture. Pedobiologia 23: 403-414.

(28) DOANE, J.F., SCOTTI, P.D., SUTHERLAND, O.R.W. and POTTINGER, R.P., 1985. Serological identification of wireworm and staphylinid predators of the Australian soldier fly (Inopus rubriceps) and wireworm feeding on plant and animal food. Entomol. exp. appl. 38: 65-72.

(29) DOBSON, R.M., 1961. Observations on natural mortality, parasites and predators of wheat bulb fly, Leptohylemyia coarctata (Fall.). Bull. ent. Res. 52: 281-291.

(30) DUNN, J.A., 1960. The natural enemies of the lettuce root aphid, Pemphigus bursarius (L.). Bull. ent. Res. 51: 271-278.

93

(31) EAST, R., 1974. Predation on the soil-dwelling stages of the winter moth at Wytham Woods, Berkshire. J. anim. Ecol. 43: 611-626.

(32) EDWARDS, E.E., 1929. A survey of the insect and other invertebrate fauna of permanent pasture and arable land of certain soil types at Aberystwyth. Ann. appl. Biol. 16: 299-323.

(33) EDWARDS, C.A., BUTLER, C.G. AND LOFTY, J.R., 1975. The invertebrate fauna of the Park Grass Plots. II. Surface fauna. Rothamsted Report for 1975, 2: 63-89.

(34) EGHTEDAR, E., 1970. Zur Biologie und Ökologie der Staphyliniden Philonthus fuscipennis Mannh. und Oxytelus rugosus Grav. Pedobiologia 10: 169-179.

(35) EKBOM, B.S. and WIKTELIUS, S., 1985. Polyphagous arthropod predators in cereal crops in central Sweden, 1979-1982. Z. ang. Entomol. 99: 433-442.

(36) EMPSON, D.V. and GAIR, R., 1982. Cereal pests. Ministry of Agriculture, Fisheries and Food Reference Book 186. HMSO, London.

(37) ERNSTING, G. and JOOSSE, E.N.G., 1974. Predation on two species of surface dwelling Collembola. A study with radio-isotope labelled prey. Pedobiologia 14: 222-231.

(38) EVANS, M.E.G., 1964. A comparative account of the feeding methods of the beetles Nebria brevicollis (F.) (Carabidae) and Philonthus decorus (Grav.) (Staphylinidae). Trans. Roy. Soc. Edinb. 66: 91-109.

(39) EVANS, M.E.G., 1967. Notes on feeding in some predaceous beetles of the woodland floor. Entomologist 100: 300-303.

(40) FAYAD, Y.H., HAFEZ, M. and EL-KIFL, A.H., 1979. Survey of the natural enemies of the three corn borers Sesamia cretica Led., Chilo agamemnon Bles. and Ostrinia nubilalis Hbn., in Egypt. Agric. Res. Rev. Cairo 57: 29-33.

(41) FINCH, S. and COLLIER, R.H., 1984. Parasitisation of overwintering pupae of cabbage root fly, Delia radicum (L.) (Diptera, Anthomyiidae), in England and Wales. Bull. ent. Res. 74: 79-86.

(42) FOX, C.J.S. and MACLELLAN, C.R., 1956. Some Carabidae and Staphylinidae shown to feed on a wireworm Agriotes sputator (L.) by precipitin test. Can. Ent. 88: 228-231.

(43) FRANK, J.H., 1967a. The effect of pupal predators on a population of winter moth, Operophtera brumata (L.) (Hydriomenidae). J. anim. Ecol. 36: 611-621.

(44) FRANK, J.H., 1967b. A serological method used in the investigation of the predators of the pupal stage of the winter moth, Operophtera brumata (L.) (Hydriomenidae). Quaest. entomol. 3: 95-105.

(45) GOOD, J.A. and GILLER, P.S., 1987. Impact of crop management on staphylinid diet variability: a preliminary evaluation of electrophoretic prey detection. In: Fourth European Ecology Symposium, Ecological Implications of Contemporary Agriculture, Wageningen.

(46) GOOD, J.A. and GILLER, P.S. (in preparation).

(47) GRADWELL, G.R., 1959. Carabus nemoralis Muell. as prey of Staphylinus olens Muell. (Col.). Ent. mon. Mag. 95: 43.

(48) GREATHEAD, D.J., 1976. A review of biological control in Southern and Western Europe. Commonwealth Agricultural Bureaux, Slough, UK.

(49) GREEN, J., 1953. The beetles of a Cheshire farm. Ent. mon. Mag. 89: 81-86.

(50) HANSKI, I. and TIAINEN, J., 1987. Population dynamics in changing agroecosystems. In: Fourth European Ecology Symposium, Ecological Implications of Contemporary Agriculture, Wageningen.

(51) HARRIS, R.L. and OLIVER, L.M., 1979. Predation of Philonthus flavolimbatus on the horn fly. Environ. Ent. 8: 259-260.

(52) HEESSEN, H.J.L. and BRUNSTING, A.M.H., 1981. Mortality of larvae of *Pterostichus oblongopunctatus* (Fabricius) (Col., Carabidae) and *Philonthus decorus* (Gravenhorst) (Col., Staphylinidae). Neth. J. Zool. 31: 729-745.

(53) HEYDEMANN, B., 1956. Untersuchungen über die Winteraktivität von Staphyliniden auf Feldern. Entomol. Blätt. 52: 138-150.

(54) HINTON, H.E. and STEPHENS, F.L., 1941. Notes on the food of *Micropeplus,* with a description of the pupa of *M. fulvus* Erichson (Coleoptera, Micropeplidae). Proc. R. ent. Soc. Lond. (A) 16: 29-32.

(55) HOLMES, P.R., 1984. A field study of the predators of the grain aphid, *Sitobion avenae* (F.) (Hemiptera: Aphididae), in winter wheat in Britain. Bull. ent. Res. 74: 623-631.

(56) JONES, M.G., 1965. The effects of some insecticides on populations of frit fly (*Oscinella frit*) and its enemies. J. appl. Ecol. 2: 391-401.

(57) JONES, M.G., 1969. The effect of weather on frit fly (*Oscinella frit*) and its predators. J. appl. Ecol. 6: 425-441.

(58) JONES, M.G., 1976a. The arthropod fauna of a winter wheat field. J. appl. Ecol. 13: 61-85.

(59) JONES, M.G., 1976b. Arthropods from fallow land in a winter wheat-fallow sequence. J. appl. Ecol. 13: 87-101.

(60) JONES, M.G., 1976c. The carabid and staphylinid fauna of winter wheat and fallow on a clay with flints soil. J. appl. Ecol. 13: 775-791.

(61) KAJAK, A., 1980. Invertebrate predator subsystem. In: Breymeyer, A.I. and van Dyne, G.M. (eds.) Grasslands, systems analysis and man. Cambridge University Press, Cambridge, UK.

(62) KASULE, F.K., 1970. The larvae of Paederinae and Staphylininae (Coleoptera: Staphylinidae) with keys to the known British genera. Trans. R. ent. Soc. Lond. 122: 49-80.

(63) KELLY, M.T. and CURRY, J.P., 1985. Studies on the arthropod fauna of winter wheat crop and its response to the pesticide methiocarb. Pediobiologia 28: 413-421.

(64) KENNEDY, T.F., EVANS, G.O. and FEENEY, A.M., 1986. Studies on the biology of *Tachyporus hypnorum* F. (Col., Staphylinidae) associated with cereal fields in Ireland. Ir. J. Agric. Res. 25: 81-95.

(65) KLIMASZEWSKI, J., 1984. A revision of the genus *Aleochara* Gravenhorst of America North of Mexico (Coleoptera: Staphylinidae: Aleocharinae). Mem. ent. Soc. Can. 129: 1-211.

(66) KOWALSKI, R., 1977. Further elaboration of the winter moth population models, J. anim. Ecol. 46: 471-482.

(67) LÄITINEN, T. and RAATIKÄINEN, M., 1981. Composition and zonation of the beetle fauna of oatfields in Finland. Ann. ent. Fenn. 47: 33-42.

(68) LIPKOW, E., 1966. Biologisch-ökologische Untersuchungen über *Tachyporus*-arten und *Tachinus rufipes* (Col., Staphyl.). Pedobiologia 6: 140-177.

(69) LIPKOW, E., 1982. Lebensweise von *Philonthus*-Arten und anderen Staphylinidae (Coleoptera) des Dungs. Drosera 1982: 47-54.

(70) LUFF, M.L., 1965. A list of Coleoptera occurring in grass tussocks. Ent. mon. Mag. 101: 240-245.

(71) LUFF, M.L., 1983. The potential of predators for pest control. Agric. Ecosyst. Environ. 10: 159-181.

(72) MILLER, K.V. and WILLIAMS, R.N., 1982. Expansion of the Holling Disc Equation to include changing prey densities. J. Georgia Entomol. Soc. 17: 404-410.

(73) MILLER, K.V. and WILLIAMS, R.N. 1983. Biology and host preference of *Atheta coriaria* (Coleoptera: Staphylinidae), an egg predator of Nitidulidae and Muscidae. Ann. Ent. Soc. Am. 76: 158-161.

(74) MORRIS, H.M., 1922. The insect and other invertebrate fauna of arable land at Rothamsted. Ann. appl. Biol. 7: 282-305.

(75) MORRIS, M.G., 1968. Differences between the invertebrate faunas of grazed and ungrazed chalk grasslands. II. The fauna of sample turves. J. appl. Ecol. 5: 601-611.

(76) MURDOCH, W.W., 1969. Switching in general predators. Experiments on predator specificity and stability of prey population. Ecol. Monogr. 39: 335-354.

(77) NABIAŁCZYK-KARG, J., 1980. Density and biomass of soil inhabiting insect larvae in a rape field and in a meadow. Pol. Ecol. Stud. 6: 305-316.

(78) NEWTON, A.F., Jr., 1984. Mycophagy in the Staphylinoidea. In: Wheeler, Q. and Blackwell, M. (eds.) Fungus-insect relationships: perspectives in ecology and evolution. Columbia University Press, New York. pp.302-353.

(79) OUAYOGODE, B.V. and DAVIS, D.W., 1981. Feeding by selected predators on alfalfa weevil larvae. Environ. Ent. 10: 62-64.

(80) PESCHKE, K. and FULDNER, D., 1977. Übersicht und neue Untersuchungen zur Lebensweise der parasitoiden Aleocharinae (Coleoptera; Staphylinidae). Zool. Jb. Syst. 104: 242-262.

(81) PIETRASZKO, R. and DECLERCQ, R., 1978. Studie van de Staphylinidae-fauna in wintertarwevelden. Parasitica 34: 191-198.

(82) PIETRASZKO, R. and DECLERCQ, R.,1980a. Etude de la population d'arthropodes epigés dans les cultures agricoles au cours de la période 1974-1978. Revue Agric. 33: 719-733.

(83) PIETRASZKO, R. and DECLERCQ, R., 1980b. Studie van de bovengrondse arthropodenfauna in landbouwgewassen gedurende de periode 1974-1978. Landbouwtijds. 33: 711-724.

(84) PIETRASZKO, R. and DECLERCQ, R., 1982. Influence of organic matter on epigeic arthropods. Med. Fac. Landbouww. Rijksuniv. Gent. 47: 721-728.

(85) POTTS, G.R. and VICKERMAN, G.P., 1974. Studies on the cereal ecosystem. Adv. Ecol. Res. 8: 107-187.

(86) QUAYLE, H.J., 1912. Red spiders and mites of citrus trees. Calif. Agr. Exp. Sta. Bul. 234: 483-530.

(87) READ, D.C., 1962. Notes on the life history of Aleochara bilineata (Gyll.) (Coleoptera: Staphylinidae), and on its potential value as a control agent for the cabbage maggot, Hylemya brassicae (Bouché) (Diptera: Anthomyiidae). Can. Ent. 94: 417-424.

(88) RIECHERT, S.E. and LOCKLEY, T., 1984. Spiders as biological control agents. Ann. Rev. Entomol. 29: 299-320.

(89) ROITBERG, B.D. and MYERS, J.H., 1978. Effect of adult Coccinellidae on the spread of a plant virus by an aphid. J. appl. Ecol. 15: 775-779.

(90) ROTH, J.P., FINCHER, G.T. and SUMMERLIN, J.H., 1983. Competition and predation as mortality factors of the horn fly, Haematobia irritans (L.) (Diptera: Muscidae), in a central Texas pasture habitat. Environ. Ent. 12: 106-109.

(91) RYAN, M.F., 1975. The natural mortality of wheat-bulb fly eggs in bare fallow soils. J. appl. Ecol. 10: 869-874.

(92) SALT, G., HOLLICK, F.S.J., RAW, F. and BRIAN, M.V., 1948. The arthropod population of pasture soil. J. anim. Ecol. 17: 139-150.

(93) SHUKLA, G.S. and UPADHYAY, V.B., 1980. Morphology of mouthparts of Pheropsophus occipitalis Mael. (Coleoptera: Carabidae). J. anim. Morphol. Physiol. 27: 302-311.

(94) SILTATALA, 1967. cited in 93.

(95) SILVESTRI, F., 1945. Descrizone e biologia del coleottero stafilinide Belonuchus formosus Grav. introdotto in Italia per la lotta contro ditteri tripaneide. Boll. R. Lab. Ent. agr. Portici 5: 312-326.

(96) SOPP, P. and CHIVERTON, P. (in press) Autumn predation of cereal aphids by polyphagous predators in Southern England: a 'first look' using ELISA. Bull. SROP/WPRS.

(97) SOPP, P. and WRATTEN, S.D., 1986. Rates of consumption of cereal aphids by some polyphagous predators in the laboratory. Entomol. exp. appl. 41: 69-73.

(98) SOTHERTON, N.W., WRATTEN, S.D. and VICKERMAN, G.P., 1985. The role of egg predation in the population dynamics of Gastrophysa polygoni (Coleoptera) in cereal fields. Oikos 43: 301-308.

(99) SOUTHWOOD, T.R.E. and JEPSON, W.F., 1962. Studies of the population of Oscinella frit L. (Dipt., Chloropidae) in the oat crop. J. anim. Ecol. 31: 481-495.

(100) SPEIGHT, M.R. and LAWTON, J.H., 1976. The influence of weed-cover on the mortality imposed on artificial prey by predatory ground beetles in cereal fields. Oecologia 23: 211-223.

(101) SPEYER, W., 1954. Aleochara laevigata Gyll. (Coleop., Staphylinidae) als Puppenparasit der Brachfliege Hylemyia coarctata Fall. Nachrichtenbl. dtsch. Pflanzenschutzdienst. 6: 6-7.

(102) STEEL, W.O., 1970. The larvae of the genera of the Omaliinae (Coleoptera: Staphylinidae) with particular reference to the British fauna. Trans. R. ent. Soc. Lond. 122: 1-47.

(103) SUNDERLAND, K.D., 1975. The diet of some predatory arthropods in cereal crops. J. appl. Ecol. 12: 507-515.

(104) SUNDERLAND, K.D., CHAMBERS, R.J., STACEY, D.L. and CROOK, N.E., 1985. Invertebrate polyphagous predators and cereal aphids. Bull. SROP/WPRS 1985/VIII/3. pp.105-114.

(105) SUNDERLAND, K.D. and SUTTON, S.L., 1980. A serological study of arthropod predation on woodlice in a dune grassland ecosystem. J. anim. Ecol. 49: 987-1004.

(106) SUNDERLAND, K.D. and VICKERMAN, G.P., 1980. Aphid feeding by some polyphagous predators in relation to aphid density in cereal fields. J. appl. Ecol. 17: 389-396.

(107) SYCHEVSKAYA, V.I., 1972. Aleocharinae (Coleoptera, Staphylinidae) as natural enemies of synanthropic flies of the family Sarcophagidae in the central Asia. Zool. Zhurn. 51: 142-144.

(108) TISCHLER, W., 1958. Synökologische Untersuchungen an der Fauna der Felder und Feldgehölze (Ein Beitrag zur Ökologie der Kulturlandschaft). Z. Morph. Oekol. Tiere 47: 54-114.

(109) TOPP, W., 1977. Einfluss des Strukturmosaiks einer Agrarlandschaft auf die Austbreitung der Staphyliniden (Col.). Pedobiologia 17: 43-50.

(110) VICKERMAN, G.P., 1974. Some effects of grass weed control on the arthropod fauna of cereals. Proc. 12th Brit. Weed Control Conf., 1974, 929-939.

(111) VICKERMAN, G.P., 1978. The arthropod fauna of undersown grass and cereal fields. Scient. Proc. R. Dubl. Soc. (A) 6: 273-283.

(112) VICKERMAN, G.P. and SUNDERLAND, K.D., 1975. Arthropods in cereal crops: nocturnal activity, vertical distribution and aphid predation. J. appl. Ecol. 12: 755-766.

(113) VORIS, R., 1934. Biologic investigations on the Staphylinidae (Coleoptera). Trans. Acad. Sci. St. Louis 28: 233-261.

(114) WALLIN, H., WIKTELIUS, S. and EKBOM, B.S., 1981. Förekomst och utbredning av skalbaggar i vårkorn. Ent. Tidskr. 102: 51-56.

(115) WALKER, M.A., 1985. A pitfall trap study on Carabidae and Staphylinidae (Col.) in County Durham. Ent. mon. Mag. 121: 9-18.

(116) WATT, A.D., 1981. Wild grasses and the grain aphid (Sitobion avenae). In: Thresh, J.M. (ed.) Pests, pathogens and vegetation. pp.299-305.

(117) WHITE, E.B. and LEGNER, E.F., 1966. Notes on the life history of *Aleochara taeniata*, a staphylinid parasite of the house fly, *Musca domestica*. Ann. ent. Soc. Am. 59: 573-577.

(118) WINDELS, C.E., WINDELS, M.B. and KOMMEDAHL, T., 1976. Association of *Fusarium* species with picnic beetles on corn ears. Phytopath. 66: 328-331.

(119) WRATTEN, S.D., BRYAN, K., COOMBES, D. and SOPP, P.I., 1984. Evaluation of polyphagous predators of aphids in arable crops. Proc. Brit. Crop Prot. Conf. - Pests and Diseases, 1984, 271-276.

(120) WRATTEN, S.D. and PEARSON, J., 1982. Predation of sugar beet aphids in New Zealand. Ann. appl. Biol. 101: 178-181.

The occurrence of spiders in cereal fields

A.M.Feeney & T.Kennedy
Plant Pathology & Entomology Department, An Foras Taluntais, Oak Park, Carlow, Ireland

Summary

Investigations carried out in Ireland 1979-83 on the effects of pesticides on some of the fauna of cereal fields suggested that spiders were among the groups affected adversely (1). These have been considered in recent years to be potentially important predators of cereal aphids (3) but detailed information is scanty on their predatory behaviour (4, 2).

1. MATERIALS AND METHODS

During the period May-September 1985, preliminary investigations were initiated using pitfall traps to determine the species occurring and the relative numbers of each species in a winter wheat crop. Five pitfall traps 6.5 cm diameter were placed at random in a field of 2 ha at Oak Park. The previous cropping was continuous cereals.

Catches were examined at fortnightly intervals and stored in 70% alcohol for later identification.

From November 1985 to November 1986 monitoring of spiders in winter wheat was carried out by pitfall traps, D-Vac suction sampling, visual inspection of soil surface and examination of webs.

Twenty pitfall traps of 10 cm diameter were placed in four rows at predetermined positions on an area 93 m x 49.5 m. D-Vac sampling was carried out at approximately weekly intervals on an adjacent area of 1.5 ha and comprised 20 sub-samples of 10 seconds duration. Visual inspections were carried out on 16 quadrats 30 cm x 30 cm, permanently placed within the 1.5 ha area and spiders captured by pooter. Another 10 quadrats were examined for web formation and presence of spiders.

2. RESULTS AND DISCUSSION

May-September 1985: Table 1 gives the results of the preliminary period of trapping, May-Septmber 1985, for the family Linyphiidae. Members of this family occurred in much greater numbers than those of any other family. Erigone atra and E. dentipalpis were the dominant species and males greatly outnumbered females. This appears normal for most species within the family, but with the genus Oedothorax, females outnumbered males two to threefold. Considerable numbers of Meioneta, Bathyphantes, Lepthyphantes and Savignia also occurred. Altogether, 24 different species were captured. All other species captured over the period are given in Table 2, which shows the relatively small numbers of non-Linyphiid members occurring. The genera Trochosa and Lycosa were most common in the grouping.

November 1985 - April 1986: Results of monitoring after November 1985 are only available up to April 1986 for pitfall and D-Vac sampling but for

TABLE 1. Relative Numbers of the Most Common Linyphiid
Spiders in Pitfall Traps from May to September 1985

Species	Male	Female
Erigone atra	516	17
E. dentipalpis	334	34
Oedothorax fuscus	66	197
O. apicatus	39	83
Meioneta rurestris	57	1
Bathyphantes gracilis	42	11
Lepthyphantes tenuis	23	5
Savignia frontata	31	12

TABLE 2. Relative Number of Non-Linyphiid Spiders Caught
in Pitfall Traps from May to September 1986

Species	Male	Female
Trocosa ruricola	22	11
Lycosa amentata	14	5
Pachygnatha degeeri	7	1

visual examination of the soil surface, results are available up to mid-May. Monitoring results of spider occurrence on webs are available up to harvest.

Pitfall Trapping: Spider catches from 26 September 1985 to 29 April 1986 comprised six families, namely Linyphiidae, Tetragnathidae, Clubionidae, Thomisidae, Dictynidae and Lycosidae. Linyphiidae comprised 99.3% of the total catch and consisted of 29 genera. The family Lycosidae was represented by three genera and all others by a single genus. As in the earlier sampling Erigone atra and E. dentipalpis were the dominant species and together account for 61% of the total linyphiid catch. Males were caught in much higher numbers than females and this was constant throughout the family except in the case of Oedothorax (represented by two species). Table 3 is a list of the total linyphiid species captured together with their relative numbers.

D-Vac Sampling: Members and diversity of species caught by D-Vac sampling (Table 4) was much smaller than by trapping and this reflects the non-continuous nature of the method. Erigone atra, Meioneta rurestris and Bathyphantes gracilis were the dominant species.

Quadrat Examination: From December to April inclusive, spider occurrence was negligible but during May there appeared to be a pronounced increase as shown in Table 5.

Web Formation: Web formation throughout the winter and early spring was very restricted but from May onwards there was a major increase. This was reflected in an increase in numbers of spiders captured in webs. The species are shown in Table 6, which suggests that females mainly frequent the webs while males were observed predominantly on the soil surface. Two species, Bathyphantes gracilis and Lepthyphantes tenuis were captured most commonly.

TABLE 3. Total Linyphiid Spiders Caught From 20 Pitfall Traps, over the Period 26/11/86 to 29/4/86

Species	Male	Female	Species	Male	Female
Erigone dentipalpis	1064	153	Dicymbium nigrum	7	12
Erigone atra	1937	238	D. tibiale	1	1
Centromerita concinna	21	4	Tapinocymba praecox	0	1
Oedothorax fuscus	26	198	Tiso vagans	1	0
O. retusus	1	0	Araeoncus humilis	1	0
Oedothorax apicatus	4	16	Troxochrus scabrichulus	6	0
Meioneta rurestris	453	62	Monocephalus fuscipes	7	5
Bathyphantes gracilis	507	125	Centromerus bicolor	1	0
Lepthyphantes tenuis	147	60	C. expertus	2	0
L. insignia	2	1	Erigonella heimalis	5	0
L. zimmermanni	0	1	Walckenaera acuminata	1	0
Savignia frontata	171	49	Leptorhoptrum robustrum	0	1
Diplocephalus latifrons	2	0	Bolyphantes luteolus	1	0
D. cristatus	0	1	Hypomma cornutum	0	1
D. permixtus	2	0	Oreonetides firmus	0	1
Porrhomma pygmaeum	4	5	Labulla thoracica	0	1
P. errans	1	3	Trachynella nudipalpis	1	0
Silometopus interjectus	7	5	Wideria antica	0	1
Gongylidellium vivum	1	3	Poeciloneta globosa	33	0
Ostearius melanopygius	5	7	Immatures	57	63

TABLE 4. Linyphiid Spiders Captured by D-Vac Suction Sampling (20 subsamples per sample) over the Period 28/11/85 to 1/5/86

Species	Male	Female
Erigone atra	6	11
Oedothorax fuscus	1	1
Meioneta rurestris	7	5
Bathyphantes gracilis	9	4
Lepthyphantes tenuis	0	4
Porrhomma pygmaeum	0	1
Porrhomma errans	1	0
Erigonella hiemalis	1	0
Immatures	12	20

TABLE 5. Total Number of Spiders Captured from 16 Quadrats over
the Period 10/12/85 to 20/5/86

| Species | 10/12/85 to 29/4/86 | | 8-20/5/86 | |
	Male	Female	Male	Female
Erigone dentipalpis	2	1	0	0
Erigone atra	0	2	1	3
Oedothorax fuscus	0	0	0	1
Meioneta rurestris	0	0	1	3
Bathyphantes gracilis	0	0	1	6
Lepthyphantes tenuis	0	0	2	3
Savignia frontata	0	0	2	2
Linyphia impigra	0	0	0	1
Immatures	1	0	2	1

TABLE 6. Total Number of Linyphiid Spiders Captured From
Webs over the Period 8/5/86 to 16/7/86

Species	Male	Female
Erigone atra	0	4
Oedothorax fuscus	0	1
Oedothorax apicatus	0	1
Meioneta rurestris	0	4
Bathyphantes gracilis	2	25
Lepthyphantes tenuis	7	19

REFERENCES

(1)   FEENEY, A.M., 1984. The effects of pesticides on some of the fauna of
       cereal fields. CEC Programme on Integrated and Biological Control.
       Final Report 1979/1983. EUR 8689 EN433-441.
(2)   NYFFELER, M. and BENZ, G., 1979. Studies on the ecological importance
       of spider populations of the field layer of cereal and rape fields
       near Zurich, Switzerland. Zeitschrift für angewandte Entomologie, 87,
       348-367.
(3)   SUNDERLAND, K.D., CHAMBERS, R.J., STACEY, D.L. and CROOK, N.E., 1985.
       Invertebrate polyphagous predators and cereal aphids. Bulletin
       SROP/WPRS VIII/3, pp. 105-114.
(4)   VICKERMAN, G.P. and SUNDERLAND, K.D., 1975. Arthropods in cereal
       crops: nocturnal activity, vertical distribution and aphid predation.
       Journal of Applied Ecology, 12, 755-766.

Study of some components of BYDV epidemiology in the Rennes basin

Monique Henry
Laboratoire d'Entomologie Fondamentale et Appliquée, Université de Rennes, France
H. Gillet
GRISP, Le Rheu, France
C.A. Dedryver
INRA, Laboratoire de Zoologie, Le Rheu, France

Summary

BYDV has been of increasing importance in the western part of France since 1982 and experiments were undertaken from 1984 onward for a better knowledge of the epidemiology of the virus disease and a better forecast of its risks in this area. The main isolate of the virus (PAV) has been found on three different crops: on barley between December and June, on maize between June and October and on some cultivated grasses (brome-grass, fescue and Italian rye-grass) in autumn and spring. It is thus obvious that the virus inoculum is present all year round in the Rennes basin and that perennial grasses could be important as permanent sources of the virus. All these crops are also reservoirs of the main virus vector, Rhopalosiphum padi L. and consequently all the conditions are combined for the disease to be endemic in this region.

1. INTRODUCTION

Barley yellow dwarf virus (BYDV) is an aphid-borne virus agent of a severe disease of Gramineae; its spread is increasing in the western part of France. In 1982, 1984 and 1985 it was considered of economic importance in Brittany, especially on early-sown winter barley.

The western and south-western parts of France seem particularly favourable to the BYDV epidemiology due to specific climatic and agronomic reasons:

1) In these oceanic regions the mild climate generally allows the anholocyclic populations of Gramineae aphids to overwinter and thus to reproduce parthenogenetically all year round without a break in their cycle (3).

2) The cropping systems ensure throughout the year very large areas cultivated with different species of Gramineae (5) which are hosts of the virus and its vectors and multiply them to different degrees. The growing periods of some of these crops follow one another during the year which makes them temporary reservoirs of virus and aphids, e.g. winter cereals, maize and Italian rye-grass. Other crops are perennial, e.g. most species of cultivated grasses; they may be considered as permanent sources of virus and vectors.

In this paper, the situation in the years 1984-1985 is given as an example to study the main sequence of annual BYDV epidemiology on winter barley and maize and the role of long-term reservoirs of mono or pluriannual cultivated grasses in the Rennes basin.

2. MATERIALS AND METHODS

At Le Rheu Research Centre, weekly or fortnightly samplings were taken (i) from December 1984 to June 1985 on early-sown winter barley (cv Capri), (ii) from June to October 1985 on maize (cv Dea).

In the departement d'Ille et Vilaine, two sets of observations were made in autumn 1984 and in spring 1985, 21 and 37 fields being sampled, respectively; each sample consisting of the following cultivated grasses: brome-grass (Bromus sp.), cocksfoot (Dactylis sp.), fescue (Festuca sp.), perennial rye-grass (Lolium perenne) and Italian rye-grass (Lolium italicum).

For the barley and grass samples, the species and number of aphids were determined visually on 100 to 200 tillers, in groups of 10. For maize samples, 25 plants per week or per fortnight were collected.

Five to 25 whole plants or leaves (maize) were randomly collected for all crops at each sampling date and 3 grams of each plant sample were tested by ELISA using the double antibody sandwich ELISA test described by Clark and Adams (2). The antiserum used was able to detect PAV isolates, otherwise termed NS2 by Lapierre and Maroquin (6). This isolate is transmitted non-specifically, mainly by Rhopalosiphum padi L., and less efficiently by Sitobion avenae F. and Metopolophium dirhodum Wlk. (7). As a matter of fact the serum ('type B, Inothec') detects mainly PAV, but more weakly another strain, MAV (or F).

3. RESULTS

3.1 Population Dynamics of Vectors and Virus Detection on Winter Barley

Figure 1 shows that R. padi represented the almost entire amount of aphids present on barley in autumn and winter 1984-1985. Populations declined gradually and disappeared due to the unusually low temperatures of January and February 1985. In late winter or early spring, the recontamination of barley was due to S. avenae (in March) and then to M. dirhodum (in April). The recontamination by R. padi occurred late and

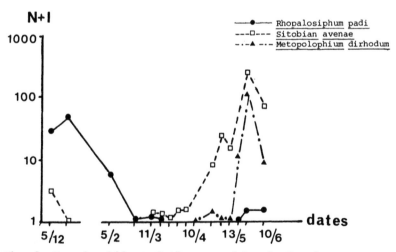

Fig. 1. Cereal aphid populations on winter barley from December 1984 to June 1985 (for 100 tillers).

weakly, at the end of May. The aphids disappeared after 10th June, essentially following plant maturation.

The PAV isolate was looked for and detected three times during the growing cycle:

1) on December 5th 1984, proving that the virus was actually inoculated during autumn by migrant aphids (almost exclusively R. padi);
2) on February 5th 1985;
3) on May 28th 1985 when alates, from the three species, left the crop and migrated to other Gramineae.

3.2 Population Dynamics of Vectors and Some Components of BYVD Epidemiology on Maize

S. avenae and M. dirhodum infested the maize fields in mid-June 1985 (Figure 2a) and were the dominant species at the beginning of the summer with a peak on July 15th. Their numbers decreased quickly under the pressure

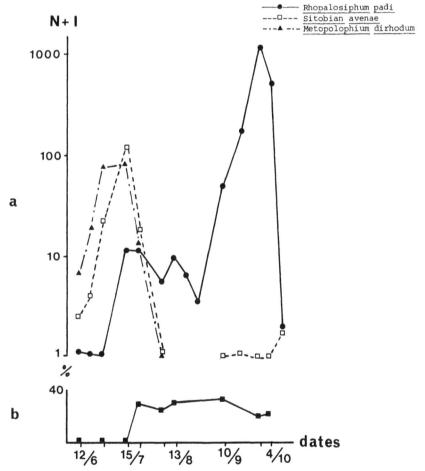

Fig. 2a. Cereal aphid populations on maize (per plant) in summer 1985.
2b. Percentage of maize plants infected by BYDV; —■—

of natural enemies at the end of July, whem some R. padi invaded the crop. R. padi maintained weak populations on maize from late July to early September, when S. avenae and M. dirhodum disappeared at the beginning of August. R. padi populations increased rapidly in September: 100% of the plants were infested on September 30th with a mean over 1000 aphids/plant. In October the aphid populations decreased according to the early maturation of maize, due to the drought of autumn 1985.

The first maize samples infected by PAV isolate were recorded on July 22nd (Figure 2b) some days after the crop contamination by R. padi. Afterwards the percentage of maize plants infected by BYVD increased quickly to a stable level of 30% from late July to early October.

3.3   Some Components of BYVD Epidemiology on Cultivated Grasses

Table 1 shows that the three species of aphids were recorded on grasses in autumn 1984 and that R. padi represented a large majority of the populations. On the other hand, among the very rare aphids recorded during the spring 1985, all were S. avenae. The levels of infestation were very variable depending:

1)   On the Gramineae species; the perennial rye-grass plants were particularly weakly infested by aphids but brome-grass plants were more heavily colonised, at least in autumn.

2)   On the season; very few aphids were found during the spring in comparison with the autumn situation, independently of the grass species.

3)   On the field location; for the same species there were, especially in autumn, large inter-field variations, e.g. from 1% to 60% of the tillers infested for brome-grass in autumn 1984.

TABLE 1. Levels of Aphid Populations on Grasses

Host plants	N	Autumn 1984			N	Spring 1985		
		R.p.	S.a.	M.d.		R.p.	S.a.	M.D.
Brome-grass	700	247	30	0	1100	0	1	0
Cocksfoot	200	26	0	2	500	0	1	0
Fescue	400	21	0	0	900	0	0	0
Italian R.G.	500	50	0	3	500	0	1	0
Perennial R.G.	300	6	0	0	700	0	0	0

(N: number of tillers sampled; R.p.: R. padi; S.a.; S. avenae; M.d.: M. dirhodum)

The PAV isolate was recorded during autumn and spring in three of the studied grasses (Table 2), fescue, brome-grass and Italian rye-grass. For these two latter grasses the percentage of infected samples decreased between autumn and spring. In 1985 the virus was not detected in perennial rye-grass nor cocksfoot but other tests designed in 1986 showed that in some cases perennial rye-grass could also be infected.

TABLE 2. Results of the ELISA Tests on Grasses

| Host plants | Autumn 1984 | | Spring 1985 | |
	Number of samples	Percentage of infected samples	Number of samples	Percentage of infected samples
Brome-grass	35	60	55	16
Cocksfoot	10	0	29	0
Fescue	20	40	41	49
Italian R.G.	25	16	25	4
Perennial R.G.	15	0	39	0

4. DISCUSSION

The above-cited examples allow a preliminary explanation of the epidemiological process of BYDV - risk zones in the oceanic parts of France.

The virus infecting winter barley (and winter wheat) from autumn to spring is inoculated to maize by migrant aphids leaving the former crops. The main vector at this period seems to be R. padi but our results have to be confirmed by the results of subsequent years. Maize appears to multiply the virus, at least at earlier growing stages and to be a reservoir from early summer to mid autumn. At this period migrant aphids (mainly R. padi too) which are mass-produced by maize (4) leave the crop and infest early-sown barley and wheat. The latter multiply the virus during the second part of autumn and the following spring. This basic explanation includes only two types of hosts for vectors and virus, the reality is certainly more complex.

Other relay-hosts certainly play a role in the annual cycle of the disease:

1) wheat and barley volunteers or regrowths growing at the end of the summer are probably 'concentrators' of virus between the maize and winter cereals (1), and
2) spring cereals are susceptible to infection in spring but mature later than winter cereals. They may be hosts for aphids and virus till mid-July.

The role of perennial cultivated grasses is more difficult to assess. Some of them are infected by PAV but are very weak producers of aphids in comparison with straw cereals in spring and maize in autumn. Our hypothesis, which has to be tested, is the following: cultivated perennial grasses are probably of importance as long-term reservoirs of virus, especially after a break of the above described annual cycle of the virus. The breaks may be due to the failure of transmission of the virus to one of its annual hosts because of the lack of vectors at the right period for virus transmission (e.g. destruction of aphids by frost in winter or by natural enemies in spring or summer) or because of the lack of a relay host when the aphids leave the previous one (e.g. because of late sowing).

It is true that components have only been described for the PAV isolate, the only one detected. By chance it seems to be the most widespread and certainly the most severe strain in the west of France. Nevertheless, in order to complete this study, it would be necessary to collect data on the epidemiology of other strains like MAV and RPV for which the host plant range and hierarchy are not necessarily the same as that for the PAV isolate.

REFERENCES

(1)   BAYON, F., AYRAULT, J.P. and PICHON, P., 1982. Epidémiologie de la jaunisse nanisante de l'orge (BYVD) en Poitou-Charentes. Mededelingen Faculteit van Landbouwwetenschappen Rijks universiteit Gent 47/3, 1039-1052.

(2)   CLARK, M.F. and ADAMS, A.N., 1977. Characteristics of the Microplate Method of Enzyme-Linked Immunosorbent Assay for the Detection of Plant Viruses. Journal of general Virology, 34, 475-483.

(3)   DEDRYVER, C.A. and GELLE, A., 1982. Biologie des pucerons des céréales dans l'ouest de la France. IV - Etude de l'hivernation des populations anholocycliques de Rhopalosiphum padi L., Metopolophium dirhodum Wlk. et Sitobion avenae F., sur repousses de céréales dans trois stations de Bretagne et du Bassin parisien. Acta Oecologica, Oecologica Applicata, 3, 4, 321-342.

(4)   DEDRYVER, C.A. and ROBERT, Y., 1981. Ecological role of maize and cereal volunteers as reservoirs for gramineae virus transmitting aphids. Proceedings 3rd Conference on Virus Diseases of Gramineae, pp. 61-66. Rothamsted Experimental Station, Harpenden.

(5)   HENRY, M., GILLET, H. and DEDRYVER, C.A., 1986. Premiers résultats sur l'épidemiologie de la jaunisse nanisante de l'orge en Bretagne. OILB IX, (in press).

(6)   LAPIERRE, H. and MAROQUIN, C., 1986. Modalités de la transmission des virus des céréales et conséquences pour la sélection. In 'Les résistances génétiques dans les systèmes de protection des cultures céréalières contre les champignons, virus et nématodes.' Ed. INRA.

(7)   ROCHOW, W.F., 1970. Barley yellow dwarf virus. CMI/AAB Description of Plant Viruses 32, 4 pp.

Can significant changes in BYDV epidemics be obtained with resistant maize cultivars?

J.P.Moreau
Station de Zoologie, Institut National de la Recherche Agronomique, Etoile de Choisy, Versailles, France
F.Beauvais & H.Lapierre
Station de Pathologie Végétale, Institut National de la Recherche Agronomique, Etoile de Choisy, Versailles, France

Summary

A number of maize varieties have been tested for their BYDV susceptibility in the field and in the glasshouse. Differences in the establishment and multiplication of Rhopalosiphum padi L., the aphid vector of BYDV, as well as the multiplication and the effects of this virus on the crop were shown. The objective of the ongoing breeding programme is to find varieties resistant to BYDV and its aphid vectors in order to avoid yield losses and to reduce viral inoculum which is transferred to winter small grain cereals.

1. INTRODUCTION

Since 1965 (13), maize has been known to be susceptible to Barley Yellow Dwarf Virus. The role of maize crops as a reservoir of cereal aphids (mainly R. padi) and BYDV in several countries is also well established.

The presence of BYDV in maize has been observed in France (7, 12), Greece (11), Italy (4, 2, 10) and Switzerland (5).

In the Paris area, natural infestations of maize fields with this virus under normal cropping conditions has been very low since 1984, and probably cannot account for any yield loss. However, early experimental infection is likely to decrease crop yield even in the absence of visible symptoms throughout the life cycle of the plant (9).

Therefore we decided to determine the effect of BYDV on the maize varieties most commonly grown in the Atlantic area and to start a comparative study on various cultivated maize varieties and a number of simple hybrids currently tested.

A complementary objective is to limit the action of maize as a BYDV reservoir by reducing the frequency of virus affected plants, lowering virus levels, and consequently decreasing the number of viruliferous aphids that leave these plants in autumn.

In addition, the successful achievement of this programme requires a suitable management of maize crops. In particular, heavy R. padi infestations, which frequently result from secondary and adverse effects of insecticidal treatments, should be avoided (6).

2. MATERIALS AND METHODS

For all our experiments, we used the non-specific BYDV strain that occurs naturally at Versailles and is reproduced in the greenhouse on winter barley. Vector aphids were collected on barley volunteers, reared in the greenhouse and in environment controlled chambers at 15 or 20°C depending on requirements. Virus-free aphid rearing was conducted on healthy plants from seeds, and from newborn larvae.

2.1 Greenhouse Tests

Maize plants were sown in small peat pots, placed in larger 14 cm pots after emergence, and watered twice a week with nutrient solution for neutrophile plants.

The aphids reared on barley were placed on maize plants at the 3-leaf stage. Their establishment and multiplication was recorded by biweekly countings. The test essentially consisted of a diallel trial with 45 simple hybrids from 10 lines (A 641, A 665, W 117, W 729, F 243, F 252, F 264, F 492, F 627, Lo 516).

The first series of plants sown (4 plants/hybrid) were infested with aphids reared on virus-diseased barley to allow an evaluation of BYDV reproduction in the greenhouse.

2.2 Field Tests

In 1985, a number of inbred lines included in the diallel were naturally infested with R. padi and BYDV in the field and submitted to ELISA tests by the end of their life cycle.

In 1986, a part of the diallel (hybrids contained A 641 and W 117) and 24 commercialised varieties were artificially infested. Sowing occurred on two dates (May 22 and June 10). We then established aphids (R. padi) previously reared on virus-diseased barley at the 4-5 leaf and 7-8 leaf stages on the plants sown in May (June 12 and 24), and only at the 4-5 leaf stage on the plants sown in June (July 1). For each variety or hybrid, we had 5, 5 and 10 plants respectively, plus 15 controls (5 in the first sowing and 10 in the second sowing), treated with insecticides after each deposit of viruliferous aphids on the neighbouring plants. Each plant to be inoculated received 30 or so aphids, larvae or wingless adults dressed with talc and placed in the funnel formed by the youngest curled leaf. ELISA tests (sampling of a fragment of the next-to-last leaf present) were made 10 and 40 days after contamination. Plants were harvested at maturity. The present data are only related to weight of the entire ears at harvest which give a good approximation of the yield as shown by tests over the two preceding years.

2.3 ELISA Tests

Diluted leaf extracts (5v/w) were tested by the ELISA sandwich method. IgG and phosphatase labelled IgG anti BYDV (non-specific strain) were purified from a rabbit polyclonal antiserum.

3. RESULTS

3.1 Greenhouse Tests

The establishment and multiplication of R. padi on young greenhouse maize varied with hybrids. The lines in diallel trial, which were evaluated on the average of the 9 corresponding hybrids, generally appeared more favourable than the control variety DEA (Figure 1). In agreement with previous results (8), line Lo 516 was the most favourable. Others, like W 117 and F 492 provided adequate aphid reproduction, in spite of a less satisfactory establishment at the 3-4 leaf stage.

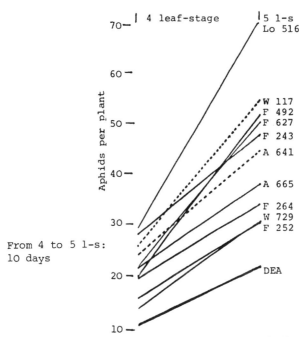

Fig. 1. <u>Rhopalosiphum</u> <u>padi</u> free establishment and multiplication on young maize hybrids grown in greenhouse (diallel trial 1986).

ELISA determinations 10 days after establishment of viruliferous aphids allowed the hybrids to be classified in three categories. Results are listed in Table 1.

TABLE 1. <u>ELISA</u> <u>Tests</u> <u>With</u> <u>Diallel</u> <u>Hybrids</u> <u>Grown</u> <u>in</u> <u>a</u> <u>Greenhouse</u>.

A A 641 x A 665	A 665 x F 627	**B** A 641 x F 264	W 729 x F 492
A 641 x W 117	W 117 x W 729	A 641 x Lo516	W 729 x F 627
A 641 x W 729	W 117 x F 252	A 665 x F 243	F 243 x F 492
A 641 x F 243	W 117 x F 627	A 665 x F 264	F 243 x F 627
A 641 x F 252	W 729 x F 252	A 665 x Lo516	F 243 x Lo516
A 641 x F 492	W 729 x Lo516	W 117 x F 243	F 252 x F 627
A 665 x W 117	F 243 x F 252	W 117 x F 264	F 252 x Lo516
A 665 x W 729	F 243 x F 264	W 117 x Lo516	F 492 x Lo516
A 665 x F 252		W 729 x F 243	F 627 x Lo516
		W 729 x F 264	
C A 665 x F 492	F 252 x F 264	F 264 x F 492	F 264 x Lo516
W 117 x F 492	F 252 x F 492	F 264 x F 627	F 492 x F 627

Category **A**: no (level 1) or low concentration of BYDV (level 2)
Category **B**: intermediate virus concentration (level 3)
Category **C**: high virus concentration (levels 4 and 5)

3.2   Field Tests

Two weeks after the aphid deposition, the numbers of observed individuals were very low. However, rather severe infestation of the hybrid W 117 x F 492 with R. padi was observed by late July/early August.

A slight decrease in plant growth was noted for the plants sown first and inoculated at the 4-5 leaf stage (Table 2).

TABLE 2. Reduced Growth (in %)

	July 2	July 18
Hybrids incl. A 641	2.0	7.6
Hybrids incl. W 117	5.2	2.8
24 varieties	11.0	9.0

Plants inoculated later at the 7-8 leaf stage showed a lesser decrease in growth.

During August various leaf symptoms appeared in inoculated plants: red spots at the margin of the limb or at the leaf tip, often symmetrically to the vein. The details of these symptoms varied with varieties, growth stages and inoculation dates, and will not be reported here.

ELISA tests confirmed that the line W 117 confers resistance to BYDV multiplication.

Yield loss in virus-diseased plants (first sowing, first and second inoculation) compared to controls, as measured by gross weight of mature ears harvested, amounted to 21.4% for the 24 commercialised varieties, 30.4% for the A 641 hybrids, and only 4.5% for the W 117 hybrids. Strong variations, however, were noticed depending on the hybrid considered (Table 3).

TABLE 3. ELISA Tests and Yield Reduction

Maize lines	1985 naturally infected lines	1986 artificial inoculation			
		hybrids inc. A 641		hybrids inc. W 117	
		ELISA level	% yield reduction	ELISA level	% yield reduction
A 665	1 / 22 pl	2	47	4	24
W 117	0 / 6	1	< 0? (+)		
W 729	0 / 1	3	55	4	32
F 243	(abs) -	4	< 0	3	∿0
F 252	0 / 19	1	37	1	<<0
F 264	8 / 28	1	37	1	<0
F 492	13 / 28	5	5	5	38
F 627	0 / 19	not tested	38	5	6
Lo516	1 / 21	2	2	5	17

(+) for W 117 x A 641, the yields were very high for the inoculated plants, but without control

The data obtained for the variety DEA were:
'3' as ELISA level and '17%' as yield reduction.

4. DISCUSSION

The varieties or inbred lines studied here can be classified differently according to whether we consider the insect feeding behaviour and capacity to establish, or the BYDV multiplication in maize and its effects upon this crop.

The number of alimentary sheaths by R. padi reaching the phloem is known to be higher, under experimental conditions, in an inbred line unfavourable to the aphid than in a favourable one (8). It cannot be inferred, however, that the virus inoculation risk in the field is different.

Undoubtedly, the best way is to breed maize lines that resist virus multiplication (e.g. W 117) and to cross them with other lines or hybrids conferring resistance to aphid multiplication at the various stages of plant growth.

Several varieties currently grown (e.g. DEA) are good for avoiding virus multiplication and damage caused to the crop. However, some of them often host large populations of R. padi during autumn.

In the north of France the association of rather tolerant varieties and a low rate of contamination has led until now to a situation without any major risk for maize crops. In warmer areas, climatic conditions favour both winter survival of virus-diseased plants and the spreading of the aphids vectors during spring and autumn, which induces a higher risk of viral epidemics. Brown et al. (3) showed that especially in the case of irrigated cropping the percentage of plants infected by BYDV may be considerable. In these conditions the risk of damage encountered by maize varieties susceptible to the virus is high.

Further studies should be aimed at:

- establishing a more accurate classification of commercialised varieties to allow both a better choice in the regions potentially affected by BYDV and a rational pest management;
- decreasing the role of maize as a 'virus and aphids' reservoir by adequate breeding, and reducing the risk encountered by the crop itself.

REFERENCES

(1) BEAUVAIS, F., 1984. Multiplication du Virus de la Jaunisse Nanisante de l'Orge (VJNO) chez l'espèce Zea mays L. Thèse d'Université - Orsay.

(2) BELLI, G., CINQUANTA, S. and SONCINI, C., 1980. Infezioni miste da MDMV (Maize Dwarf Mosaic Virus) e BYDV (Barley Yellow Dwarf Virus) in piante di maïs in Lombardia. Riv. Pat. Veg. S IV, 16, 83-86.

(3) BROWN, J.K., WYATT, S.D. and HAZELWOOD, D., 1984. Irrigated corn as a source of barley yellow dwarf virus and vector in Eastern Washington. Phytopathology 74, 46-49.

(4) CONTI, M., 1978. Osservazioni su gravi danni da virus del nanismo giallo dell'orzo (BYDV) su frumento ed orzo in Italia. L Inf. Agrario 37, 2997-2998.

(5) GUGERLY, P. and DERRON, J., 1981. L'épidémie de Jaunisse de l'orge dans le Bassin Lémanique. Revue Suisse Agric. 13, (5), 207-211.

(6) MOREAU, J.P., 1985. Surveillance des pucerons (Rhopalosiphum padi L.) en cultures de maïs protégé contre la Pyrale (Ostrinia nubilalis Hbn). EC Experts' Group Meeting - Montpellier, France. 7-9 May (in press).

(7) MOREAU, J.P. and LAPIERRE, H., 1977. Essai de caractérisation de la Jaunisse nanisante de l'Orge (JNO) en cultures céréalières. Ann. Phytopathol., 9, (3), 343-345.

(8) MOREAU, J.P., BOULAY, C. and KONE, S., 1984. Etude histologique de la piqûre de Rhopalosiphum padi L. (Homoptères Aphididae) sur deux lignées de maïs. Bull. SROP 1984/VII/4 OILB Capbreton - April 1983 - 19-20.

(9)   MOREAU, J.P. and LAPIERRE, H., 1986. Rôle des pucerons du maïs dans le cycle et les conséquences du virus de la Jaunisse nanisante de l'Orge (VJNO). Bull. SROP. Groupe de travail 'Integrated Control of Cereal Pests' - Gembloux, Belgique. 10-12 Feb., 1986 (in press).

(10)   OSLER, R., LOI, N., LORENZONI, C., SNIDARO, M. and REFATTI, E., 1985. Barley yellow dwarf virus infections in maize (Zea mays L.), inbreds and hybrids in Northern Italy. Maydica XXX, 285-299.

(11)   PANAYOTOU, P.C., 1977. Effects of barley yellow dwarf on several varieties of maize. Plant Dis. Reptr. 61, 815-819.

(12)   SIGNORET, P.A. and ALLIOT, B., 1981. A new virus disease of maize in France. Proc. 3rd Conf. on Virus Diseases of Gramineae in Europe. Rothamsted Exp. Sta., Harpenden, Herts., 28-30 May, 1980, 1-4.

(13)   STONER, W.N., 1965. Studies of transmission of barley yellow dwarf virus to corn (Abstr) Phytopathology 55, 1978.

Untreated headlands next to hedges in Denmark

L.Samsøe-Petersen

Danish Research Centre for Plant Protection, Zoological Department, Lyngby

Summary

The effects of a 6 m wide untreated strip on natural enemies of pests
are being recorded continuously during a three year period, starting
in 1985. Pitfall methods, D-vac samples, emergence traps and a removal
trapping method are used to investigate the effects of untreated
headlands on populations of natural enemies of insect pests and their
penetration into the field. Results from the first year's pitfall
trappings and experience with the trapping-out design are presented.

1. THE PROJECT

A three year project was started in 1985 to investigate effects of a
6 m wide untreated strip next to hedges in cereal fields. The investigations
are taking place in two cereal fields, both of which are surrounded by old
hedges, several metres wide, consisting of trees and bushes. One field is
situated in the south eastern part of Zealand (Stevns, 'Gjorslev') and the
other near the east coast in the middle of Jutland, close to Århus ('Kalø').
In both fields a hedge (about 1 km long), oriented more or less north-south
has been selected, and along the northern half of this, the 6 m closest to
the field do not receive any pesticide treatments, whereas the rest of the
field is treated by the grower following the usual farming practice. In 1985
at Gjorslev this consisted of two treatments: on June 3rd with DPD 667 (a.i.
2,4-D + dichlorprop), Sportak 45EC (prochloraz) and Maneb (maneb), and on
June 23rd with Tilt 250EC (propiconazol), Cerone (ethephon) and Dimethoate
(dimethoate).
The following investigations are being conducted:

- Analyses of soils.
- Continuous climatic measurements during the growing season.
- Measurements of deposits of chemicals at spraying, in the field, in
  the untreated zone, in the hedge, and in the neighbouring field.
- Analyses of vegetation at points from the hedge to 14 m into the field
  before and after treatments with herbicides.
- Pitfall trapping of ground-living arthropods in the hedge and at
  varying distances into the field - the traps are emptied weekly from
  April to October-November (only at Gjorslev).
- D-vac samples from hedge and field - the first year these were taken
  weekly, in 1986 six times during the growing season.
- Emergence traps in hedge and field from April to June (1986).
- Weekly assessments of aphids and fungal diseases of the crop during
  the season,
- Yield in the 6 m zones and in the field.

The part of the investigations to be described here is the pitfall trapping of ground-living arthropods, of which many are important natural enemies of aphids. The traps were placed at Gjorslev on Zealand only.

The purpose of this part of the project is to investigate possible effects of the untreated strip on the populations of natural enemies of aphids and their penetration into the field.

2. TRAPPING

2.1 Construction

The traps used are flower pots (11 cm diameter, 18 cm deep) with a plastic inner cup that fits tightly 1 cm below the rim of the flower pot. The inner cup has little holes 1 cm beneath the upper rim for overflow of excess water as no shelters are used. The trap liquid (water with 0.5% formaldehyde and 0.05% unscented dishwashing agent) is filled in to about 1 cm depth.

2.2 Arrangement

The first year 228 traps were placed in rows at different distances from the hedge: 0 m (in the hedge), 3 m, 14 m, 35 m, 65 m and 106 m from the hedge. In each part of the field a row consisted of 18 traps - except for the 3 m rows with 24 traps.

This arrangement has been changed in 1986 to: 5 rows (0, 3, 14, 35 and 80 m) with 12 traps in each of the two parts of the field, giving a total of 120 traps.

2.3 Emptying

The traps were emptied once every week from April 30th to November 5th in 1985 and from April 8th to September 30th in 1986.

2.4 Species Recorded

Only adults of four species of carabids and one genus of staphylinids were identified in the samples. These are known to be important natural enemies of aphids in Denmark, with different biology. The species are: Agonum dorsale (active in spring, mostly active at night, to some degree specific aphid predator, climbs plants), Bembidion lampros (spring, day, polyphagous, no climbing), Pterostichus melanarius (autumn, night, poly-phagous, no climbing), Trechus quadristriatus (autumn, night, polyphagous, no climbing). The rest of the carabid species in the samples are counted as 'other carabids'. The staphylinid genus selected is Tachyporus; all members of this genus are simply counted as 'Tachyporus sp.'.

2.5 Results and Discussion

As the project is still running, there is no result concerning possible changes during the three year period.

The catches from 1986 have not been fully analysed yet, so at present only catches from the first year can be presented. As the field has been cultivated for years and treated all over every year, the results from the first year can be taken as a recording of the activities of the species 'before' the untreated zone was established. In the case of the spring-active species (A. dorsale and B. lampros) that appear again in the autumn, an effect of the 1985 untreated strip may be observable in the late catches.

The phenology of the species investigated was as expected, except for the fact that B. lampros was trapped in low numbers compared with other Danish investigations.

The results were analysed by means of two-way analyses of variance on $\log(n+1)$, evaluating differences caused by the three variables 'part of the field', 'distance from the hedge', and 'date/time with respect to dimethoate treatment'. $P=0.05$ was used as significance level.

Analyses showed changes in catches between the two parts of the field before and after the treatment 3 m from the hedge. In most of the field no such change was detectable. Three metres from the hedge A. dorsale was most numerous in the southern part of the field before the dimethoate treatment. After spraying, total catches of A. dorsale in this row were greater in the northern part than in the southern (Figures 1 and 2). A similar picture was seen for B. lampros, the only difference being that catches in the two parts of the field were equal before spraying.

With a few exceptions most of the species were caught in the highest numbers in the southern part of the field all through the season - this is the part where all of the field is treated, so that there is no untreated strip. This is illustrated in Figures 3 and 4 showing catches of T. quadristriatus in the two parts of the field during the season. This means that the fauna in the two parts of the field was not homogeneous at the start of the experiment.

3. TRAPPING-OUT

As the results from pitfalls do not give a measure of population densities, the investigations were supplemented by a trapping-out design in 1986.

3.1 Materials

A 30 cm high piece of steel drum, diameter 60 cm, was used to enclose the trapping-out area.

Two pitfalls were placed in each drum about 10 cm from the drum side with a small barrier leading from the drum to the pitfall.

A fine net was used to cover the drum; it was held in position by a rubber band.

D-vac, pooter, forceps, etc.

3.2 Method

The drum was pushed hard into the soil, and immediately after, the interior was sucked clean with the D-vac. Then the vegetation inside the drum was cut off, shaken inside the drum and examined. The soil in the drum was ground-searched. All arthropods visible during these exercises were collected in the pooter.

Then the drum was pushed/dug as deep into the soil as possible (at best 20 cm) whereafter the pitfalls and barriers were placed inside it, and the drum was covered with the net.

Pitfalls were emptied weekly for three to four weeks, until catches were close to zero.

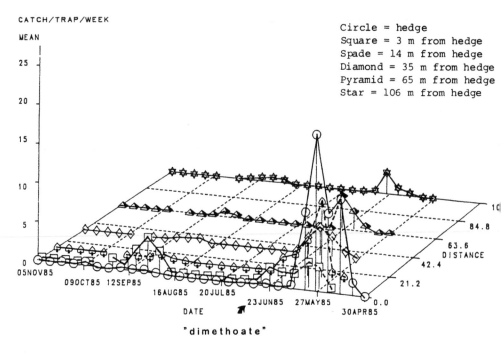

Fig. 1. **Agonum dorsale**, 1985. Southern part of field.

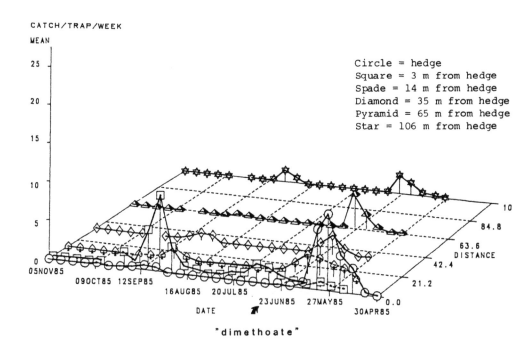

Fig. 2. **Agonum dorsale**, 1985. Northern part of field.

118

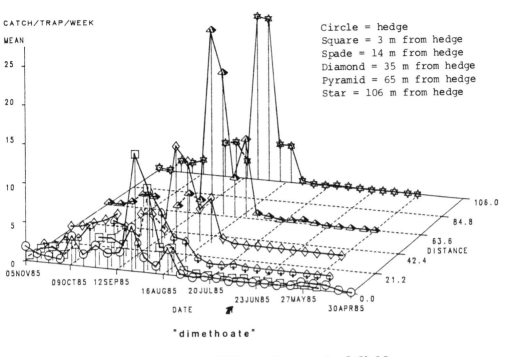

Fig. 3. Trechus quadristriatus, 1985. Southern part of field.

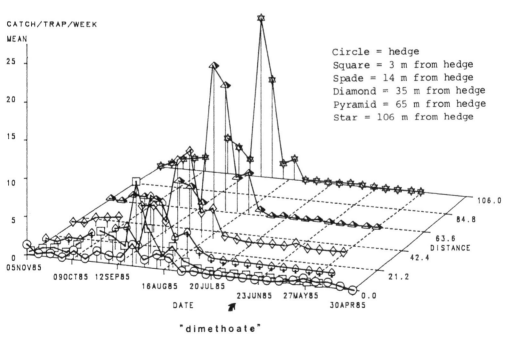

Fig. 4. Trechus quadristriatus, 1985. Northern part of field.

3.3   Number of Drums

Drums were placed in the hedge and at distances of 3, 35 and 80 m from the hedge. In each part of the field, two drums were placed at every trapping-out site.

3.4   Times of Sampling

The drums were set up the day before and two days after every pesticide treatment in the field.

3.5   Results

These results have not yet been analysed, but the preliminary experience is:

1)   The D-vac gave no catches of the species of interest in this study.
2)   Plant examination and ground search gave almost no catches.
3)   The pitfalls seemed to empty the enclosed area, as catches were negligible after three weeks and after four weeks.
4)   The method has one weak point: when the soil is dry and hard, it is almost impossible to get the drum into it. This means that digging is necessary, and the violent efforts involved give beetles a chance to escape.

Thus all results from this project are at the moment preliminary, and conclusions about effects of the untreated strip on beetle populations and their penetration into the field can be drawn after the third year, i.e. 1987, at the earliest.

Experimental options for developing selective pesticide usage tactics within cereals IPM

P.C.Jepson
Departments of Biology and Chemistry, University of Southampton, UK

Summary

Pathways via which chemically or ecologically selective pesticide usage tactics can be developed in cereals are reviewed. Current protocols for examination of pesticide side-effects on non-target invertebrates are considered in this context and the range of questions appropriate to commercial and research/advisory objectives discussed. An improved framework for the evaluation of pesticide side-effects is developed which emphasises the selection of key species, the integration of laboratory, semi-field and field methods and the measurement of insect exposure to pesticides. This framework meets two of the basic criteria required for insecticide testing by being quantitative and repeatable.

1. Introduction

The technology of crop protection within European small grain cereals is becoming increasingly advanced. For example, decision making can now be guided by sophisticated on-line, interactive advisory services which integrate yield loss relationships with economic data to give precise recommendations concerning the justification for and timing of pesticide application (5). In parallel with these developments, there have been considerable advances in our understanding of the role and importance of beneficial insects within the crop (1,6,7). These advances are however, still to be extended into practical advice concerning the conservation or enhancement of natural control. One of the most important topics in this respect is the selective management of pesticide use in a way that is compatible with natural control by predators and parasites. Until the present time, this area has been neglected and few experimental procedures or research methodologies exist to investigate such topics as screening for selectivity, dose optimisation or improved spray targeting. This paper discusses the range of factors which may be manipulated to achieve selectivity and reviews the potential for developing experimental procedures that may test ideas concerning both chemical and ecological ways of achieving it.

2. Experimental questions concerning side-effects of pesticides on natural enemies

One of the major difficulties in attempting to establish new experimental methodologies in this area is the breadth of objectives which industrial, governmental and research scientists may have. One approach to overcoming this difficulty has been the production of standard testing methods for pesticide side-effects which are approved by all parties concerned (2). The protocols devised tend however to be generalised in nature and to lack the necessary flexibility to be adapted to more specific questions. Alternatively therefore, a broader methodological framework might permit a greater diversity of factors concerned with selectivity to be tested but still retain a common system for interpretation and cross reference. To clarify this argument, some of the questions, concerning pesticide selectivity, posed by commercial and research/advisory scientists are listed below.

A) Commercial
   Objectives: concerned with the development, marketing and registration of pesticides.
   Constraints: short-term, cost-effective methods are required, concentrating on direct, rather than indirect effects of treatment. Research is confined to the crop habitat.
   Topics:
   e.g. 1. Product development, compound selection.
        2. Marketing, potential for selectivity.
        3. Registration, testing for approval.
        4. Environmental, general hazard assessment.
B) Research/Advisory
   Objectives: concerned with potential impact of new and established compounds, ranking of selective materials and development of selective management tactics.
   Constraints: short and long-term effects investigated including indirect consequences of use on a large scale. Research also into effects within non-crop habitats.
   Topics:
   e.g. 1. Environmental impact, economic species and wildlife.
        2. Regulatory procedures, realistic and sensitive test development.
        3. Development of IPM.

There is considerable research and development activity in areas A (3,4) and B (2) at present however, some of the other topics are largely ignored because of the complexity of the problems on a research scale. All the topics however, contain certain common features. They require that data can be interpreted in terms of potential effects on predation rate and mortality whether they derive from laboratory, semi-field or field based evaluations. They differ in terms of the resolution required and the time scale over which they operate.

3. A critique of current methodologies

To examine the potential for developing a new experimental framework, current attempts to develop tests of direct effects for the purposes of registration requirements are examined. A more general discussion of this topic can be found elsewhere (3).

Successful insecticide tests, whether based in laboratory or field, share a series of common features, briefly these are:

A) Quantitative expression of effects.
   Achieved in the laboratory by establishing quantal response criteria and deriving dose-response statistics.
   Achieved in the field via direct measurement of mortality or economic benefit in comparison with untreated, control areas.

B) Repeatability.
   Achieved in the laboratory by use of homogeneous populations
   of test organisms, treated in a standard way within rigorously
   controlled conditions.
   Achieved in the field by control over experimental parameters.

C) Relevance to testing criteria.
   Commonly, acceptability thresholds are established whereby
   effects with one compound or treatment are compared with a
   toxic standard.

D) Cost effectiveness.
   In all cases, testing procedures must be developed to
   minimise cost and technical requirements.

These basic criteria are satisfied by the battery of stepwise testing
procedures used in the screening and development of insecticidal products
against phytophagous pests. In this case, the correlation between
laboratory bioassay results and subsequent field tests is sufficient for
all the initial stages of compound selection to be carried out in the
laboratory. Field tests are carried out at a later stage, with only the
most promising compounds. Does this same stepwise progression apply
therefore, to tests for side-effects on beneficial species? There are two
factors in particular which illustrate the fundamental differences between
phytophagous insect pests and their natural enemies as subjects for pesticide testing programmes.

A) The level and predictability of exposure to pesticides:
   The degree to which a mobile insect is exposed to pesticides
   is a complex function of the distribution and physicochemical
   properties of the compound and the distribution and behaviour
   of insects with respect to these. During the period of
   economic damage, phytophagous insects are often predictably
   associated with specific plant parts such as leaves, stems
   or fruits and this distribution may be relatively stable over
   days and even weeks. In contrast, natural enemies offer
   a more varied target for pesticides. Many species are
   dispersive and their position within a crop or canopy may
   show large short-term variation and diel periodicity. In
   addition, a large number of species will be associated
   with the crop or crop boundary for the whole season and will
   thus be exposed to a far greater number of pesticide treatments than pest species.

B) Taxonomic and ecological diversity:
   The selection of pests as targets for insecticide application is governed by their relative economic importance.
   In temperate arable crops, few pest species will be active
   within a given crop and therefore, this selection procedure is relatively simple. In contrast, the taxonomic
   and ecological diversity of natural enemies far exceeds
   the levels associated with pests. Over 300 predatory
   invertebrate species from a wide-range of arthropod
   orders and families, have been described within European
   cereal crops and although these vary in their relative
   importance, each one represents a different and variable
   target for pesticide application. The ecological
   diversity of natural enemies is also wide. For example
   the spectrum of predatory habit varies from polyphagous
   to pest specific. Many parasitic species are also

found. Predators may be univoltine or multivoltine
with highly dispersive phases or life cycles that
remain associated with the field or field boundary
throughout the year.

The task of developing realistic tests of pesticide side-effects against
natural enemies is therefore complicated by the breakdown in the simple
association between laboratory and field based tests which exists for
phytophagous pest species. In addition, the large numbers of predatory
species and ecological groups makes the selection of organisms a far more
complex process.

What techniques have therefore been developed for pesticide side-
effects against beneficial species? The most significant and rigorous
work in this area is reported by Hassan et al. (2). The rationale of
this approach is for testing to progress in a stepwise manner from
laboratory to field with progress to the succeeding step being determined by criteria of mortality. This rationale has been reviewed by
Jepson (3), the most important features of the approach with notes on
limitations with respect to data interpretation are summarised in Table 1.
The most important of these testing procedures is the field programme,
designed to quantify the effects of pesticides on cereal non-target
invertebrates. This currently forms the basis of draft protocols for
registration testing in several European countries.

The protocol as described, forms a complete methodology for field
testing and evaluation of compounds and can therefore be assessed in
terms of the basic criteria for insecticide testing methods, listed above.

A) Is the protocol quantitative?
The lack of specific statistical recommendations associated
with the experimental design is an important limitation to
the quantitative nature of this protocol. Problems are also
caused by the use of activity dependent trapping methods
which cannot be interpreted in terms of mortality or
predation rate. A further limitation is the lack of
guidance given concerning the selection of key species or

TABLE 1.   CURRENT APPROACH TO PESTICIDE/NON-TARGET
                  INVERTEBRATE TESTING, DEVELOPED BY HASSAN, (2)

		FEATURES	LIMITATIONS (eg)
1)	LABORATORY DOSE: RESPONSE RELATIONSHIPS	- Easily cultured species. - Topical application. - Controlled conditions.	Relevance to field effects not as clear as for phytophagous species.
2)	FIELD EXPERIMENTS	- Standard application practices. - Trapping of ground and plant active fauna. - No specialised analyses.	No site or species selection criteria; no direct measurement of mortality; use of activity dependent trapping methods; plot size problems; no criteria of harmfulness.
3)	LAB-SEMI-FIELD PROGRESSION	- Rigid, stepwise program governed by mortality thresholds.	As above; no feedback between testing methods.

124

the division of species into particular ecological or behavioural groupings.

B) Is the protocol repeatable?
As designed, the protocol is site and season dependent.
No guidance is given concerning site selection and the
composition, diversity and abundance of the non-target
invertebrate community is therefore likely to vary widely
between selected sites.

C) Is the protocol relevant to specific test criteria?
The absence of direct records of mortality or predation
rate, limits the options for development of specific
criteria of harmfulness. No guidance is given concerning
the type or level of effect that might be considered deleterious.

D) Is the protocol cost-effective?
As outlined, the protocol entails 60-90 man-days work (around
0.4 man-years) for each chemical/toxic standard combination
tested. In particular, sample sorting and identification is
a time consuming procedure and dependent upon specialised
knowledge.
It is concluded therefore that there is a requirement for further
development of experimental procedures in this area of research and that
several specific questions deserve particular attention.These include:
A) The selection of key organisms.
B) The quantification of insect exposure to pesticide application.
C) The measurement of indirect and sub-lethal effects and the
separation of these from direct mortality.
D) The establishment of more rational laboratory, semi-field and
field methods which permit feedback between the different sites
of testing and which exploit the particular advantages of each.

4. The development of new testing procedures
This section is divided into two parts. The first considers if there
are ways of improving methodology immediately, based on current knowledge.
The second, more speculative section, considers options for research
towards this goal.

A) Proposal for improved testing protocols based on current knowledge.
This proposal outlines the basic procedure carried out on the
Agrochemical Evaluation Unit at Southampton University to determine direct
effects of pesticides on the key natural enemies of cereal aphids. These
methods were developed from basic principles and contain the following
important features:
1) The integrated use of laboratory, semi-field and field testing
methods with feedback between the different methodologies.
Mortality data is therefore an important component.
2) The selection of a range of key natural enemy species in the
tests.
3) An emphasis on quantitative techniques, hence the use of large-
scale, replicated experimental designs in the field with
sophisticated statistical support.
In summary the procedures are as follows (detailed reports will be
published elsewhere).

LABORATORY TESTS.   The procedures, based on topical
                               or residual dose tests, utilise the range
                               of key species under test in the semi-

125

field and field experiments. Data is
used only to support and aid interpretation
of the field tests.

SEMI-FIELD TESTS.   These specially designed tests aim to yield
high quality dose-response data from the
field. Their key feature is the release
of marked organisms into cages or arenas
at the correct crop growth stage and the
direct assessment of mortality at intervals following controlled dose spraying.
The technique can be adapted for all the
major groups of insect natural enemy
and is flexible enough for a wide-range
of experimental questions to be asked.

FIELD TESTS.   These experiments feature Latin-square
designs with plot sizes exceeding 5 ha.
Sites with important species present are
selected and specific sampling methods
for these are used. Sampling frequency
is varied to be particularly intense
prior to and following spraying.
Statistical analysis is based on comparisons of treated and control areas
and of trends in individual plots in
comparison with pre-treatment levels.
Laboratory and semi-field toxicological
data are used to support the interpretation of results and finally, sampling
may continue for up to 12 months following
treatment.

In terms of meeting the basic test criteria cited above, these
procedures give a more quantitative and repeatable basis for chemical
testing. The semi-field methodology in particular, meets all the basic
test requirements and is especially inexpensive to carry out, given the
availability of suitable cages and containment facilities for collected
or cultured insects. The testing methods still have basic limitations
however, including firstly the undefined relationship between field and
laboratory exposure to pesticides, which limits the usefulness of the
laboratory data and prevents comparative analysis of effects on different
species. Secondly, none of the tests record sub-lethal or indirect
effects or effects on predation rate; these factors limit the interpretation of field results. Finally, it is still not possible to define
criteria of harmfulness or hazard other than to indicate relative
effects in comparison with the toxic standard or untreated control.

B) Research options for development of improved testing methods.
This paper concerns the development of selective pesticide management tactics. It is therefore relevant to outline the pathways via which
selectivity may be obtained.
The toxic effect of a compound is a complex function of its intrinsic
toxicity and the level of uptake by a particular organism. The discussion
above has so far, concerned tests which may detect hazardous compounds and
thus effect chemical selectivity, it has not considered the factors which
may influence the eventual uptake of a compound which could be manipulated
to give ecological selectivity. The aim of current research at Southampton
University is to examine the potential for manipulating the bioavailability

of compounds to non-target organisms to achieve ecological selectivity.
A major component of this study is the development of new methods to
examine these factors which may themselves lead to new, rigorous pesticide
screening methods.

The schematic diagram below represents the steps via which pesticide
is tranferred to non-target invertebrates. It also indicates the way in
which manipulation of some of the parameters could achieve the objectives
of chemical or ecological selectivity.

	POTENTIAL FOR SELECTIVITY	
	CHEMICAL	ECOLOGICAL
1. DECISION TO SPRAY		REDUCED NUMBER OF SPRAYS
2. CHOICE OF (a.i., formulation, CONCENTRATION)	MINIMISE INTRINSIC TOXICITY	
3. CHOICE OF APPLICATION TECHNIQUE		*IMPROVE TARGETTING OF SPRAY. OPTIMISE DROP SIZE DISTRIBUTION
4. TIMING OF APPLICATION		AVOID 'SENSITIVE' PERIODS
5. DISTRIBUTION IN CROP AND NON CROP HABITAT*		
6. RELEASE OF a.i. FROM FORMULATION	MANIPULATE TOXICITY VIA PHYSICOCHEMICAL PROPERTIES	
7. REDISTRIBUTION AND TRANSFER OF a.i. †		
8. DOSAGE UPTAKE BY ORGANISM		
9. BIOLOGICAL EFFECT		

The pathway of pesticide dosage transfer to non-target invertebrates

Selectivity can therefore be achieved via a variety of routes involving
on-farm decision making, compound selection, compound formulation, spray
technology, spray distribution and timing. Before these can be achieved,
the basic interactions between pesticides and beneficial insects must be
described in more detail. Current research at Southampton is attempting
to develop quantitative descriptions of components 5,8 and 9 (in the
schematic diagram) to derive predictive models of biological effect
in different conditions.

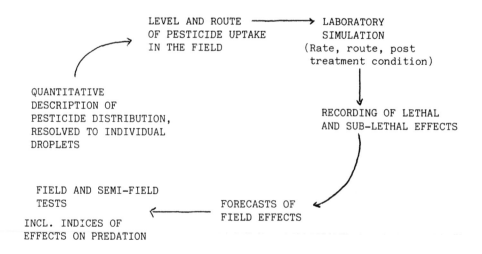

LEVEL AND ROUTE ——————→ LABORATORY
OF PESTICIDE UPTAKE   SIMULATION
IN THE FIELD   (Rate, route, post
                          treatment condition)

QUANTITATIVE
DESCRIPTION OF
PESTICIDE DISTRIBUTION,   RECORDING OF LETHAL
RESOLVED TO INDIVIDUAL   AND SUB-LETHAL EFFECTS
DROPLETS

FIELD AND SEMI-FIELD
TESTS   FORECASTS OF
                           FIELD EFFECTS
INCL. INDICES OF
EFFECTS ON PREDATION

Flow diagram of current research components.

The advantages of this approach are that it permits the testing of understanding via simple predictive models and that it attempts to record the
factors which affect uptake of compounds in different natural enemy groups.
The disadvantages lie in the inherent complexity of the system which means
that all the approach can do is indicate the relative importance of
different factors and determine the potential extent of variation in
particular effects. Some components of this research are summarised
diagramatically below. The most advanced component at present is the
system which has been developed to give a quantitative description of
pesticide distribution (Fig. 1). This has, for the first time, provided
a description of pesticide deposition on a biologically relevant scale;
distribution of droplets over individual plant parts. Large differences
in mortality of insects placed at various positions in the crop canopy
have been detected and the consequences of different natural distributions
recorded for several species currently being evaluated.

5. Speculative frameworks for revised testing methods to achieve
    selective chemical tactics
      The research outlined above will result in the development of a new
methodological framework however, the composition of this framework must
at present, be highly speculative. The summary below gives an example
of how such a package might be constructed and the uses to which
individual components might be put.

A)   SELECTION OF THE ORGANISM

            Organisms selected using ecological, behavioural
            and economic criteria to represent the range of
            important groups, active within the crop.

        eg.1) Plant active, aphid specific predator
                    eg. Syrphid or coccinellid larvae
            2) Plant active, polyphagous predator
                    eg. Staphylinid or carabid beetle
            3) Ground active, polyphagous predator
                    eg. Carabid beetle

128

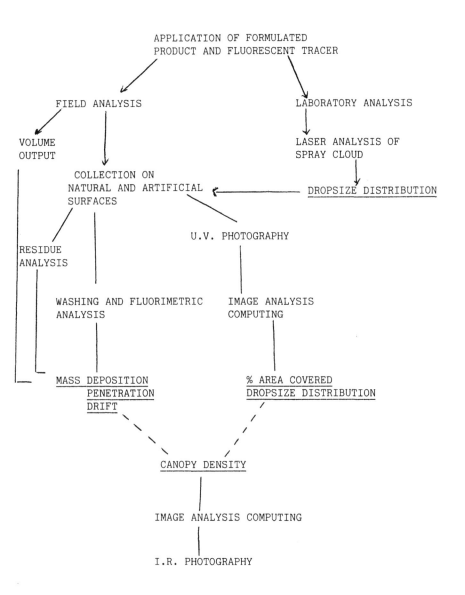

APPLICATION OF FORMULATED
PRODUCT AND FLUORESCENT TRACER

FIELD ANALYSIS

LABORATORY ANALYSIS

VOLUME
OUTPUT

LASER ANALYSIS OF
SPRAY CLOUD

COLLECTION ON
NATURAL AND ARTIFICIAL
SURFACES

DROPSIZE DISTRIBUTION

U.V. PHOTOGRAPHY

RESIDUE
ANALYSIS

WASHING AND FLUORIMETRIC
ANALYSIS

IMAGE ANALYSIS
COMPUTING

MASS DEPOSITION
PENETRATION
DRIFT

% AREA COVERED
DROPSIZE DISTRIBUTION

CANOPY DENSITY

IMAGE ANALYSIS COMPUTING

I.R. PHOTOGRAPHY

Fig. 1.   QUANTITATIVE approach to the description of pesticide
distribution through the crop canopy and field boundary
zone.

    4) Ground active, non-insect predator
            eg. Liniphiid spider
    5) Parasitoid
            eg. parasitic hymenopteran

    Evaluation data exist for cereal aphid natural
    enemies in the above categories. Other classi-
    fications could include phenological characters
    and dispersiveness between fields.

B)   SELECTION OF TEST PROCEDURE

        OBJECTIVE   TEST PROCEDURE

    Basic chemical selection/   LABORATORY TESTS (key
    ranking by toxicity   features)
                                    - Simulation of rates and routes
                                       of field exposure and
                                       environmental conditions
                                       post treatment.
                                    - Calculation of dose response
                                       surfaces, specific to the
                                       above conditions with
                                       estimates of levels of
                                       field uptake.

                                       SEMI-FIELD TESTS IN CAGES
                                       OR ARENAS
    - Screening formulations   - Release of marked insects
    - Improved spray targetting   in presence of prey, at
    - Basic registration tests   correct growth stage.
    - Spray timing etc.   - Controlled dose application.
                                    - Measurement of spray
                                       distribution.
                                    - Recording of environmental
                                       conditions.
                                    - Recapture of insects for
                                       mortality estimate.
                                   *- Estimate of effects on
                                       predation.

                          FIELD PROCEDURES

- Specialised topics   - Large-scale, fully repli-
                                      cated experimental design.
eg.*Long-term effects   - Site characterised by key
   Community effects   species abundance criteria.
   *Interactions between   - Specialised, absolute
    trophic levels   sampling methods.
   *Interactions between crop   - Recording of spray distri-
    and non-crop habitats.   bution.
   Settling arguments   - Recording of key environ-
    Registration tests   mental parameters,
   IPM Development   *- Independent index of
   *Selective practices   predation rate.
    integrated with crop   *- Measurement of effects on
    production   population processes of key
                                      species.
                                      ie. Fecundity
                                          Survival
                                          Dispersal

Several of the topics listed above, have not been previously discussed. This may indicate the range of factors that need to be examined in the development of selective chemical usage tactics.

With reference to registration testing, it is proposed that semi-field methods are employed to provide the basic data on compound toxicity to key beneficial species. Laboratory screening methods may also offer scope for rapid assessments and screening if the rate of routes of possible field exposure are simulated. These modifications exploit the most advantageous aspects of laboratory and semi-field tests and avoid some of the disadvantages discussed above. Some combination of these testing methods will also meet the four basic criteria for establishment of successful testing methods.

The field test has now been elevated to a far more specialised function, the measurement of effects on community processed and inter-actions between tropic levels. It is also on the field scale that progress will be made towards integrating selective spraying practices with crop production methods. A considerable amount of basic research is required before significant progress can be made towards truly selective pesticide usage tactics, this paper argues that this can only be achieved via an improved understanding of the nature of the interaction between beneficial invertebrates and pesticides.

This work is variously funded by the Ministry of Agriculture, SERC, ODA, the University Committee for Advanced Studies, the Game Conservancy, NERC, the British Council and several chemical companies. We are also grateful for the support of ICI, (Plant Protection) Ltd. The research area was inspired by the work of Dr. N.W. Sotherton and the Game Conservancy, Cereals and Gamebirds Research Project.

Current research into pesticide natural enemy interactions at Southampton (4).

A) Measuring the distribution of pesticides through crop canopies

Artificial targets such as MgO slides can be used to collect droplets impacting at different levels of the canopy. Image analysis techniques are then used to measure coverage and droplet sizes.

Application Technique	Target	Reference Area μm^2	Area occupied by droplets μm^2	Number Droplets Sampled	% Cover	T	T*
Tractor Mounted Sprayer	WSP	4.9×10^9	4.1×10^8	2301	8.13	P<0.029	
	MgO	6.4×10^9	5.9×10^8	2592	9.15		P>0.32
Knapsack Sprayer	WSP	3.3×10^9	3.0×10^8	1782	9.08	P<0.016	
	MgO	7.4×10^9	8.0×10^8	4131	10.96		

Target Position	Number Droplets Analysed	Droplet size um Max	Min	Mean	Number um NMD	84/	16/	Volume um VMD	84/	16/	NMD/VMD
60cm	1620	30.69	636.9	164.3	119	225	44	251	327	171	0.474
40cm	1178	29.95	622.4	191.5	138	259	64	288	380	190	0.497
20cm	2075	31.38	703.9	191.0	134	242	75	308	433	182	0.435
Ground	1328	30.62	785.2	248.6	189	323	81	353	462	246	0.535

The upper table shows a comparison of two types of collecting surface and two types of sprayer. The lower table, a breakdown of droplet size statistics throughout the canopy following treatment.

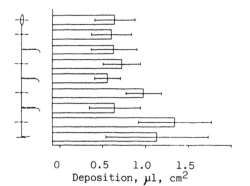

Deposition, μl, cm^2

Alternatively fluorescent wash-off techniques can be used to determine mass deposition rates throughout the canopy.

This data is being used to calculate contact and residual dose exposure of natural enemies within the crop.

B) Toxicological analysis

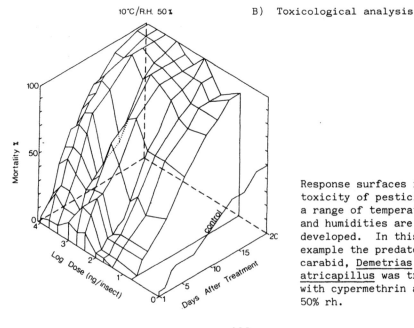

10°C/R.H. 50%

Response surfaces for the toxicity of pesticides in a range of temperatures and humidities are being developed. In this example the predatory carabid, <u>Demetrias atricapillus</u> was treated with cypermethrin at 10°C, 50% rh.

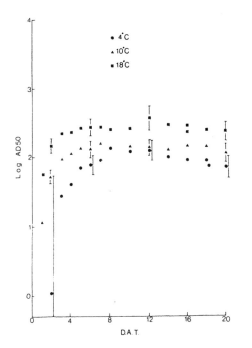

These data can be simplified to LD_{50} or AD_{50} (combined mortality and knockdown) progress curves which form the basis for predictive Models of toxic effect in field conditions.

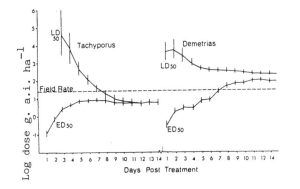

Residual exposure methods are also used to compare knockdown and mortality of different predatory groups in simulated field conditions.

REFERENCES

1. CHAMBERS, R.J., SUNDERLAND, K.D., STACEY, D.L. and WYATT, I.J. (1986). Control of cereal aphids in winter wheat by natural enemies: aphid-specific predators, parasitoids and pathogenic fungi. Annals of Applied Biology, 108, 219-231.
2. HASSAN, S.A. (1985). Standard methods to test the side-effects of pesticides on natural enemies of insects and mites developed by the IOBC/WPRS Working Group 'Pesticides and Beneficial Organisms'. Bulletin OEPP/EPPO, 15, 214-255.
3. JEPSON, P.C. (in press). An experimental rationale for the quantitative evaluation of pesticide side-effects on beneficial insects in cereal crops. IOBC/WPRS Bulletin.

4. JEPSON, P.C., CUTHBERTSON, P., DOWNHAM, M., NORTHEY, D., O'MALLEY, S., PETERS, A., PULLEN, A., THACKER, R., THACKRAY, D., THOMAS, C. and SMITH, C. (In press). A quantitative ecotoxicological investigation of the impact of synthetic pyrethroids on beneficial insects in winter cereals. IOBC/WPRS Bulletin.

5. WRATTEN, S.D., MANN, B., and WOOD, D. (1986). Information exchange in future crop protection - interactions within the independent sector. BCPC Conference proceedings, 1986, VOL. 2, 703-711.

6. WRATTEN, S.D., BRYAN, K., COOMBES, D., SOPP, P. (1984). Evaluation of polyphagous predators of aphids in arable crops. BCPC Conference proceedings, 1984. VOL. 1, 271-276.

7. VORLEY, W.T. (1986). The activity of parasitoids (Hymenoptera: Braconidae) of cereal pahids (Hemiptera:Aphididae) in winter and spring in Southern England. Bulletin of Entomological Research 76, 491-504.

Control of cereal aphids in winter wheat with reduced dose rates of different insecticides with special respect to side effects to beneficial arthropods

H.M.Poehling
Institut für Pflanzenkrankheiten und Pflanzenschutz, Universität Hannover, FR Germany

Summary

The influence of insecticide treatments on cereal aphids and aphid specific predators was investigated in field and laboratory studies from 1983-1986. In the majority of years an increase of aphid density above the economic threshold could be observed late in the season at growth stage 69. Under these circumstances spraying with Pirimor (Pirimicarb) and Sumicidin (Fenvalerate) in reduced dose rates was sufficient to prevent yield losses and to achieve a better economic return. In addition, the selective properties of Pirimor and Sumicidin for coccinellid and syrphid larvae could be further increased by this modified application procedure. Apart from direct toxic effects often a long term reduction of aphid-specific predators occurred even after treatment with selective aphicides as a result of the complete elimination of aphids. Starvation then became a main mortality factor. These detrimental effects can be avoided only to a certain degree if the efficacy of used aphicides is reduced. Again, the choice of appropriate dose rates offers the possibility to maintain limited aphid populations in the field as an essential food source for these predators.

1. INTRODUCTION

The increasing abundance of cereal aphids, particularly in the northern parts of Germany with a high degree of intensively cultured winter wheat in the rotation, has resulted in an almost regular treatment of large areas with insecticides to prevent yield losses. Many farmers spray prophylactically at growth stage 55 (1), but a more reliable aphid control with a better return in economic terms can be achieved if the treatment is more exactly timed according to economic thresholds (2, 3). Often this threshold - in Germany 1 aphid/shoot before the end of flowering - is reached late in the season, when the invasion of cereal aphids in the fields has nearly stopped. At this time the risk of a second critical increase of aphid population is very low and possibly the high efficiencies of commonly used insecticides would not be necessary to get an effective aphid control. Therefore a better return in economic terms could be expected from dose rate reduction, which had to be examined in the field investigations described here.

However, this was only one of the major points of interest in our studies. Other aspects, like the development of cereal aphid control to a 'real' integrated pest management system, and the reduction of environmental loading, in general could be of greater importance, depending on the point of view. Since integrated cereal aphid pest management should be more than optimised chemical control it is desirable to use also the often demonstrated regulatory potential of aphid antagonists (4, 5, 6, 7). In certain years the complex of different species of predators and parasitoids alone cannot prevent aphid population growth from exceeding the very low thresholds and an additional fast aphid reduction by insecticides is necessary.

For an integration of both regulation factors, 'moderate' chemical control methods have to be evolved to reduce the negative side effects on aphid antagonists and non-target organisms in general, if the above mentioned ecological aspects of low loading of the environment are taken into consideration. The first important condition is a high selectivity of active ingredients. With reduced efficacy, sometimes an increase in the selective properties of insecticides was observed (8, 9, 10, 11). But selectivity alone is not the 'magic formula' to guarantee survival, development and effectiveness of aphid antagonists. Most important are the so-called secondary or indirect side effects of cereal aphid control, caused by the often dramatic changes in prey/predator or host/parasite relationships. This problem is often discussed but until today without any consequence for the classification of pesticides by standard test procedures as done by the IOBC - Working Group (12) or even for practical pest management (10, 13, 14).

Therefore we investigated, mainly in field, but also in some additional laboratory studies from 1983-1986, whether a reduction in insecticide efficacy by using reduced dose rates for aphid control could either increase selectivity of different insecticides or help to maintain limited, small aphid populations in the field to improve the conditions (food supply) for survival and development of stenophagous aphid antagonists. In this paper some selected studies are described and discussed, concentrated on the larvae of syrphids and coccinellids, which were of major importance as regulating factors in winter wheat during the last years in the observed areas.

2. MATERIALS AND METHODS

The field studies were done from 1983-1986 on two farms in Lower Saxony in Ahnsen and Hess. Oldendorf. During each season about 12 ha winter wheat were used for the experimental studies. Fields of 3-5 ha were divided into plots of about 2700 m^2, four replicates were treated with different insecticides or different dose rates or left untreated (control). Fungicide and herbicide treatment as well as fertilisation and application of growth regulators were carried out according to common practice.

Insecticide sprays were carried out at growth stage 55 or 69. Here only experiments with a late application are described. The names of the commercial preparations are used in this paper; for a more complete characterisation the commercial preparation names and the corresponding active ingredients are listed below:

Commercial preparation		Active ingredient	Dose rates used (com. prep.)
Pirimor	W.P.	50% Pirimicarb	50, 100 and 200 g/ha
Sumicidin 30	E.C.	30% Fenvalerate	40, 80 and 100 ml/ha
E 605	E.C.	50% Parathion	210 ml/ha
Metasystox	E.C.	25% Demeton-S-Methyl	500 ml/ha

136

Aphid density was recorded continuously from the middle of June until population breakdown at the end of July or early August. Aphids were counted and identified on ears and flag leaves on 4 x 100 plants/plot. The density of syrphid eggs, larvae and pupae and coccinellid larvae was estimated visually. Individuals were counted on 4 x 100 plants/plot. For a determination of the apparent abundance (15, 10) of arthropods on upper leaves and ears of wheat plants a sweep net was used (10 strokes, going 5 m ahead, 10 replicates). Additionally, insects were collected using a vacuum net sampler, but this method gave satisfactory results only for Hymenoptera and Diptera. (These data will be published in a separate paper).

The exact determination of direct effects (mortality, altered behaviour) can be performed only under clearly arranged experimental conditions. Apart from the standard test procedure according to the guidelines of the IOBC/WPRS working group 'Pesticides and beneficial organisms' (12) a simple test system was used. Wheat plants in pots infested with aphids were sprayed before or after syrphid or coccinellid larvae were added. Mortality or alterations in feeding behaviour (weight increase) of the larvae, which were prevented from leaving the contaminated plant or soil by special barriers, were recorded.

Calculations in terms of economic returns were done by the University of Southampton using a simulation model of a computer-based advisory system for cereal aphid control (16, 17).

3. RESULTS

The influence of the insecticides Pirimor, Sumicidin 30 and E 605 in different dose rates on cereal aphid density is shown for three selected fields in 1985 and 1986 in Figures 1-3.

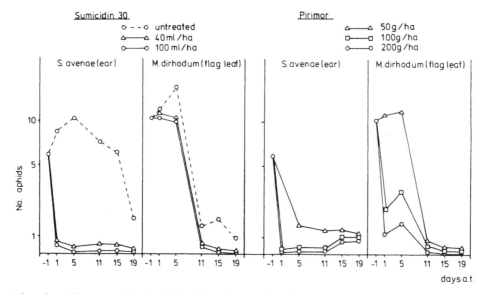

Fig. 1. **Effects of Sumicidin (Fenvalerate) and Pirimor (Pirimicarb) in different dose rates on cereal aphids in winter wheat. Field Hess. Oldendorf I (1985) with high aphid peak density, particularly large colonies of M. dirhodum**

137

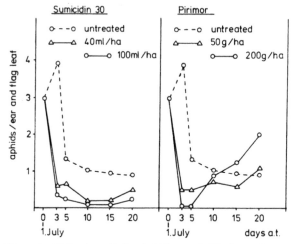

Fig. 2. Effects of Sumicidin and Pirimor in different dose rates on cereal aphids in winter wheat. Field Hess. Oldendorf II (1985) with a relatively low aphid peak density.

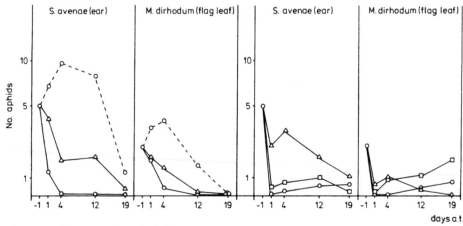

Fig. 3. Effects of Sumicidin and Pirimor in different dose rates on cereal aphids in winter wheat. Field Ahnsen (1986). Very low aphid peak density.

In 1985 and also 1986 (as well as 1984, not described here) the increase of aphid density up to the critical threshold occurred late in the season, immediately before the end of flowering, but anyway the very low economic threshold in West Germany was reached in both years. The further development of the aphid population however was quite different: 1985 a peak density of more than 30 aphids per shoot was reached at the milky ripe stage, whereas in 1986 a sudden unexpected decrease of aphid density could be noticed in early July after the decision to spray, due to the extreme

138

weather conditions with a very hot and dry period. Under these circumstances the lowest dose rate certainly gave the best return in economic terms as shown by the calculation done by the University of Southampton (Table 1).

TABLE 1. Calculation of Yield and Economic Returns from Three Fields Treated with Reduced Dose Rates of Pirimicarb and Sumicidin for Cereal Aphid Control

		Profit from spraying* (£/ha)			Yield (t/ha)	
		A	B	C	A	B
Control - untreated					6.6	7.2
Pirimor	50 g/ha	34.5	85.2	5.5	6.9	7.6
	100 g/ha	38.3	100.5	n.a.	7.1	7.9
	300 g/ha	37.3	98.7	-1.3	7.2	7.9
Sumicidin	40 ml/ha	29.6	87.8	5.5	7.0	8.1
	80 ml/ha	38.7	93.5	n.a.	7.0	8.3
	100 ml/ha	n.a.	n.a.	-8.4	7.0	8.4

* Calculation by University of Southampton (computer-based advisory system)
Fields: A - Hess. Olendorf I, expected yield 6.5 t/ha, 1985
        B - Hess. Oldendorf II, expected yield 6.0 t/ha, 1985
        C - Ahnsen, expected yield 8.0 t/ha, 1986
        All calculations at 1986 values of £118/t.

In 1985, however, insecticides were sprayed at the time of maximal population growth. Only Pirimor at dose rates of 300 and 100 g/ha and E 605 reduced aphid density on ears and flag leaves for a short period nearly completely whereas at a dose rate of 50 g Pirimor/ha limited aphid populations survived on both plant organs. Sumicidin on the other hand removed Sitobion avenae very efficiently from the ears but at all tested dose rates small colonies of Metopolophium dirhodum remained underneath the flag leaves. In Table 2 comparison of the efficiencies of the different insecticides and dose rates is given. A long-term decrease in efficacy could often be observed in high dose rates plots with an initial complete elimination of aphids in contrast to those treatments with lower aphicidal activity but also with higher densities of stenophagous predators (Table 2). The detrimental effects of surviving aphids from the 1985 trials were rather low, as documented by the yield data from the corresponding plots (Table 1). A calculation in economic terms showed the same tendency with highest profit from sprays at dose rates of 100 g Pirimor/ha or 80 ml Sumicidin/ha.

At the time of insecticide application normally relatively high numbers of syrphid larvae, mainly Episyrphus balteatus and Metasyrphus corollae, could be found on the wheat plants. The larval population increased simultaneously with the increasing aphid density. Therefore, initial and long-term effects could be studied. The results of field sampling for two selected fields (1985) with quite different aphid densities are shown in Figures 4 and 5.

Considering that an exact evaluation of mortality of syrphid larvae is nearly impossible in the field, the changes in relative abundance immediately after spraying may give an indication of the initial toxic effects. Pirimor and particularly E 605 reduced the density of syrphid larvae if used at the recommended dose rates, whereas the effect of Sumicidin was less harmful. Distinct graduations of efficacy of Pirimor, and

139

TABLE 2. Efficacy of E 605 (Parathion), Pirimor (Pirimicarb) and Sumicidin (Fenvalerate) in Different Dose Rates. Mortality of Cereal Aphids Calculated According to Henderson and Tilton (18). Treatment of all Three Fields at Growth Stage 69.

Days a.t.		1			5			10			15			20	
Field	A	B	C	A	B	C	A	B	C	A	B	C	A	B	C
Aphids/shoot (control)	11.4	30.5	3.9	14.0	36.4	1.3	12.5	n.a.	1.1	5.2	30.3	1.0	1.4	4.2	0.9
E 605 (Parathion) 210 ml/ha	100.0	96.7	96.1	96.0	80.4	84.0	80.0	n.a.	-176.2	10.4	30.3	-240.2	-89.8	2.4	-322.2
Pirimor (Pirimicarb) 50 g/ha	70.2	90.1	87.2	65.0	85.7	60.0	73.4	n.a.	33.3	75.2	75.0	38.1	12.4	35.7	-22.2
100 g/ha	86.8	95.2	n.a.	87.1	92.1	n.a.	78.0	n.a.	n.a.	20.4	83.1	-71.4	n.a.	26.1	n.a.
300 g/ha	98.2	97.0	98.7	97.1	92.0	92.0	92.6	n.a.	4.8	7.6	90.0	-34.0	-38.6	30.9	-122.2
Sumicidin (Fenvalerate) 40 ml/ha	43.9	88.2	83.3	73.6	81.3	44.0	76.2	n.a.	80.9	64.4	95.0	53.6	59.8	80.9	-8.2
100 ml/ha	67.5	89.1	89.7	95.4	90.1	80.0	98.1	n.a.	85.7	95.1	98.8	79.4	92.7	97.6	44.4

Fields: A - Hess. Oldendorf II (Spray 11.7.1985); B - Ahnsen (Spray 10.7.1985); C - Ahnsen (Spray 1.7.1986)

140

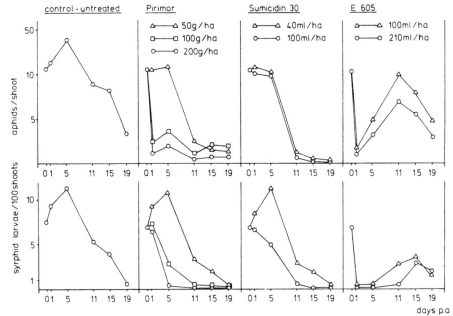

Fig. 4. Effects of E 605, Pirimor and Sumicidin 30 on cereal aphids and
syrphid larvae in winter wheat. Field Hess. Oldendorf I (1985),
treatment 13 July at growth stage 69, high aphid peak density.

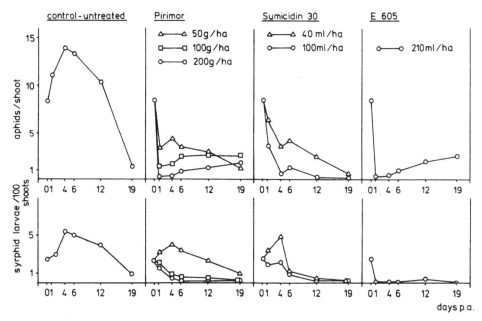

Fig. 5. Effects of E 605, Pirimor and Sumicidin 30 on cereal aphids and
syrphid larvae in winter wheat. Field Hess. Oldendorf II (1985),
relatively low aphid density, treatment 12 July, growth stage 69/71.

even Sumicidin, could be observed in relation to the amount of active ingredient applied. The selectivity of both aphicides was enhanced at lower dose rates. Nearly the same effects could be shown if the larval mortality was evaluated under more controlled conditions in a plant test system in climate rooms, but completely different results were obtained with the standard glass plate test system, especially for the pyrethroid Sumicidin (Table 3, Figure.6).

TABLE 3. Mortality of Larvae (L3) of Episyrphus balteatus on Dry Residues of Different Insecticides on Glass Plates (Standard Test System - IOBC Guidelines (12)) or Plants (Wheat) and Soil.

		Glass plate	Plant
E 605 (Parathion)	0.05%	100.0	100.0
Pirimor (Pirimicarb)	0.05%	100.0	100.0
	0.01%	78.5	52.2
	0.005%	46.0	26.0
Sumicidin (Fenvalerate)	0.05%	54.2	0.0
	0.025%	38.2	0.0
	0.01	32.8	0.0

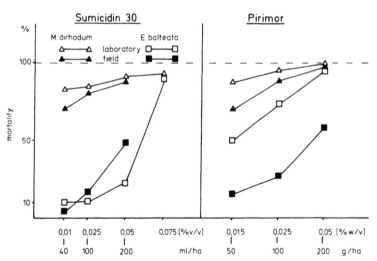

Fig. 6. **Relation of syrphid larvae (L3 E. balteatus) and cereal aphid (M. dirhodum) mortality on different concentrations of Pirimor and Sumicidin. Aphids and syrphid larvae were directly treated on the plants (treatment with fresh insecticide preparations).**

Besides the initial toxicity alterations in the prey/predator relationships, the elimination of the cereal aphids strongly influenced the abundance of those syrphid larvae which had survived the insecticide treatment. Only in plots with remaining aphids could a further development of parts of the larval population to the pupal stage be observed.

Apart from syrphid larvae in 1983 and 1985, larger densities of coccinellid larvae could be found in the wheat fields. The two dominant species were <u>Propylea quatuordecimpunctata</u> and <u>Coccinella septempunctata</u>. Maximal abundance of the larval stages (L3, L4) could be monitored one or even two weeks after insecticide application (Figure 7).

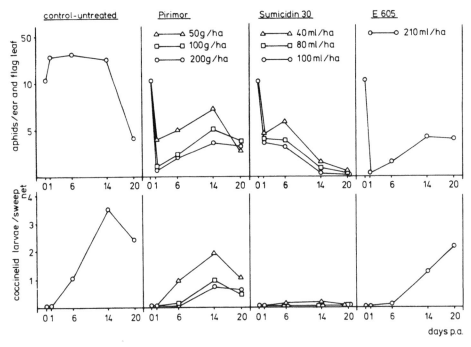

Fig. 7. Effects of Pirimor, Sumicidin 30 and E 605 on cereal aphids and coccinellid larvae in winter wheat (Ahnsen, 1985). Treatment: 10 July, growth stage 69.

However, it was not possible to register initial effects of the insecticides in the field, because at the time of treatment only eggs and small larval stages were present in the field, which (the larvae) were very difficult to count exactly. But from laboratory tests it could be assumed that Pirimor has nearly no initial toxicity for the coccinellid larvae in all tested dose rates, in contrast to Sumicidin and E 605. In Pirimor plots the surviving larvae showed a strong relation to fluctuations in prey density as documented in Figure 7. Only in those Pirimor plots with low dose rates were many coccinellid larvae able to finish the development. It seems - detailed studies are in progress today - that coccinellid larvae have a higher searching activity and prey capture efficiency at low aphid densities in the field compared to the syrphid larvae, but their sensitivity to starvation periods is much higher.

4. DISCUSSION

The experiments described here show that in seasons with late aphid invasion spraying of insecticides like Pirimor or Sumicidin in reduced dose rates could prevent yield losses like a 'regular' treatment. Considering economic terms, better returns can often be achieved. Particularly when the aphid population development after an insecticide application is lower than expected from theoretical calculations (simulation models), in our area in 1984 and 1986, the advantages of dose rate reductions are even more obvious.

On the other hand, it seems that this modification of aphid pest management may be less effective and safe - from the economic point of view - if an earlier insecticide treatment is necessary or if a prophylactic insecticide spray is the method of choice. However, today this conclusion is without any experimental background. It has to be kept in mind that the observed, often distinct, differences in efficacy of aphid control between high and low dose rate plots, two or three weeks after insecticide application, may be caused by the more or less intensive negative side effects on aphid antagonists by a destabilisation of prey/predator or host/parasite interactions at higher spray levels. Therefore, it may be interesting in future to study also the effects of low dose rates for early treatments, perhaps followed by a second 'soft' regulation if aphid population growth increases again above a tolerable density. Hopefully (for research purposes only!) one or two years with heavy, early aphid attack will occur soon.

The question remains open whether only an optimal economic benefit should be the main driving force for research projects on cereal aphid pest management. Ecological aspects will be and have to be of increasing importance. This means that evaluation of side effects on non-target organisms, and the development of strategies to keep these effects as low as possible, will be a central question for a modern integrated cereal aphid control. The often-observed concentration in different research studies on only a few species of such a complex ecosystem as a wheat field is a very unsatisfactory simplification, but these species may be regarded as more or less representative indicators of possible negative influences to the whole fauna.

For some species the initial toxic effects resulting from contamination with active ingredients or contact with residues could be avoided to a certain degree if insecticides with selective properties like Pirimor or Sumicidin were used instead of broad spectrum ones like E 605 or Metasystox. The selectivity of the interesting aphicides mentioned can be further enhanced by the choice of the 'minimal' essential (for aphid control) dose rate. But it is very doubtful (German situation) today whether these ecological advantages of selective aphicides will be arguments strong enough to compete with the economic 'value' of low cost broad spectrum insecticides.

The use of selective aphicides alone only reduces the direct effects which are indeed of major importance for polyphagous species like carabids, staphylinids or spiders. Stenophagous antagonists are severely affected by the removal of prey. This is an aspect which has to be more critically discussed in future. Again here a decrease of insecticide efficacy by dose rate reduction can improve the conditions for these species.

In summary, 'limited' or 'soft' aphid control may be not only of ecological but also of economic importance. The data of four years of field and laboratory experiments available up to now are not sufficient for a final conclusion yet but demonstrate some interesting aspects for a modification of the common practice of cereal aphid pest management.

ACKNOWLEDGEMENT

The experiments were carried out with help from the Agricultural Department of Niedersachsen.

REFERENCES

(1)   ZADOKS, J.C., CHANG, T.T. and KONZAK, C.F., 1974. A decimal code for the growth stages of cereals. Weed Research 14, 415-421.

(2)   BASEDOW, T., BAUERS, C. and LAUENSTEIN, G., 1983. Zur Bekämpfungsschwelle der Getreideblattläuse an Winterweizen. Nachrichtenblatt Deut. Pflanzenschutzd. 35, 141-142.

(3)   POEHLING, H.M. and DEHNE, H.W., 1986. Mehrjährige Untersuchungen zur Bekämpfung von Getreideblattläusen in Winterweizen unter besonderer Berücksichtigung direkter und indirekter Nebenwirkungen auf Nutzarthropoden. Med. Fac. Landbouww. Rijksuniv. Gent 51/3a, 1131-1145.

(4)   CARTER, N., GARDNER, S., FRASER, A.M. and ADAMS, T.H., 1982. The role of natural enemies in cereal aphid population dynamics. Ann. appl. Biol. 101, 190-195.

(5)   CHAMBERS, R.J., SUNDERLAND, K.D., WYATT, J.J. and VICKERMAN, G.P., 1983. The effects of predator exclusion and caging on cereal aphids in winter wheat. J. appl. Ecol. 20, 209-224.

(6)   CHAMBERS, R.J., SUNDERLAND, K.D., STACY, D.L. and WYATT, J.J., 1986. Control of cereal aphids in winter wheat by natural enemies: aphid specific predators, parasitoids and pathogenic fungi. Ann. appl. Biol. 108, 219-231.

(7)   SCHIER, A., 1986. Untersuchungen zur Populationsdynamik der Getreideblattläuse unter Berücksichtigung ihrer natürlichen Feinde. Mitt. Biol. Bundesanstalt. 232, 128.

(8)   HELLPAP, C. and SCHMUTTERER, H., 1982. Untersuchungen zur Wirkung verminderter Pirimorkonzentrationen auf Erbsenblattläuse (Acyrthosiphon pisum) und natürliche Feinde. Z. angew. Ent. 94, 246-252.

(9)   BODE, E., 1981. Begrenzung der Massenvermehrung von Getreideblattläusen durch Spritzbrühen mit vermindertem Aphzidgehalt als ein Beitrag zum Konzept des integrierten Pflanzenschutzes. Mitt. Biol. Bundesanstalt 203, 80-81.

(10)   POEHLING, H.M., DEHNE, H.W. and PICARD, K., 1985. Untersuchungen zum Einsatz von Fenvalerate zur Bekämpfung von Getreideblattläusen in Winterweizen unter besonderer Berücksichtigung von Nebenwirkungen auf Nutzarthropoden. Med. Fac. Landbouww. Rijksuniv Gent 50/2b, 539-554.

(11)   STORCK-WEYERMÜLLER, S., 1984. Untersuchungen über die Wirkung niedriger Dosierungen selektiver Insektizide auf Getreideblattläuse und deren Feinde. Mitt. Biol. Bundesanstalt 223, 273.

(12)   HASSAN, S., 1985. Standard methods to test the side-effects of pesticides on natural enemies of insects and mites. Bull. OEPP/EPPO 15, 214-255.

(13)   POWELL, W., DEAN, J.G. and BARDNER, K., 1985. Effects of Pirimicarb, Dimethoat and Benomyl on natural enemies of cereal aphids in winter wheat. Ann. appl. Biol. 106, 235-242.

(14)   CAVALLORO, R., (ed.), 1983. Aphid antagonists. Proc. Meeting EC Experts Group. Portici 23-24.11.1982.

(15)   FREIER, B. and WETZEL, Th., 1984. Abundanzdynamik von Schadinsekten im Winterweizen. Z. angew. Ent. 98, 483-494.

(16)   MANN, B.P., WRATTEN, S.D. and WATT, A., 1986. A computer-based advisory system for cereal aphid control. Computers and Electronics in Agriculture (in press).

(17)   WRATTEN, S.D., HOLT, J. and WATT, A., 1984. Conversion of research models to micro computer advice on cereal aphid control. British Crop Protection Conference - Pests and Diseases. 609-614.

(18)   HENDERSON, C.F. and TILTON, E.W., 1955. Tests with acaricides against the brown wheat mite. J. Econ. Ent. 48, 157-161.

The effects of intensive pesticide use on arthropod predators in cereals

A.J.Burn
Department of Applied Biology, University of Cambridge, UK

SUMMARY:
Three years of differential pesticide treatments have been
completed in the long-term, large-scale M.A.F.F. study at
Boxworth EHF, on the side effects of intensive pesticide use.
Strongly contrasting levels of pesticide input have been
achieved in the low input (supervised and integrated areas) and
full insurance treatment areas. A range of ecological effects
are being investigated, and this paper presents interim results
of the effects on cereal pests and their natural enemies.
Populations of a range of polyphagous predators have shown
significant reductions in the full insurance area. Although
effects on predation of artificial prey have been variable, in
the most recent year of treatments predators had a significantly
lower impact on cereal aphid population development in the full
insurance treatment area than in the low pesticide input areas.

1. INTRODUCTION:

The use of pesticides on cereals in the United Kingdom has risen
strikingly over the past 10 years (6). Whilst foliar fungicides have
contributed largely to this increase (5.25 million treated hectares in
1982), both insecticides + molluscicides (1.1 million treated hectares) and
herbicides (7.4 million treated hectares) are used widely and often
routinely as part of a pre-determined spray programme. The effects on
non-target organisms of many of these chemicals have been the subject of
increasing concern, and much recent research has aimed primarily to
investigate the effects of usually single pesticide applications on a range
of non-target fauna (e.g. 1,2,5,9). Attention has focussed particularly on
the side-effects of such pesticides on predatory arthropods, especially in
the light of extensive evidence for the role of natural enemies in
preventing outbreaks of cereal aphids (8).

The Boxworth Project was initiated in Cambridgeshire, England, in 1982
with the aim of examining the long-term effects of intensive pesticide use
on winter wheat. The treatments, which will continue until 1988, comprise
four fields recieving a complete package of insurance pesticide treatments,
covering 67 ha; three fields (45 ha) recieving pesticide treatments only
when pests, weeds or diseases exceeded a treatment threshold (the
"supervised" area); and four fields (21 ha) under "integrated" control, in
which the effects of pesticides were further reduced by, among others, the
use of selective chemicals and resistant cereal varieties. A full
description of the aims of the project (which include the effects of such
farming systems on management, yield and a range of non-target organisms),
and of the treatments is published elsewhere (4). Of the range of

ecological effects being studied in this project, it is recognised that the most immediate impact on the farming community is that concerning the effects of the treatment regimes on predators of cereal pests, with possible consequences for pest resurgence, and hence further commitment to intensive pesticide use. In this paper, results are presented for changes in the polyphagous predator fauna over the first three years of differential treatments. Changes in predation levels (using artificial prey) are presented, and the consequent effects on cereal aphid population development are examined for the most recent crop year. As this represents the mid-way position for the study as a whole, these results should be seen as a preliminary indication of the likely trends, rather than a final assessment of the long-term effects of pesticides.

2. MATERIALS AND METHODS:

The polyphagous predator fauna was monitored for two baseline years (during which time all treatment areas recieved the same low pesticide input) and for the three subsequent treatment years. Pitfall traps, at densities of 8 - 24 per field, were opened for periods of one week at irregular intervals during winter and weekly during peak aphid populations in summer. These were supplemented with D-vac suction samples (2.4 m^2 per field) and soil searches to 5cm depth (1.25 - 2.25 m^2 per field) at irregular intervals during summer, to provide independent population estimates for the more abundant polyphagous predators.

Predation levels were measured on 5 - 6 occasions throughout summer, using Drosophila pupae as artificial prey (7). Twelve pupae were glued onto square cards (2.5 X 2.5cm), 24 squares were placed within a selected field in each treatment area and exposed to predation for 24h on each occasion. The effects of predation on cereal aphid population development during summer 1986 was measured using three predator exclusion plots (9 x 9m) within a selected field in each treatment area. Polythene barriers were used (3) and predator numbers were further reduced within the plots by soil applications of carbaryl (7 l of 0.52 % a.i. per plot) between cereal rows. Predator exclusion and control plots recieved identical pesticide and herbicide treatments to the surrounding fields, but were protected from summer aphicide applications to allow a comparison of aphid population development between the three treatment areas. Aphids were sampled using a D-vac (0.9m^2 per plot), or counted on tillers (30 per plot) within predator exclusion and control plots at weekly intervals up to the time of the aphid population peak.

3. RESULTS AND DISCUSSION:

Since the aim of the Boxworth project is to show long-term trends, and a further two crop years remain before the completion of the five year treatment phase of the study, the results presented here should be interpreted only as interim findings. However there have been consistent differences in the levels of pesticide application to the high and reduced input areas (Table 1). Total pesticide use (presented in Table 1 as the number of different applications of commercial products) has consistently been 2.2 - 2.5 times greater in the full insurance treatment area, whilst only in one year has pest monitoring indicated the need for insecticide applications in the integrated and supervised areas. Pesticide application has varied surprisingly widely in the full insurance treatment area, largely due to difficulties in spraying to a fixed programme during prolonged periods of adverse weather. Despite the use of, for example, mixed cultivars and varietal choice for increased disease resistance in the integrated area, there has been no significant reduction in pesticide use here relative to the supervised area over the first three years of

148

TABLE 1. Pesticide Use in the Boxworth Study Site
Average Numbers of Different Pesticides used per Treatment Area

| | 1983/4 | | 1984/5 | | 1985/6 | |
	Total pesticides	Insecticides only	Total pesticides	Insecticides only	Total pesticides	Insecticides only
Insurance	18.8	6.0	17.7	5.0	14.8	4.5
Supervised	8.0	2.0	8.0	0	6.7	0
Integrated	8.3	2.0	8.0	0	6.0	0

differential treatments, and these two areas are considered equivalent in the following discussion of changes in predator populations.

Full details of changes in the predator fauna will be published elsewhere, however Table 2 summarises the present status of the most abundant groups of polyphagous predators in the full insurance treatment area relative to their occurrence in the low pesticide input areas. Over three years there has been a reduction in species richness in the full insurance area, and populations of certain groups of predators given a high ranking as cereal aphid predators (10) have been reduced (e.g. A.dorsale, Linyphiid spiders) or eliminated altogether (Bembidion spp.). Levels of predation measured using artificial prey, in contrast, showed little evidence of reduction during the first two years of differential treatments (Fig 1, 1984 and 1985). At that stage, the only significant difference between the full insurance and other two treatment areas, was a higher level of predation in the insurance area on two occasions, possibly related to the low availability of alternative prey. However, in 1986 predation levels were significantly lower in the full insurance area in trials made at the beginning and towards the end of summer, and at no stage was predation higher in the insurance than in the low pesticide treatment areas. A fuller analysis of changes in predator populations over the complete period of the study is necessary before possible reasons can be stated with confidence, but it seems probable that the greater range of predators affected at this later stage of the programme may have been responsible.

TABLE 2. Status of Predator Taxa in 1986

Absent from full insurance area:	Bembidion lampros, B. obtusum
Very low frequency in insurance area:	Notiophilus biguttatus, Trechus quadristriatus (spring only) Linyphiid spiders (spring only), Agonum dorsale, Pterostichus madidus, P. melanarius, predatory Prostigmatid mites
Similar/higher density in full insurance area, relative to low pesticide input areas:	Pterostichus macer, Tachyporus spp., Xantholinus spp., Lathrobium spp., Cantharidae

During 1984 and 1985, predators were unable to prevent outbreaks of cereal aphids in any of the treatment areas. In 1986, however, aphids appeared late in the season; <u>Metopolophium dirhodum</u> reached 1 per tiller at GS 75 - 79, and <u>Sitobion avenae</u> at GS 80 - 85. The effects of predator exclusion on the subsequent development of aphid infestations by both species are shown in Figs. 2a and b. Levels of <u>S. avenae</u> stayed below five per tiller in control plots in all areas, whilst <u>M. dirhodum</u> reached outbreak levels only in the insurance area. Predator exclusion allowed a significant increase in the population development of both aphid species in the two low pesticide input areas, whereas it had no effect in the full insurance treatment area, although <u>S. avenae</u> did not reach damaging levels there.

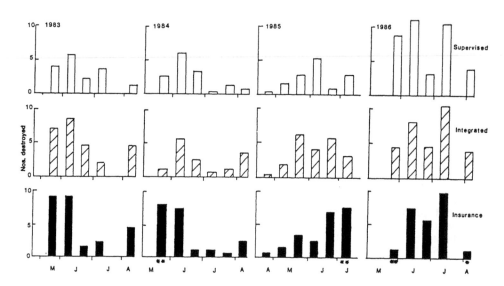

Fig. 1. **Predation of artificial prey (Drosophila pupae) in three treatment areas during 1983-1986. Significant differences between treatments are indicated: * P < 0.01, ** P < 0.001**

4. CONCLUSIONS:

Over the first three years of differential treatments in the Boxworth project, populations of a range of polyphagous predators of cereal pests have shown significant reductions in the full insurance treatment area. Effects of this reduction on predation of artificial prey have been variable, and only in the third treatment year was any significant reduction shown in predation levels in the full insurance area. The consequences of this fall in predation rate on pest population dynamics have yet to be fully explored, but exclusion experiments have indicated that predators had little effect on the late season population development of both <u>M. dirhodum</u> and <u>S. avenae</u> in the full insurance area in 1986, relative to their evident role in the two low pesticide input areas. It should be emphasised that the project is still two years from completion and that the summary of results presented here should not be interpreted as a final indication of any long-term trends which the programme hopes to investigate.

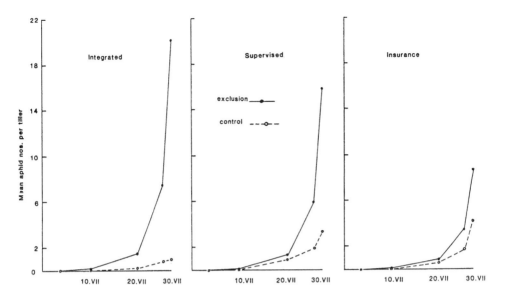

Fig. 2a. Population growth curves for <u>Sitobion</u> <u>avenae</u> in predator exclusion
and control plots in three treatment areas, 1986

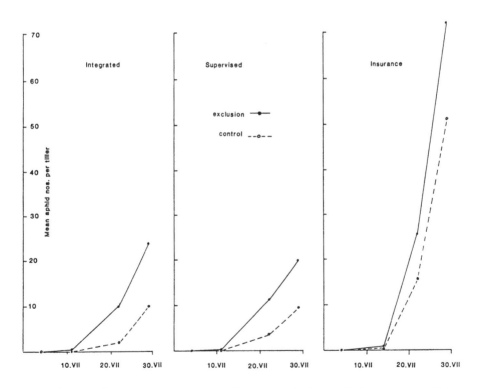

Fig. 2b. Population growth curves for <u>Metopolophium</u> <u>dirhodum</u> in predator
exclusion and control plots in three treatment areas, 1986

5. REFERENCES:

1. CHIVERTON,P.A. (1984). Pitfall trap catches of the carabid beetle _Pterostichus melanarius_ in relation to gut contents and prey density, in insecticide treated and untreated spring barley. Ent. Exp. et Appl. 36, 23-30.

2. COLE, J.F.H. and WILKINSON, W. (1984). Selctivity of pirimicarb in cereal crops. 1984 B.C.P.C. Pests and Diseases. 1, 311-316.

3. EDWARDS, C.A., SUNDERLAND, K.D., GEORGE, K.S. (1979). Studies on the polyphagous predators of cereal aphids. J. Appl. Ecol. 16, 811-823.

4. HARDY, A.R. (1986). The Boxworth Project: a progress report. 1986 B.C.P.C. Pests and Diseases (in press)

5. MATCHAM, E.J. and HAWKES, C. (1985). Field assessment of the effects of deltamethrin on polyphagous predators in winter wheat. Pestic. Sci. 16, 317-320.

6. SLY, J.M.A. (1984) Arable farm crops and grass 1982. (Prelim. Pestic. Usage Survey Report No.35) MAFF.

7. SPEIGHT, M.R. and LAWTON, J.H. (1976). The influence of ground cover on the mortality imposed on artificial prey by predatory ground beetles in cereal fields. Oecologia 23, 211-223

8. SUNDERLAND, K.D., CHAMBERS, R.J., STACEY, D.C., CROOK, N.E. (1985). Invertebrate polyphagous predators and cereal aphids. Bull. SROP/WPRS VIII/3, 105-114.

9. VICKERMAN, G.P. and SUNDERLAND, K.D. (1977). Some effects of dimethoate on arthropods in winter wheat. J. Appl. Ecol. 14, 767-777.

10. WRATTEN, S.D., BRYAN, K., COOMBES, D., SOPP, P., (1984). Evaluation of polyphagous predators of aphids in arable crops. 1984 BCPC. Pests and Diseases. 1, 271-276.

Observations on the effects of an autumn application of a pyrethroid insecticide on non-target predatory species in winter cereals

G.Purvis
ADAS, Rothamsted Experimental Station, Harpenden, UK
N.Carter & W.Powell
AFRC, Rothamsted Experimental Station, Harpenden, UK

Summary

Preliminary results are presented from the first year of an investigation of the effects on polyphagous predator populations of a pyrethroid insecticide applied to winter cereals in the autumn. At nine sites throughout England and Wales, the numbers of Carabidae caught in pitfall traps on insecticide treated plots (2-5 ha) were reduced on average by about seventy percent compared with untreated control plots during a two month period following spray application. Complete recolonization of treated areas occurred by the end of this two month period. At one site, where trapping continued, subsequent summer populations of some autumn-breeding Carabidae and Staphylinidae emerging from pupation on the study plots appeared to be reduced by fifty percent or more but very rapid dispersal across the site following emergence quickly masked this effect. Activity of Linyphiidae on treated plots was initially reduced by about seventy five percent and this reduction persisted until at least early summer across the range of sites. Where trapping continued, spider activity was only restored by a large influx of the two commonest species in May, presumably achieved by ballooning from considerable distances, whilst the activity of other species remained depressed well into summer on the treated area. Difficulties of interpretation in field studies involving highly mobile species on even quite large plots are discussed and the danger of erroneously ascribing greater insecticide susceptibility to less mobile groups is pointed out.

1. Introduction

There has been a trend in England and Wales in recent years towards earlier sowing of winter cereals which has increased the risk of infestation by cereal aphids carrying Barley Yellow Dwarf Virus (BYDV) (7, 2). This has resulted in the widespread application of insecticides, principally synthetic pyrethroids, to crops in October and early November. As pyrethroids are broad spectrum in their effect there is a need to investigate their potential impact on non-target arthropods, especially polyphagous predators such as carabid and staphylinid beetles and spiders which are present in fields in the autumn. These predators are potentially important components of the natural enemy complex which can limit the development of cereal aphid populations in the summer (6, 11, 3).

Previous field studies of the influence of pyrethroid insecticides on predatory species have concentrated on applications aimed at summer pest populations (9, 1, 10) and knowledge of the effects of late autumn applications for BYDV control is limited (8).

The aim of this investigation is to assess any immediate effects of an autumn pyrethroid application on beneficial non-target species active in the crop at this time and in the following spring when aphid populations are becoming established.

2. Methods

Nine trial sites were established on newly sown winter cereal crops in autumn 1985 throughout five of the regional divisions of the Agricultural Development and Advisory Service (ADAS). At each site, one of two plots each measuring at least 2 ha was given a single application of a synthetic pyrethoid insecticide (either deltamethrin or cypermethrin) at 5 g a.i. per ha in late October/early November as recommended for control of aphid vectors of BYDV, whilst the second plot was left untreated as a control (Table I).

Table I   Site locations and treatment details

ADAS region	County	Crop	Insecticide	Application date
Northern	Northumberland	Winter wheat	Deltamethrin	7 November
Northern	N. Yorkshire	Winter wheat	Deltamethrin	13 November
Midland & Western	Leicestershire	Winter wheat	Cypermethrin	4 November
Midland & Western	Shropshire	Winter wheat	Deltamethrin	1 November
Eastern	Cambridgeshire	Winter wheat	Deltamethrin	13 November
Eastern	Hertfordshire	Winter wheat	Deltamethrin	7 November
South Eastern	Hampshire	Winter wheat	Deltamethrin	30 October
Wales	Dyfed 1	Winter barley	Cypermethrin	12 November
Wales	Dyfed 2	Winter barley	Cypermethrin	29 October

At all sites except that in Hertfordshire, twelve pitfall traps were operated on each plot in a 3 x 4 grid at least 20 m apart. Each trap consisted of a 400 ml aluminium can with a 5 cm diameter hole cut in its lid suspended in a piece of buried plastic drain pipe. Traps were filled with Gault's solution and operated for approximately seven day periods from two weeks prior to insecticide application until four weeks after application, and thereafter, for similar periods at monthly intervals until at least March 1986.

On the Hertfordshire site at Rothamsted Experimental Station,

twenty-five traps containing water and a little detergent were operated
almost continuously on each plot in a 5 x 5 grid for seven day periods from
five weeks prior to insecticide application until the beginning of August
1986. Short breaks in this programme occurred mainly during periods of
frost and snow in winter.

Total numbers of carabid and staphylinid beetles and of spiders were
recorded in trap catches at all sites. The most common beetle taxa were
determined to specific level and, at the N. Yorkhsire, Hertfordshire and
two Dyfed sites, the dominant linyphiid spiders were also separately
identified.

3. Results

3.1 Immediate effects on winter-active populations

A wide variation between trial sites was found in the surface activity
of carabid and staphylinid beetles and spiders during the late autumn and
winter (Table II).

Table II   Total numbers of Carabidae, Staphylinidae and Araneae caught
during the period October-March (pre and post treatment)

	Total trap days per plot	Carabidae treated	untreated	Staphylinidae* treated	untreated	Araneae treated	untreated
Northumber-land	588	78	143	216	294	322	640
N. Yorkshire	552	112	189	166	254	87	147
Leicester-shire	444	538	714	447	671	273	402
Shropshire	540	22	23	70	66	106	130
Cambridge-shire	924	180	155	109	71	96	81
Hertford-shire	1875	263	387	168	231	243	534
Hampshire	540	336	482	38	16	94	95
Dyfed 1	744	137	244	39	39	170	277
Dyfed 2	636	12	81	48	44	120	204
TOTAL	6,843	1,678	2,418	1,301	1,686	1,551	2,510

*includes predatory and non-predatory species.

A comparison of relative activity on treated and untreated plots throughout
this period, revealed marked contrasts in the immediate influence of

155

insecticide application on these different groups (Fig. 1).

Carabid activity declined immediately following spray application to levels, on average, between 30 and 40 percent of those on untreated areas. This decline was short-lived with an apparently rapid recolonization of treated areas being effected within eight weeks of application. Carabid catches throughout this early post-treatment period, with the exception of those at the Dyfed sites, consisted largely of the same single species, *Trechus quadristriatus* (Schrank) (Table III).

Table III   Relative abundance of carabid species caught during the period from October to March; percentage of total carabids caught on treated/untreated areas

Site	*Trechus quadristriatus*	*Bembidion* spp.	*Nebria brevicollis*	*Agonum dorsale*	Others
Northumberland	99/97	-/0.5	-/0.5	-/-	1/2
N. Yorkshire	85/77	10/15	1/2	-/-	4/6
Leicestershire	98/98	2/2	-/-	-/-	-/-
Shropshire	52/60	22/8	11/16	-/-	15/16
Cambridgeshire	79/79	?/?	?/?	-/-	21/21
Hertfordshire	89/81	2/4	2/3	-/-	7/12
Hampshire	56/53	31/37	13/10	-/-	-/-
Dyfed 1	13/13	?/?	65/75	6/7	16/5
Dyfed 2	19/7	?/?	58/67	19/15	4/11

Staphylinid activity seemed little influenced by spray application with comparable trap catches being made on treated and untreated plots at the majority of sites throughout the immediate post-treatment period (Fig. 1). The composition of the staphylinid fauna varied substantially, both between sites and trapping occasions. *Xantholinus linearis* (Olivier) was generally the commonest predatory species trapped but, at northern sites (Northumberland, N. Yorkshire and Leicestershire), the non-predatory species, *Lesteva longoelytrata* (Goeze) greatly predominated throughout the winter.

On all sites, linyphiid species accounted for virtually all the spiders trapped between October and March and of these the vast majority were male. Their relative activity on treated plots declined after treatment, rather slowly initially, to a level less than 30 percent of that on treated plots at about eight to twelve weeks after spray application. Thereafter, their numbers in traps on the treated plot remained markedly depressed until the end of sampling in March (Fig. 1).

3.2 Subsequent summer activity

At the Hertfordshire site, trapping continued after the winter period into spring and summer, by which time, general activity levels of all groups markedly increased with the appearance of many additional species in trap catches (Figs. 2, 3 and 4).

Much of this increased carabid activity was due to spring breeding species re-entering the field after overwintering at field margins, e.g. *Bembidion lampros* (Herbst.), *Loricera pilicornis* (F.). No treatment difference was noted in the abundance of such species or in overall carabid activity. Of the commonest autumn-breeding carabids which would have been present on the field as larvae at the time of spray application, *Harpalus rufipes* (Degeer) showed substantially greater adult activity on the control plot at three peaks in occurrence during May-July (Fig. 5). Between these peaks, plot differences in activity disappeared suggesting a rapid redistribution of beetles across the site after pulses of emergence from pupation. A second autumn-breeding species, *Calathus fuscipes* (Goeze), displayed lower activity on the treated plot in the majority of trapping periods during peak incidence in July (Fig.5). Neither of two large *Pterostichus* spp. (*P. madidus* (Fab.) and *P. melanarius* (Illiger)) showed any apparent treatment effect on their numbers (Fig. 5).

Many of the staphylinids caught in the summer were small non-predatory Aleocharinae. Of the commonest predatory species, *Tachyporus* spp. (mainly *T. hypnorum* (Fab.)), *Philonthus* spp. (mainly *P. cognatus* Stephens and *P. laminatus* (Creutzer)) and *Xantholinus* spp. (largely *X. linearis* (Olivier)), were all caught in greater numbers on the untreated plot at times of peak occurrence and showed an apparent redistribution across plots at intervening periods (Fig. 6).

A substantial suppression in overall spider activity on the treated plot persisted until early May when activity greatly increased on both plots (Fig. 4). Linyphiids greatly predominated throughout trapping and the commonest species, *Erigone atra* (Blackwall) and *E. dentipalpis* (Wider), which together accounted for more than 90 percent of all Araneae caught, both exhibited a marked increase in activity in early summer which overwhelmed previous plot differences in spider catches (Fig. 7). *Oedothorax* spp. (mainly *O. fuscus* (Blackwell)) were the most abundant of the other linyphiids but they remained less active on the treated plot even during peak abundance in June (Fig. 7). Of other spiders, lycosids became more numerous in catches during May and June at which time considerably fewer were caught on the treated plot (Fig. 7).

4. Discussion

In a previous study of the effects of a deltamethrin application for BYDV control on winter wheat, the general level of predator activity was reduced by about thirty percent (8). Preliminary results from the current studies suggest that this reduction may be more substantial but great mobility on the part of the populations involved makes clear demonstration of such effects in the field difficult. Initially, carabid activity was reduced, on average by about seventy percent, on treated plots across a wide range of study sites. Although recolonization of the 2-5 ha plots appeared to occur within two months, the initial decline following application, if interpreted as mortality, suggests that application to entire crops may have a much more persistant effect as the general predator reservoir is depleted. This would be especially important in areas where a high proportion of land is given over to winter cereal production.

Treatment effects on the larval populations of autumn-breeding species, which in the following summer prey as adults on establishing aphid populations, are perhaps of greatest consequence in affecting aphid incidence (6). Trap catches at the Hertfordshire site suggest that the numbers of some species emerging from pupation (e.g. *Harpalus rufipes* and staphylinids of the genera *Tachyporus, Philonthus* and *Xantholinus*) were reduced substantially on treated areas although post-emergence mobility appears to mask this reduction rapidly on trial plots. The life style, and particularly the surface activity, of larvae is undoubtedly of great importance in determining susceptibility to autumn-applied insecticides. In this respect, the large *Pterostichus* species appeared unaffected by the spray application possibly because their larvae are quite deep in the soil and so avoid contact with autumn spray applications.

Linyphiid spiders are now recognized as important aphid predators being present in arable habitats in considerable numbers early in the season during the establishment of aphid populations (12). Their susceptibility to insecticide application is, therefore, of particular interest.

Pitfall traps are not an efficient means of estimating total linyphiid activity as females are very rarely caught, probably because they largely remain associated with webs and are thus much more sedentary than males which, presumably while seeking out females, are caught in traps in considerable numbers. *Oedothorax* spp. are exceptional in this context in that both sexes are readily caught, possibly because they are less reliant on webs to catch their prey (13). Bearing in mind this limitation, trapping records across the range of sites showed a considerable reduction in at least male linyphiid activity which persisted well into the spring and was only restored at the Hertfordshire site in May by a large immigration of the commonest species, *E. atra* and *E. dentipalpis*, probably achieved by ballooning from considerable distances (4).

Some other linyphiid species did not undergo such an influx and like the lycosid spiders, continued to show an apparent reduction in numbers well into the summer on the treated plot. The effect of autumn spray application on spider predation on early summer aphid populations would, therefore, seem to depend strongly upon the timing and extent of aerial dispersal by the dominant species present which may vary in different regions. Although *Erigone* spp. were common at all sites studied, the relative abundance of other genera e.g. *Oedothorax, Lepthyphantes* and *Meioneta* varied considerably over the winter period. Different genera vary in their efficiency as aphid predators (12) and more information on the extent and timing of their aerial dispersal is required.

In any further field work on the effects of insecticide application on polyphagous predators it will be necessary to recognize the difficulties presented by the great mobility of the species concerned. Many predatory beetles are very active making even large plot experiments difficult to interpret. Spider movement, although achieved by ballooning over much greater distances at certain seasons, is probably much less within fields. Consequently, it is perhaps relatively easier to demonstrate insecticidal effects on spider populations and to draw the conclusion, possibly erroneous, that spiders are more susceptible to insecticides than are other polyphagous predators (9, 1). It is planned that, in continuing the present studies, use will be made of barriered plots (5) at the Rothamsted site in an attempt to minimize recolonization of treated areas and so to determine more accurately the mortality of surface active predators following insecticide application.

5. Acknowledgements

The collaboration of ADAS entomologists who undertook regional trials and made available their data for this account is gratefully acknowledged. It is a pleasure to thank Jim Ashby and Ann Wright for assistance with the identification of samples at Rothamsted and to Dr. P. Jepson who advised in the early stages of this work.

REFERENCES

1. BASEDOW, T., RZEHAK, H. and VOSS, K. (1985). Studies on the effect of deltamethrin sprays on the numbers of epigeal predatory arthropods occurring in arable fields. Pesticide Science, 16, 325-331.
2. CARTER, N. (1984). Cereal pests : present problems and future prospects. Proceedings British Crop Protection Conference - Pests and Diseases, Brighton, 1984, 151-158.
3. DE CLERCQ, R. and PIETRASZKO, R. (1982). Epigeal arthropods in relation to predation of cereal aphids. In "Aphid Antagonists", Proceedings of a meeting of the EC Experts' Group, Portici, 1982, 88-92. A.A. Balkema, Rotterdam.
4. DUFFEY, E. (1956). Aerial dispersal in a known spider population. Journal of Animal Ecology, 25, 85-111.
5. EDWARDS, C.A. and THOMPSON, A.R. (1975). Some effects of insecticides on predatory beetles. Annals of Applied Biology, 80, 132-135.
6. EDWARDS, C.A., SUNDERLAND, K.D. and GEORGE, K.S. (1979). Studies on polyphagous predators of cereal aphids. Journal of Applied Ecology, 16, 811-823.
7. GAIR, R. (1981). A review of cereal pests in the United Kingdom. Proceedings British Crop Protection Conference - Pests and Diseases, Brighton, 1981, 3, 779-785.
8. MATCHAM, E.J. and HAWKES, C. (1985). Field assessment of the effects of deltamethrin on polyphagous predators in winter wheat. Pesticide Science, 16, 317-320.
9. RZEHAK, H. and BASEDOW, T. (1982). Die Auswirkungen verschiedener Insektizide auf die epigaischen Raubarthropoden in Winterraps-feldern. Anz. Schadlingsk. Pflanzenschutz Umweltschutz, 55, 71-75.
10. SHIRES, S.W. (1985). A comparison of the effects of cypermethrin, parathion-methyl and DDT on cereal aphids, predatory beetles, earthworms and litter decomposition in spring wheat. Crop Protection, 4, 177-193.
11. SUNDERLAND, K.D., STACEY, D.L. and EDWARDS, C.A. (1980). The role of polyphagous predators in limiting the increase of cereal aphids in winter wheat. Bulletin SROP/WPRS, III/4, 85-91.
12. SUNDERLAND, K.D., FRASER, A.M. and DIXON, A.F.G. (1986). Field and laboratory studies on money spiders (Linyphiidae) as predators of cereal aphids. Journal of Applied Ecology, 23, 433-447.
13. THORNHILL, W.A. (1983). The distribution and probable importance of linyphiid spiders living on the soil surface of sugar-beet fields. Bulletin of the British Arachnological Society, 6, 127-136.

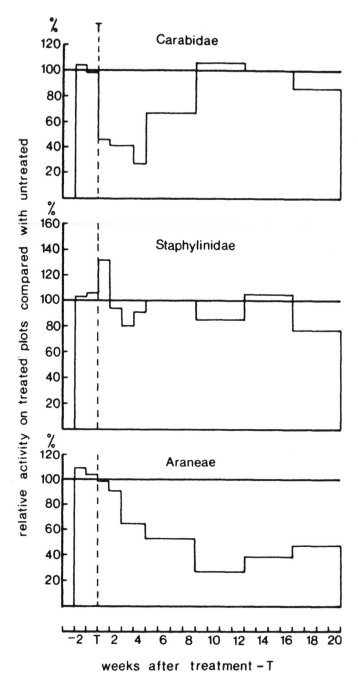

Fig. 1. Relative activity of carabid and staphylinid beetles and Araneae
on treated and untreated plots at nine sites throughout the period
October, 1985 – March, 1986.

160

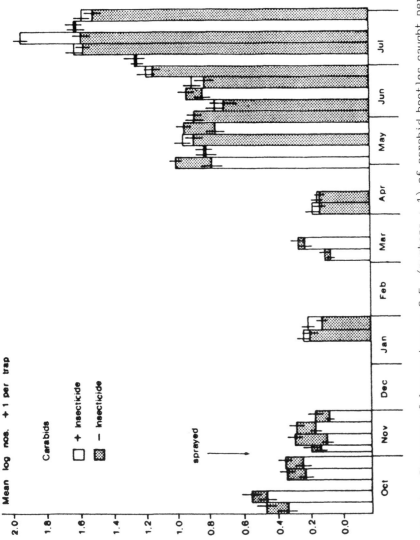

Fig. 2. The mean of log numbers ± S.E. (numbers + 1) of carabid beetles caught per trap at Rothamsted during seven-day trapping periods from October, 1985 – August, 1986

161

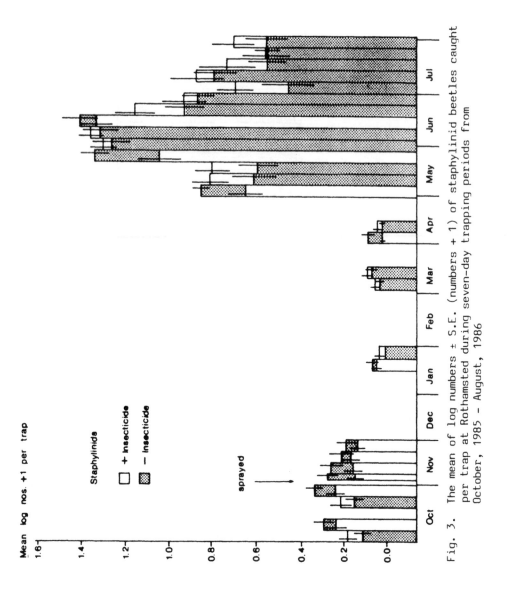

Fig. 3. The mean of log numbers ± S.E. (numbers + 1) of staphylinid beetles caught per trap at Rothamsted during seven-day trapping periods from October, 1985 – August, 1986

162

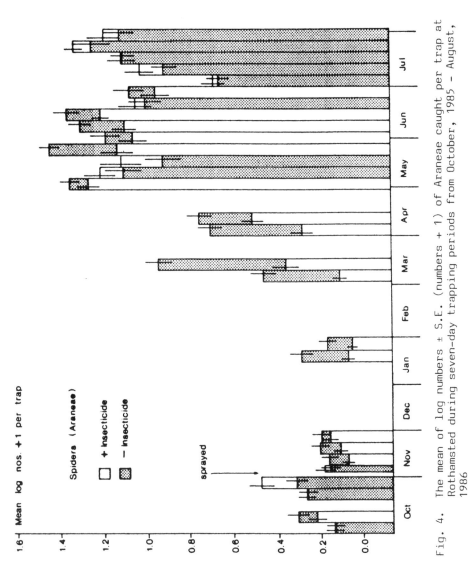

Fig. 4. The mean of log numbers ± S.E. (numbers + 1) of Araneae caught per trap at Rothamsted during seven-day trapping periods from October, 1985 – August, 1986

163

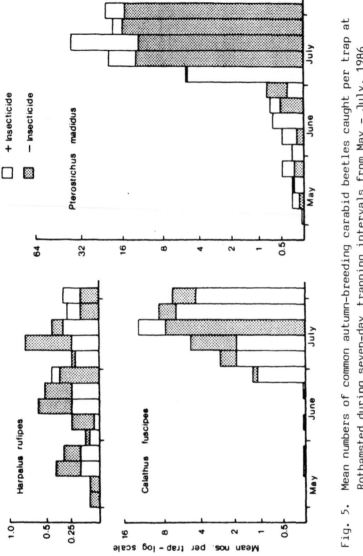

Fig. 5. Mean numbers of common autumn-breeding carabid beetles caught per trap at Rothamsted during seven-day trapping intervals from May – July, 1986

164

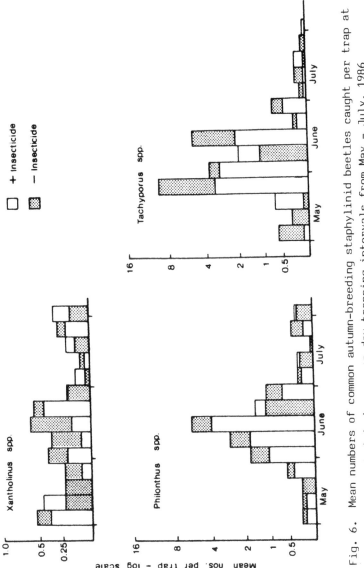

Fig. 6. Mean numbers of common autumn-breeding staphylinid beetles caught per trap at Rothamsted during seven-day trapping intervals from May – July, 1986

165

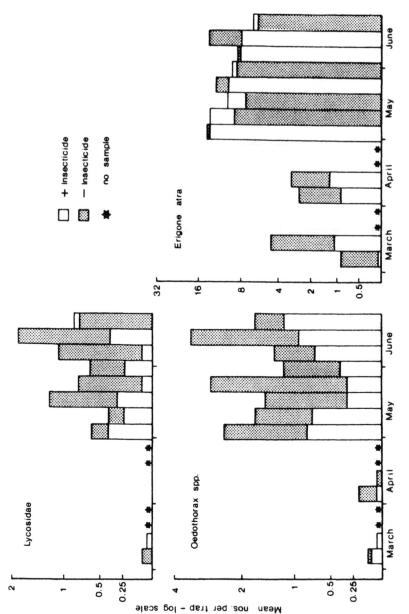

Fig. 7. Mean numbers of some common Araneae caught per trap at Rothamsted during seven-day intervals from March – June, 1986

166

On the influence of some insecticides and fungicides on the epigeal arthropod fauna in winter wheat

R.De Clercq & H.Casteels
State Nematology and Entomology Research Station (CLO-Gent), Merelbeke, Belgium

Summary

In 1986 the influence of the insecticides phosalone and fenvalerate and the fungicides fenpropimorph, prochloraz and propiconazole has been investigated on the soil predators (Carabidae, Staphylinidae and Araneae) in winter wheat. Our field experiments were carried out in 3 replicates on square barriered plots of 25 m^2. All the pesticides were applied at the end of May. The predator fauna was examined by means of pitfall traps.

From these field experiments it can be concluded that the insecticides phosalone and fenvalerate have little toxic effect on some polyphagous predators, whereas the fungicides fenpropimorph, prochloraz and propiconazole had no influence on the epigeal arthropod fauna.

1. Introduction

Since 1960 aphid attacks have become more of a problem in winter wheat. Research has shown that the presence of certain predatory insects, especially carabids, rove beetles and spiders, may reduce the possibility of such outbreaks (4, 6). These so-called polyphagous predators are very important in annual crops, because they can feed on a wide variety of foods but will switch to eating aphids, whenever they have the opportunity. In addition, there are the specific predators, such as ladybirds and the larvae of hoverflies, which feed almost entirely on aphids.

Since 1960 the control of pest and diseases in winter wheat depends very heavily on chemical pesticides. Research is needed to establish the specificity of insecticides, as well as the insecticidal action of some fungicides and herbicides.

In 1980, 1981 and 1982 the influence of the insecticides parathion, dimethoate and pirimicarb, and of the fungicide benomyl has been investigated on the epigeal predator fauna (Carabidae, Staphylinidae and Araneae) in winter wheat fields. From these experiments it can be concluded that the insecticides dimethoate and parathion are very toxic for the dominant Carabidae- and Staphylinidae spp., but only little for the spiders, whereas pirimicarb and the fungicide benomyl had no influence on this fauna (5). Analogous results are obtained by BASEDOW et al. (1) and VICKERMAN and SUNDERLAND (7).

In 1984 and 1985 the influence of the insecticides phosalone and fenvalerate has been investigated on the soil predators (Carabidae, Staphylinidae and Araneae) in winter wheat. From these field experiments it can be concluded that the insecticides phosalone and fenvalerate, applied half June, have none or little toxic effect on the soil predators in winter wheat (2, 3).

This paper describes field experiments carried out in 1986 in order to investigate the possible influence of the insecticides phosalone and fenvalerate and the fungicides fenpropimorph, prochloraz and propiconazole, all applied at the end of May, on the epigeal arthropod fauna in winter wheat.

2. Methods

The experiments were carried out in 1986 in 3 replicates on square plots of 25 m^2. Winter wheat was sown in November, using seed cultivar, seed rate, fertilizer treatment and herbicides typical of locality.

All the plots were barriered end May with 50 cm high corrigated plastic sheets which were buried 20 cm into the soil to delineate the plots and to prevent movement of the various predators from one treatment area to another. The pesticides were applied on 21 May. The following objects were compared :

- untreated
- phosalone at 900 g a.i./ha
- fenvalerate at 25 g a.i./ha
- fenpropimorph at 1500 g a.i./ha
- prochloraz at 1000 g a.i./ha
- propiconazole at 500 g a.i./ha.

Four days after the application of the pesticides 8 plastic beakers with a diameter of 10 cm were buried within each plot with their open end level to the soil surface.

The beakers, filled with a 1 % formol solution to which a spreading agent was added, were weekly emptied, beginning a week after the application of the pesticides until the first half of June. The effect of these applications on carabids, staphylinids and spiders and on the dominant subfamilies of carabids and staphylinids is dealt with consecutively.

3. Results

The total number of predators caught in 1986 within the different plots is given in table I. These results indicate that there is no toxic effect of the pesticides on the total number of polyphagous predators.

Table I. - Influence of some pesticides on the polyphagous predators in winter wheat.

Total catches in 3 x 8 pitfalls during the first half of June

Treatment	Carabidae	Staphylinidae	Araneae	Total
Untreated	703	1822	722	3247
Fenpropimorph	645	2262	716	3623
Prochloraz	704	2108	769	3581
Propiconazole	710	2600	813	4123
Fenvalerate	640	3092	441	4173
Phosalone	805	2337	821	3963

Table II. – Influence of some pesticides on the concerned subfamilies of the Staphylinidae.

Total catches in 3 x 8 pitfalls during the first half of June

Treatment	Aleocharinae	Tachyporinae	Remaining subfamilies(*)
Untreated	502	832	488
Fenpropimorph	613	1054	595
Prochloraz	569	981	558
Propiconazole	834	1134	632
Fenvalerate	992	1369	731
Phosalone	645	1165	527

(*) Remaining subfamilies : Staphylininae, Steninae, Paederinae, Oxytelinae and Micropeplinae.

Table III. – Influence of some pesticides on the concerned subfamilies of the Carabidae.

Total catches in 3 x 8 pitfalls during the first half of June

Treatment	Scaritinae	Loricerinae	Bembidiinae	Pterostichinae	Remaining subfam.(*)
Untreated	86	44	434	110	29
Fenpropimorph	91	40	402	88	24
Prochloraz	144	42	382	116	20
Propiconazole	156	33	405	93	23
Fenvalerate	126	35	370	84	25
Phosalone	127	34	374	246	24

(*) Remaining subfamilies : Nebriinae, Notiophilinae, Demetrinae, Badistrinae, Zabrinae, Stenolophinae, Harpalinae, Anisodactylinae and Trechinae.

Taking into account the three groups separately, we can conclude the staphylinids were the most abundant group, while the spiders and the carabids have the same importance.

The Araneae consisted mainly of : Erigone atra (Blackwall), Oedothorax apicatus (Blackwall), Bathyphantes gracilis (Blackwall) and Leptyphantes tenuis (Blackwall) (Linyphiidae). The results demonstrate that only fenvalerate caused a reduction of the number of spiders by 40 % during the observed post-treatment period, whereas the other pesticides had no toxic effect on the polyphagous predators.

The staphylinids consisted mainly of : Aloconota gregaria (Er.) from the Aleocharinae and Tachyporus hypnorum (L.) from the Tachyporinae. The results in table II indicate that the fungicides and both the insecticides had no effect on the rove beetles population.

The carabidae were less abundant than the former predator group; the cómmon species were : Bembidion lampros Herbst, B. ustulatum L. (Bembidiinae), Pterostichus melanarius (Illiger), Platynus dorsalis Pont. (Pterostichinae), Lorocera pilicornis F. (Loricerinae) and Clivina fossor L. (Scaritinae). In some cases a reduction of the number of 10 % to 15 % is noticed (table III).

4. Conclusion

These experiments have shown that the fungicides fenpropimorph, prochloraz and propiconazole are not toxic for carabids, staphylinids and spiders, while the insecticides phosalone and fenvalerate are little toxic to this fauna.

These results are analogous to those obtained in the experiments in 1985 (2) and to those of CHEROUX et DEBRAY (3).

Because of their low detrimental effect on the polyphagous predators, these broader spectrum insecticides can be used for cereal aphid control in winter wheat. Damage to the epigeal fauna can also be minimized when using selective pesticides, such as pirimicarb (5).

Insecticides such as phosalone, fenvalerate and pirimicarb should be applied only when the aphid infestation exceeds the economic treshold level.

REFERENCES

1. BASEDOW, Th., BORG, A. und SCHERNEY, F. (1976). Auswirkungen von Insektizidbehandlungen auf die epigaïschen Raubarthropoden in Getreide-feldern, inbesondere die Laufkäfer (Coleoptera, Carabidae). Ent. exp. and appl., 19, 37-51.
2. CASTEELS, H. and DE CLERCQ, R. (1986). On the influence of phosalone and fenvalerate on the epigeal arthropod fauna in winter wheat. Bulletin OILB/SROP, 1986, in print.
3. CHEROUX et DEBRAY, Ph. (1985). Sumicidin 10, incidence des pulvérisa-tions sur la faune auxiliaire en culture de maïs et blé. La défense des végétaux, n° 233, 31-37.
4. DE CLERCQ, R. and PIETRASZKO, R. (1982). Epigeal arthropods in relation to predation of cereal aphids. E.E.C. Report "Aphids antagonists, 1983, 88-92.
5. DE CLERCQ, R. and PIETRASZKO, R. (1984). On the influence of pesticides on Carabidae and Staphylinidae in winter wheat. Les colloques de l'INRA, n° 31, 273-278.
6. SUNDERLAND, STACEY, D. and EDWARDS, C. (1980). The role of polyphagous predators in limiting the increase of cereal aphids in winter wheat. Bulletin OILB/SROP, III/4, 85-87.
7. VICKERMAN, G. and SUNDERLAND, K. (1977). Some effects of dimethoate on arthropods in winter wheat. J. Appl. Ecol., 767-777.

The corn stalk borer, *Sesamia nonagrioides:* Forecasting, crop-loss assessment and pest management

J.A.Tsitsipis
Department of Biology, 'Democritos' National Research Center, Aghia Paraskevi, Greece

Summary

The corn stalk borer, Sesamia nonagrioides (Lef.) (Lepidoptera, Noctuidae), is an important corn pest causing severe damage to the late crop. Adults of the overwintering generation appear in April-May and last catches occur in October-early November. Three to four generations are completed within the year. Preliminary results allow estimation of insect appearance and issue of warnings for timely insecticidal applications based on adult captures and plant examination for oviposition. Damage is due to larval feeding inside the stalks and ears. Crop loss is usually 0-10% in early crop and it can reach to 100% in late crop. High stalk infestation resulted in reduced protein content of seeds. Highly infested ears developed secondary fungal infestation in which aflatoxin B_1 was detected at half the tolerance level concentration. For the late crop pest management, an insecticidal spray programme was tested in which two to three spray applications were necessary to achieve reasonable protection. Investigations on the relative resistance of 12 simple corn hybrids to borer infestation showed variation in the degree of susceptibility under natural infestation conditions. Some hybrids were very sensitive, some showed a medium susceptibility and some were relatively resistant.

1. INTRODUCTION

Maize has become an important crop in Greece during the last several years. It is the introduction of improved simple hybrids as well as the improvement of cultural practices (fertilisation, herbicide control, irrigation) that have increased corn yields remarkably. While the cultivated area increased by 52%, yields increased by 160% during the years 1975-1986 as shown in Table 1. Corn is, however, often susceptible to important insect pests. The most important is the corn stalk borer, Sesamia nonagrioides (Lef.) (Lepidoptera, Noctuidae), which causes serious damage especially to the late-sown crop, that is the one sown in July after the harvest of small cereals. Observation on the biology and ecology of the species have been done in certain south European and north African countries (7, 10, 9, 8, 6, 4). Data on this species in Greece has been very limited (11, 14, 5, 12). Information available on crop loss assessment caused by the corn stalk borer and the management of this pest is also very limited.

In this paper work is reported on the seasonal appearance of the insect, the assessment of the crop losses, the evaluation of the relative

TABLE 1. Cultivated Area with Maize in Greece and Corn Production During the Years 1975-1986

Year	Cultivated area (Hectares x 10³)	Total production (Tonnes x 10³)	Production (Tonnes/hectare)
1975	135	540	4.0
1976	128	505	4.0
1977	130	541	4.2
1978	112	536	4.5
1979	122	731	6.0
1980	163	1233	7.6
1981	161	1360	8.5
1982	164	1448	8.9
1983	171	1639	9.6
1984	205	1992	9.7
1985+	203	1800	8.8
1986+	200	2070	10.4

Source: Anonymous 1985 (2), +Ministry of Agriculture, Greece.

resistance of certain simple hybrids, the development of a spraying programme and the formulation of an approach towards developing a forecasting system.

2. MATERIALS AND METHODS

The seasonal appearance of the corn stalk borer was studied in eight different major corn producing areas in Greece (N. Greece: Kavala; Central Greece: three locations in Larissa, Lamia, Kopaïs, Aitoloakarnania; S. Greece: Ilia). Adult insects were trapped in light traps supplied with a 15 watt BL or an 8 watt Daylight fluorescent lamp. Ethyl acetate or a plaquette of DDVP were used as killing agents. Traps were serviced three times a week. Crop losses, caused by the stalk borer, were estimated by random sampling of 100 plants per field at harvest time. Plants were dissected and stalk damage, caused by the larvae, was expressed as a percentage of the total volume of the stalk tissue. Damage evaluation in the spadices was similarly expressed as a percentage of the total seed area (14). Records were made of the infestation level of each plant associated with its spadix. The effect of the plant infestation intensity on the chemical composition of the seeds of uninfested ears was also examined. A chemical analysis of the seeds was made on the total protein, lipid, ash, fibrous material and nitrogen-free extractable matter content, according to a methodology reported elsewhere (14). Heavily infested corn ears by the corn stalk borer were examined for presence of aflatoxin B_1 as a result of secondary fungal infection of the seeds. The method used was the one reported in the EEC Gazette (3).

The relative resistance or sensitivity of 12 corn hybrids, late crop sown in mid-July, was tested against natural infestation of the corn stalk borer in a field experiment in Kopaïs. A randomised complete block design was set up in four blocks. Each hybrid in each block was sown in two lines of 25 plants each. The distance between the lines was 80 cm and on the line the plants were 20 cm away from each other. All experimental plants were treated three times with a combination of trichlorfon (Dipterex) and methamidophos (Tamaron) at concentrations of 0.1 and 0.3% (a.i.)

respectively. The timing of applications was decided upon when adult activity was noticed in the field and egg plaques were ready to hatch. Plants were irrigated two days before the insecticidal applications to avoid rinsing the insecticide with subsequent water sprinkling of the plants. The interval between two consecutive sprayings was approximately three weeks and the first application occurred on August 23. Final evaluation of the hybrids was done at harvest, where ten randomly collected ears per block of each hybrid were examined.

In an experiment in Kopaïs, the degree of protection to the corn plants from a natural infestation of the corn stalk borer by a spraying programme of a combination of trichlorfon (0.1%) and methamidophos (0.03%) was tested in a late-sown corn field in mid-July. In the programme there were included: one early (Aug.23), one late (Sept.14), two early (Aug.23, Sept.14), two late (Sept.14, Oct.5), and three (Aug.23, Sept.14, Oct.5) spraying applications. The experiment was set up in a randomised complete block design in four blocks. Each treatment in the block was composed of four lines of 25 plants each. Distances between the lines and on the line were similar to the previous hybrid experiment. Final evaluation of the experiment was done at harvest and only the two central lines from each treatment block were considered. All ears were evaluated.

Data evaluation was submitted to analysis of variance and means were compared by the Duncan's multiple range test.

3. RESULTS AND DISCUSSION

The generalised curve of the seasonal appearance of the corn borer adults is shown in Figure 1. During the years 1982-1985, data from different areas around Greece were used to draw the line. The first adults usually appear from late April to June and those are the adults of the previous generation that overwinters in the larval stage. A second peak appears in July and this is the adult population of the first generation. A subsequent population rise occurs in mid to late August that gives very high insect numbers in September. From the end of September, the population decreases and the last captures usually take place at the end of October to early November. During this period three generations are completed and there is

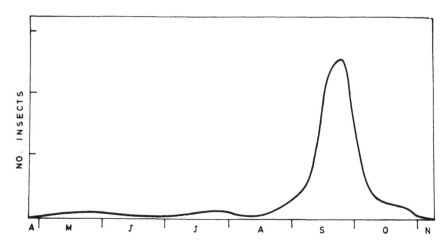

Fig. 1. **Generalised curve of seasonal appearance of Sesamia nonagrioides adults caught in light traps in South, Central and North Greece during the years 1982-1985.**

evidently a partial fourth generation. What is of great interest is the remarkable difference in the insect numbers between the first and the second generation, where differences of a hundred fold or more have often been observed. The above pattern of seasonal appearance has shown certain deviations from area to area and from year to year. It is possible that the first adults are caught two to four weeks later and the peaks of the first and second generation appear a little later. Thus, it is possible to observe slightly higher population numbers in the first generation, or a delay in the decline of the population of the second generation, or the appearance of a small peak during the decline phase of the second generation (early to mid October). Data from Italy (Sardegna) and France show much higher population levels in May and August than in Greece.

Evaluation of the impact of the corn stalk borer on the corn crop is shown in Table 2 (14). The infestation levels on the plant stalks varied considerably. Levels ranged from 2% in the Evros area, where there is practically no corn borer activity, to 97% in the Ilia area in the Peloponnese. The respective values for the spadices were 0% in Evros and 50% in a late crop in Phthiotis. The infestation intensity of the infested stalks ranged from 1.5 to 27.3% for the Evros and Ilia area respectively, while the values for spadices were 3.7% for Voiotia, 15% for Ilia and 14.1% for the late crop in Phthiotis. It is evident from these data that damage can be serious in the late crop. In a late-sown field in Voiotia, where no measures against the corn borer were taken, a complete loss of the crop was recorded. Heavily infested plants had collapsed to the soil surface and all seeds were devoured by the larvae.

TABLE 2. Infestation of Stalks and Ears of Corn by Sesamia Nonagrioides in 1984. Sample size: 100 plants

| District | % Infestation | | Degree of infestation (%, $\bar{x}$ Sd) | |
	Stalks	Ears	Stalks	Ears
Ilia	97	29	27.3 ± 21.0	15.0 ± 24.0
Larissa	43	9	12.6 ± 14.2	9.0 ± 16.0
Voiotia	37	5	9.5 ± 14.0	3.7 ± 4.7
Evros	2	0	1.5	-
Phthiotis	65	25	15.5 ± 16.7	7.2 ± 8.3
Phthiotis+	49	50	9.1 ± 12.2	14.1 ± 18.3

Source: Tsitsipis et al. 1985 (14)
+ Late crop

The effect of infestation intensity of the stalks on the yield obtained is shown in Table 3 (14). These are preliminary data from one sampling only, but they give an indication of the relationship between the two parameters measured. Stalk infestation at levels abbove 25% gave yields significantly lower than at levels ranging from 0-10%. It was of interest to see whether there is any adverse effect of the infestation level of corn stalks compared with the quality of seeds produced by uninfested ears. While there were no differences found in the lipid and the ash content in seeds deriving from plants with very high or very low infestation levels, there were statistical differences in the protein and the nitrogen-free extractable matter between low level (28-40%) and high level (68-72%) stalk infestation in late-sown corn in Kopaïs (Voiotia). The respective values were for protein 10.68 ± 0.4 and 12.49 ± 0.7 and for nitrogen-free extractable matter 82.4 ± 0.3 and 80.9 ± 0.6 (per cent values of dry weight). It seems that high infestation levels affect protein metabolism of the plant (14).

Seeds coming from ears heavily infested by the corn stalk borer (>70%) were tested for aflatoxin B_1. The test proved positive and the toxin was found at a concentration of 25 ± 2 ppb, well below the tolerance level for animal feeds (50 ppb). It showed, nevertheless, that the respective fungi producing the toxin were present and could produce it at certain levels (14).

TABLE 3. Relationship Between Stalk Infestation Intensity of Maize by Sesamia nonagrioides and Seed Production. Sampling in Larissa 1984

	Stalk infestation intensity (%)	Ears (No.)	Mean ear seed weight (g)	Compa- rison	Student's t value
1.	0-10	78	195.6 ± 71.5	(1,2)	2.03*
2.	11-25	13	163.4 ± 84.5	(1,3)	3.52*
3.	> 25	7	112.2 ± 54.7	(2.3)	1.45

Source: Tsitsipis et al. 1985 (14)
*P< 0.05

A field experiment aimed at testing the possibility of protecting the crop from the corn stalk borer by insecticidal treatments showed that three insecticidal applications could give an acceptable degree of protection, and so could two late sprayings (Table 4).

TABLE 4. Mean Ear Infestation by Sesamia nonagrioides in a Late Maize Crop Treated with One, Two or Three Insecticidal Applications. Four Blocks in a Randomised Complete Block Design

Applications	Ear infestation intensity (%)
Three	23.00 a
Two late	29.25 ab
Two early	46.25 c
One late	38.00 cd
One early	60.25 d
Control	62.50 d

Figures with same letter are not different by Duncan's multiple range test (P < 0.05)

One late or two early sprays gave almost similar protection, while an early one alone did not provide any protection at all.

Finally, from 12 different hybrids tested for relative resistance to the corn stalk borer (Table 5) some were found to be relatively resistant, some were susceptible and there were some that showed an intermediate degree of sensitivity.

A knowledge of the seasonal appearance of the corn stalk borer in Greece, the number of generations per year, the thermal constant (590 days-degrees; (6)) and the lower threshold of development (10.5°C; (6)) make possible work towards the development of a forecasting model. It is understood that more information is needed on (i) the biology of Greek geographical races of the corn stalk borer, (ii) the study of the relationship between temperature and development, (iii) comparison between the method of thermal summation and that of the logistic equation, (iv) the collection of data on trap efficiency and correlation of trap catches with infestation levels. More information is also needed on whether the low adult

175

TABLE 5. Mean Ear Infestation by Sesamia nonagrioides in 12 Maize Hybrids, Sown Late, Treated Three Times Organophosphorous Insecticides. Four Blocks in a Randomised Complete Block Design

Hybrid No.	Ear infestation intensity (%)
112	12.80 a
140	15.20 abc
105	15.50 abc
130	15.80 abc
104	17.57 abc
148	20.75 bc
101	21.35 bc
114	22.70 cd
131	23.07 cd
129	30.02 d
149	46.10 e
151	49.10 e

Figures with same letter are not different by Duncan's multiple range test ($P < 0.05$)

captures during July represent the real population levels. To date the data available are used to give limited warnings for timely insecticidal applications based on trap catches and examination of egg masses in the field.

For the pest management of the corn stalk borer the data of the present report showed the effect of a spraying programme suitable for the late crop. It is to be investigated, if a fourth application or three late ones give an even higher degree of protection. It was noted that despite low stalk infestation, ear infestation was high (probably caused by late oviposition on the ears, when insecticide activity had ceased). The role of certain parasites has also to be clarified. A tachinid, Lydella thompsoni, has been found to parasitise larvae. It is, however, an egg parasite, the scelionid Telenomus busseolae (1) that was found to be very effective during the months of August and September 1986 in the area of Istiaea, Evoia (Alexandri and Tsitsipis, unpublished data). Finally the role of Sesamia nonagrioides pheromone (Mazomenos, personal communication) has to be viewed within the framework of a pest management system.

ACKNOWLEDGEMENTS

Thanks are expressed to Mrs. Philippopoulou and Mr. Kontoghiannis for drawing the graph and photographing it respectively.

REFERENCES

(1)   ALEXANDRI, M.P., 1986. Bioecology and parasitization of Sesamia nonagrioides Lef. (Lepidoptera-Noctuidae) in the district of Istiaea, Evoia. Dissertation, College of Agriculture, Athens, Greece. (In Greek).
(2)   ANONYMOUS, 1985. Maize. Factors affecting its productivity. Modern Agric. Techn. 25:16-24. (In Greek).
(3)   EEC GAZETTE, 1976. No.372/15-4-1976, L 102/9-14pp. (Greek Edition).

(4) GALICHET, P.F., 1982. Hibernation d'une population de Sesamia nonagrioides Lef. (Lep., Noctuidae) en France méridionale. Agronomie 2:561-566.

(5) GLIATIS, A., 1983. Report on a project to study the life cycle of Sesamia sp. in the district of Larissa. Report to the Hellenic Ministry of Agriculture. Mimeo. 13pp. (In Greek).

(6) HILAL, A., 1981. Etude de développment de Sesamia nonagrioides et établissement de modèles pour la prévision de ses population dans la nature. Bull. OEPP 11:107-112.

(7) LESPES, L. and JOURDAN, M.L., 1940. Observations sur la biologie de la Sésamie du Maïs (Sesamia vuteria Stoll.) au Maroc. Rev. Zool. Agr. Appl. Bordeaux 7-8:49-58.

(8) MELAMED-MADJAR, V. and TAM, S., 1980. A field survey of changes in the composition of corn borer populations in Israel. Phytoparasitica 8:201-204.

(9) NUCIFORA, A., 1966. Appunti sulla biologia di 'Sesamia nonagrioides' (Lef.) in Sicilia. Tec. Agr. 18:395-419.

(10) PROTA, R., 1968. Osservazioni sul' etologia di Sesamia nonagrioides (Lefebvre) in Sardegna. Studi Sass. Ann. Fac. Agr. Sassari 13:336-360.

(11) STAVRAKIS, G.N., 1967. Contributions a l'étude des éspèces nuisibles au maïs en Grèce du genre Sesamia (Lepidoptères-Noctuidae). Anns. Inst. Phytopath. Benaki 8:20-23.

(12) TSITSIPIS, J.A., GLIATIS, A. and MAZOMENOS, B.E., 1984. Seasonal appearance of the corn stalk borer, Sesamia nonagrioides, in Central Greece. Med. Fac. Landbouww. Rijksuniv. Gent, 49:667-674.

(13) TSITSIPIS, J.A., MAZOMENOS, B.E., CHRISTOULAS, C., MOULOUDIS, S., STEFANAKIS, M., PAPAGEORGIOU, G., GLIATIS, A. and SINIS, D., 1983. Report on the lepidopterous insects attacking corn in Greece with emphasis on the corn stalk borer, Sesamia nonagrioides. 9th Interbalcanic Plant Protection Conference, Athens, November 7-11, 8pp.

(14) TSITSIPIS, J.A., GLIATIS, A., SALTZIS, B., STATHOPOULOS, F., MOULOUDIS, S., STEFANAKIS, M., PAPASTEFANOU, S., CHRISTOULAS, C., PAPAGEORGIOU, G., KATRANIS, N. and ECONOMOU, D., 1985. Seasonal appearance of Sesamia nonagrioides and Ostrinia nubilalis in different areas in Greece and estimation of crop losses caused to corn. 2nd National Entomological Meeting, Athens, November 6-8. (In press) (In Greek with extended English summary).

Current status of the *Sesamia nonagrioides* sex pheromone and its potential use in the suppression of the insect population

B.E.Mazomenos
Department of Biology, 'Democritos' National Research Center, Aghia Paraskevi, Greece
D.Bardas
Agricultural Cooperatives, Servota, Trikala, Greece

Summary

The synthetic sex pheromone blend of the Sesamia nonagrioides Lef. was evaluated under field conditions. An efficient trap and the optimum pheromone concentration for field attraction were established. A pilot project to suppress insect population by the mass trapping method was initiated during 1986. First year results indicated that ten pheromone traps per ha suppressed insect populations and crop damage in the early-sown crop remained at low levels. In the late corn crop, although male captures in the pheromone traps were low, the infestation of the stems and spadices increase and crop damage reached 30%.

1. INTRODUCTION

The corn stalk borer, Sesamia nonagrioides Lef., is considered one of the most serious pests of corn. The insect has been reported to occur in most of the countries surrounding the Mediterranean basin and in many countries in Africa. In Greece, since the introduction of improved single hybrids, the cultivated area has been increased and also the growing season has been greatly extended. However, S. nonagrioides has become the most serious pest (6, 7) and crop losses due to this pest, presumably in the late corn crop, (maize sown in the beginning of July) are very high.

Field and laboratory tests demonstrated that 2-3 day old virgin females produce and release an airborne sex attractant which attracts males (3). Preliminary chemical work on the collection and purification of the sex pheromone showed that when the last two abdominal segments of 2-3 day old virgin females were extracted in ether, the ether extracts were attractive to males. Partial chemical isolation indicated that the sex pheromone produced is a multicomponent pheromone (3).

In this paper we discuss the current status of the corn stalk borer S. nonagrioides sex pheromone and its potential use in an integrated programme aiming to control this pest.

2. MATING BEHAVIOUR

Preliminary studies on the mating behaviour of the S. nonagrioides showed that female mating and calling behaviour and male courtship behaviour were similar to that described for other noctuid species (1, 2, 5). Females began exhibiting calling behaviour 48 h after eclosion; they were resting on a vertical position in the cage. The last abdominal segments were extruded and a white sac was exposed; while the sac was exposed they were fanning their wings facilitating pheromone evaporation from the gland surface. The female began calling two hours into the dark phase and the peak of calling was observed between the fourth and the sixth hour.

The resting males, upon pheromone perception, were waving their antennae, vibrating their wings for a few minutes and then they were flying toward the calling female. When the male approached the calling female he landed a short distance away, courted her for a few minutes and then mated. Copualation lasted for about 1.5 hours. Most of the matings occurred four to six hours into the dark phase. Females were mated 48 hours post-emergence, while males were mated from the first day of emergence.

3. PHEROMONE IDENTIFICATION

Gas liquid chromatographic and mass spectral analysis of pheromone gland volatiles extracted in ether indicated the presence of Z-11-hexadecenyl acetate (Z-11-16:Ac) and dodecyl acetate (12:Ac), both were attractive to males (Mazomenos, in preparation). Further chromatographic studies revealed the presence in minute quantities of two components which showed some degree of attraction in laboratory bioassays. Comparison of GC retention data for these two components with that of Z-11-hexadecenol (Z-11-16:OL) and Z-11-hexadecenal (Z-11-16:ALD) on high resolution columns indicated that they were identical (Mazomenos, in preparation). Confirmation that the proposed structures for the four pheromone components were correct was provided by field tests carried out during August to October 1985. The tests were conducted in the Kopais area of central Greece. The traps were baited with rubber septa (H. Thomas, Co Phila P.A. No. 1780 J 12 stopper 7 x 11 mm), which were impregnated with 250 µg of each component. The data obtained indicated that traps baited with Z-11-16:Ac were attractive to males, while traps baited with the other components were not attractive. Few males were caught in traps baited with rubber septa impregnated with 250 µg of 12:Ac. Z-11-16:Ac is the major sex pheromone component (4). All possible combinations of the three secondary components with Z-11-16:Ac revealed various degrees of male attraction. Traps baited with 200 µg of Z-11-16:Ac + 20 µg Z-11-16:OL + 20 µg Z-11-16:AL attracted significantly more males than traps baited with Z-11-16:Ac alone. A higher number of males was attracted to traps baited with 200 µg Z-11-16:Ac + 20 µg Z-11-16:OL + 20 µg Z-11-16:ALD + 40 µg 12:Ac. The average number of males captured per trap during the experimental period in traps baited with 250 µg Z-11-16:Ac, 200 µg Z-11-16:Ac + 20 µg Z-11-16:OL + 20 µg Z-11-16:ALD and 200 µg Z-11-16:Ac + 20 µg Z-11-16:OL + 20 µg Z-11-16:ALD + 40 µg 12:Ac was 22.9, 62.4 and 97.5 males respectively (4).

4. TRAP DESIGN

Pheromone trap efficiency is expected to be affected by such parameters as shape, size and colour of the trap, trap position in the field, pheromone concentration, trap density, etc. Four different types of trap were tested during June 25 to July 26, 1985. The type of traps tested were:

a) Delta trap (INRA).
b) Cylindrical trap made from PVC drainpipe; the length of the trap was 18 cm with an internal diameter of 10 cm. A poster paper insert coated with sticky material was placed at the bottom of the trap to kill or immobilise the attracted males.

c)   Funnel-type traps made of aluminium sheet 0.2 mm thick. The large end of the funnel was 14 cm in diameter and reduced to 6 cm in diameter at the small end. The large end was covered with an aluminium sheet 16 cm in diameter. The distance between the funnel and the cover was 2.5 cm. A plastic bag was fastened at the small end to collect the trapped males.

d)   Moth trap (Pheromone International Ltd.).

A piece of slow release formulated dichlorvos (DVP) was used in traps c) and d) to kill the males entering the traps. The pheromone was suspended from rubber septa loaded with 250 µg of the pheromone blend. The traps were placed 30 m apart at the edge of a corn field on stakes 1.5 m above the ground.

Table 1 shows the number of males captured during the experimental period (a period with low population density). Of the traps examined, the moth trap was clearly more efficient than the other traps. Several general inferences were possible from this test:

(i)   Field observation during the night revealed that the males began responding to the pheromone between 00.10 to 00.30 h.

(ii) The majority of the males entering the pheromone plume did not fly directly to the pheromone source, but they landed on the lower part of the trap and then after a few seconds they walked to the pheromone source. This type of male behaviour was suited to the moth traps, which gave the insects enough space to walk.

TABLE 1. The Effect of Trap Design on Male S. nonagrioides Captures; Traps Were Baited with 250 µg of the Synthetic Sex Pheromone Blend. Kopais Greece June 25 to July 25 1985.

Trap	Total moths	Males/trap/night
Delta	4	0.13 a
Cylinder	15	0.50 a
Funnel	28	0.93 ab
Moth trap	63	2.10 c

Means followed by the same letter are not significantly different. (Duncan's multiple range test P=0.05)

Moth traps were selected for further studies, such as effect of pheromone concentration on trap catches, monitoring insect population and mass trapping studies.

5. EFFECT OF PHEROMONE CONCENTRATION

The test of concentration of pheromone was designed to determine the most effective dose to be used for monitoring or mass trapping studies. The test was conducted at Kopais between September 16 and October 15, 1985. Traps were baited with rubber septa loaded with 100, 150, 200 and 250 µg of pheromone. The traps were suspended randomly 1.5 m above the ground, 50 m apart. Three replications were used for each concentration. The traps were cleaned once a week and the males captured were recorded.

181

All the concentrations tested attracted males (Table 2). Although there is no significant difference, the concentration of 200 µg/trap attracted more males than the other three concentrations tested. Pheromone trap catches were high from the first night and remained high up to the end of the experiment. The lower and upper threshold of male response to the pheromone seems to be below and above the range of the concentrations tested.

TABLE 2. Effect of the Synthetic Female S. nonagrioides Sex
Pheromone Concentration on Male Captures.
Kopais Greece September 16 to October 15, 1985.

Concentration (µg)	Total males	Males/trap/night
100	1584	63.4
150	2421	96.8
200	2751	110.0
250	1909	76.4

6. MASS-TRAPPING

The damage caused by the corn stalk borer in the early corn crop (maize sown in the beginning of May) is low and no special efforts are taken for control. In some corn growing areas one insecticide treatment is applied during July. The effectiveness of the treatment is questioned. For the late corn crop, crop damage due to corn stalk borer is high and 3 to 5 insecticide treatments are applied, without effective control. The lack of an efficient method to prevent corn damage caused by the corn stalk borer is the main reason that farmers tend to abandon the late corn cultivation.

A pilot test was designed during 1986 to study the effectiveness of the mass-trapping method with pheromone traps on the suppression of the corn stalk borer population, especially for the late corn crop. The tests were conducted in Servota, Trikala, central Greece. The area which was covered with early-sown corn was approximately 200 ha, while the area covered with late-sown corn was less than 5 ha. An 11 ha field was selected for the experiment. The field was partially isolated from the other corn fields. Corn was sown at the beginning of May.

Another corn field of 4 ha about 2 km away and sown on the same date in May was used as a control. Near the two fields 1.5 ha of corn was sown late in July. In each field 12 traps (Phytophyl AE, 16 Averof Street, Athens, Greece), baited with 200 µg of the sex pheromone blend, were installed on April 20 to follow the insect population. Ten traps per ha were used for the mass-trapping experiment. The traps were installed on May 12 and the distance between the traps was 30 m; trap density was higher along the borders. The traps were baited with polyvinyl septa loaded with 200 µg of the synthetic pheromone blend. The pheromone dispensers were replaced once during the experiment. Twelve traps in both mass-trapped and control fields were inspected every three days to monitor the insect population. Corn plants were also sampled regularly and the number of egg masses and degree of larval infestation were recorded.

7. RESULTS

7.1 Early-sown Corn

The number of males captured in the entire field during the insect-flying period was 25 000 males. Traps which were suspended inside the field captured less males than traps suspended along the borders of the field. The first males were captured in the last week of April. These males originated from insects which hibernated as larvae. The population density was very low and flight activity lasted until the middle of July. The second flight activity was observed from the middle of July to the middle of August. At the beginning of September when the third generation emerged, the population increased and trap catches also increased significantly. The number of males captured in the control and the mass-trapped field was not significantly different for the first two generations. During the third generation, however, the male captures in the control field were significantly higher than the male captures in the mass-trapped field (Figure 1A). The results showed that pheromone traps suppressed insect populations in the mass-trapped field.

The plant infestation was low in the mass-trapped field, while plant infestation in the control was higher. The overall infestation is shown in Figure 1B. Fifty one per cent of the plants examined were infested in the control compared to 17% infested in the mass-trapped field. The infestation recorded for the spadices was 10% for the control and only 2% in the mass-trapped.

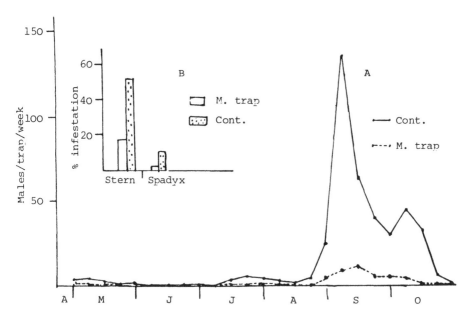

Fig. 1A. Number of males S. nonagrioides captured during the insect flying period in the control and mass-trapped field.

1B. Total stems and spadices of the early-sown crop infested by S. nonagrioides larvae. Servota, Trikala, 1986.

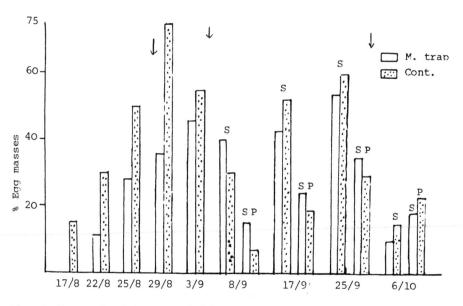

Fig. 2. Per cent of S. nonagrioides egg masses found on the stems and spadices of the late-sown corn. Servota, Trikala, 1986.

7.2   Late-sown Corn

The insect population density in the mass-trapped field was low compared to that of the control field. However, egg masses recorded during regular sampling, especially at the perimeter of the field, were high (Figure 2). Comparing the number of egg masses recorded in the mass-trapped field to those found in the control, more egg masses were found in the control field during August. In September, where the insect population density was high the number of egg masses found was approximately the same in both fields. Egg masses were also found on the developed spadices during September and October. Three insecticide treatments were applied in the control field on August 27, September 6 and October 1. The second and third treatments were also applied in the mass-trapped field. The final levels of larval infestation found in the stems and spadices are shown in Figure 3.

Thirty five per cent of the stems and 30% of the spadices in the mass-trapped field were not infested compared to 28% of the stems and 16% of the spadices in the control field.

The larval population density found in the infested stems and spadices was as follows: 51% of the stems and 45% of the spadices of the mass-trapped field were found infested with 1-5 larvae and 14% with more than 5 larvae. In the control field 61% of the stems and 46% of the spadices were found with 1-5 larvae and 25% with more than 5 larvae.

The overall crop damage due to S. nonagrioides infestation was estimated to be approximately 27% for the mass-trapped and 33% for the control fields.

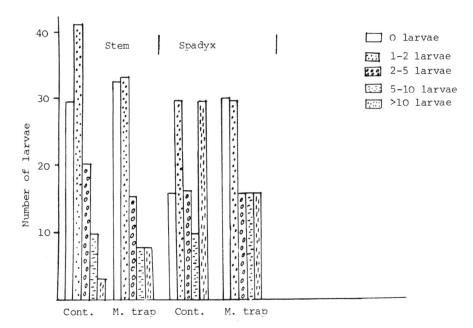

Fig. 3. Total stems and spadices of the late-sown corn infested by S. nonagrioides larvae. Servota, Trikala, 1986.

8. CONCLUSIONS

The corn stalk borer S. nonagrioides sex pheromone is a very potent attractant to males and the pheromone traps will be useful for future behavioural and other field studies. It is only two years since the corn stalk borer female sex pheromone was identified and our knowledge of the effect on insect behaviour is limited.

For the use of the pheromone for monitoring and suppression of the insect population, studies are needed on the flight activity and dispersal behaviour of the insect; also the distance of response to the pheromone source under various climatic conditions should be studied.

The first year mass-trapping data showed that in the future the female S. nonagrioides sex pheromone will be a main component in an integrated programme aiming to suppress the insect population.

ACKNOWLEDGEMENT

The mass-trapping experiment in Servota was partially supported by the Greek Department of Agriculture (GDA).

REFERENCES

(1)   BARTELL, R.J., 1977. Behavioral response of Lepidoptera to pheromones. In, Chemical Control of Insect Behavior Theory and Application. Ed. H.H. SHOREY and J.J. McKELVEY Jr., New York, John Wiley & Sons, pp. 201-230.
(2)   CASTROVILLO, P.J. and CARDE, R.T., 1979. Environmental regulation of the female calling and male pheromone response periodicities in the codling moth (Laspeyresia pomonella). J. Insect Physiol. 25, 659-667.

(3)   MAZOMENOS, B.E.,   1984.   A sex attractant of the corn stalk borer
       Sesamia nonagrioides Lef.:   Partial   chemical   purification   and   its
       biological activity under laboratory conditions.   Med. Fac. Landbouww.
       Rijksuniv. Gent, 49/3a, 675-681.

(4)   MAZOMENOS, B.E.,   1985.   Field   evaluation   of   the   sex   pheromone
       components of Sesamia nonagrioides Lef.   2nd National Entomological
       Meeting,   Athens,   November 6-8 (In press)   (In Greek with extended
       English summary).

(5)   SHOREY, H.H.,   1970. Sex pheromones of Lepidoptera. In, The Control of
       Insect Behavior by Natural Products. Ed. WOOD, D.C., SILVERSTEIN, R.M.
       and NAKAJIMA, M. New York, Acad. Press pp. 249.

(6)   STAYRAKIS, G.N., 1967. Contribution à l'étude des espèces nuisibles au
       maïs en Grèce du gerne Sesamia (Lepidoptères-Noctudae).   Annls.   Ins.
       Phytopath. Benaki 8, 20-23.

(7)   STAYRAKIS, G.N.,   1977.   Lepidopterous insects of corn.   4th Balkan
       Conference of Plant Protection. Athens, 24-27 Sept. 1973.

Session 2
Diseases and weeds

Chairman: P.Lucas

Wheat and barley diseases in Italy and prospects for integrated control

L.Corazza
Istituto Sperimentale per la Patologia Vegetale, Rome, Italy

Summary

Among cereals cultivated in Italy, at present, bread wheat, durum wheat and barley are of particular interest. Fungal and some viral diseases (SBWMV, BYDV) seem to be major problems for wheat and barley. Particular attention was given, during the last years, to varietal reaction to fungal parasites in different agro-climatic conditions. Genotypes of wheat stem rust, brown rust and powdery mildew were recorded and their potential epidemic value to cultivated varieties was tested.

Varietal reaction to the main pathogens, knowledge of the variability of obligate pathogens and choice of adequate farming practices are fundamental for integrated control of wheat and barley diseases.

1. INTRODUCTION

Winter and spring wheats (durum and bread wheat) and barley are of relevant interest to Italian agriculture. Achievements of breeding for high-yielding cultivars are definitely and rapidly changing the varietal composition of both barley and wheats and, in recent years, also modifying the area of cultivation, particularly for durum wheat and barley.

Improvement and intensification of cultural practices (tillage techniques, fertilisation, weed control), the increased monoculture, the management of crop residues and, in some arid areas of the South, the use of irrigation, are all playing a significant role in the incidence and frequency of infection by several pathogens.

2. MAIN FUNGAL DISEASES OF WHEAT

Yellow rust (Puccinia striiformis West.) is normally considered of minor importance in Italy. It is, however, more common at higher altitudes in the North, even if it causes sporadic severe epidemics also in the South (37).

It should be mentioned that, under conditions favourable to yellow rust epiphytotics, durum wheats generally show a higher level of resistance, while several bread wheat varieties, largely grown in Italy, can be badly damaged (41). Although strong epidemics of yellow rust rarely occur, it seems useful not to underestimate its potential damage to production. In 1977 a serious epidemic was destructive to high-yielding bread wheat cv 'Irnerio', released in 1973 and widespread in Central Italy (28). In the same year a strong negative correlation was found between severity of yellow rust infection and yield; high levels of nitrogen enhanced damage to moderately susceptible varieties (27).

In releasing new varieties, resistance or tolerance to this pathogen should be carefully considered.

Leaf rust (_Puccinia recondita_ Rob. ex Desm. f.sp. _tritici_ Eriks. et Henn.) rarely causes strong epidemics, but it is constantly present in our cereal fields as compared with yellow and stem rust. In fact, stem rust develops at the end of spring, when optimal conditions for the development and spreading of the uredial stage are normally satisfied. Furthermore, many varieties, both of _durum_ and bread wheats, including the most extensively cultivated ones, are susceptible (30, 42) (Tables 1 and 2).

TABLE 1. Field Reaction of Several Wheat Varieties to Rusts, Powdery Mildew and Septoria Tritici

Variety	P.recondita	P.graminis	E.graminis	S.tritici
ANIENE	S	MS	MR	S
ARGELATO	S	MR	R	MS
AUTONOMIA B	S	MR	MR	MS
CHIARANO	MS	R	MR	MS
CONCORDIA	S	R	R	S
IRNERIO	S	MS	S	MS
MARZOTTO	S	MR	MR	MS
MEC	S	MS	MR	MS
ORSO	S	MR	MR	MS
S.PASTORE	S	MS	S	S

TABLE 2. Field Reaction of Several Durum Wheat Varieties to Rusts and Powdery Mildew

Variety	P.recondita	P.graminis	E.graminis
APPULO	S	S	S
BERILLO	S	MS	S
CAPEITI 8	S	S	S
CAPPELLI	MS	S	S
CRESO	R	MR	MS
KAREL	MS	MS	S
PATRIZIO 6	MS	S	MS
PRODURA	MS	S	S
VALNOVA	MS	MS	R

Physiological specialisation of this rust is routinely analysed in a controlled environment in order to get information about the variability of rust genotypes (12, 26).

Field reaction of cultivars, lines and selections, related species, differential varieties and isogenic and near-isogenic lines are recorded yearly under different agro-climatic conditions for this rust, as well as for yellow rust, stem rust and powdery mildew. This is done in order to obtain useful information on the behaviour of varieties to be released, the usefulness of related species as sources of resistance and the effectiveness of genes for tolerance or resistance. In recent years, Lr 3ka, Lr 9, Lr 13,

Lr 14a, Lr 15, Lr 17, Lr 19, Lr 24 showed a higher degree of effectiveness, although virulent strains active on cultivars carrying some of these factors were recorded in some localities (30, 42).

Stem rust (Puccinia graminis Pers. f.sp. tritici Eriks. et Henn.) is the last rust species to appear during the wheat season. It is normally more frequent in the South, since the uredial stage is more thermophilic than that of brown rust. Despite the fact that most of our varieties are susceptible to this rust (Tables 1 and 2), as found both by seedling infection with selected biotypes in controlled conditions (11), and from the records of field behaviour in different localities (30), severe epidemics are impaired particularly by two factors:

- drought in the last period of cultivation, unless host-parasite relationship is affected by irrigation, if applied in the arid areas of the South, thus inducing a complete susceptibility in varieties generally lightly affected (15);
- earliness of recently released cultivars, which escape infection.

In the period from 1953 to 1983, 43 different physiologic races of stem rust were isolated in Italy. Many of these genotypes can develop on the most widely cultivated varieties, especially of bread wheat, and some of these genotypes were prevailing in different periods (13). As a consequence of the evolution of factors of virulence in natural populations of stem rust, in recent years, several genes for resistance have lost their effectiveness (e.g. Sr 5, Sr 6, Sr 13). On the contrary, Sr 11, Sr 22 (this last one transferred from Triticum monococcum) and to a lower degree, Sr 7b, Sr 25, Sr 32, still demonstrate a certain level of effectiveness towards the most frequently occurring biotypes of this rust (11).

Another obligate pathogen, very frequent in our wheat growing areas, is Erysiphe graminis DC.f.sp. tritici Marchal. Epidemics are more common on dwarf and semi-dwarf varieties, also in relation to higher levels of nitrogen fertilisation.

Field tests carried out yearly in different agro-climatic conditions revealed the frequency of biotypes able to match genes Pm 1, Pm 2, Pm 3a, Pm 3b, Pm 3c, Pm 5. In recent years sporadic attacks on isogenic lines carrying Pm 4 as well as on recently released durum wheat varieties, which are likely to owe their resistance to this gene transferred from Yuma, have been recorded (43).

Among durum wheat cultivars, 'Belfuggito' (with Agropyron intermedium-derived resistance) and 'Belvedere' (with Triticum timopheevii-derived resistance) combine adult plant and specific resistance.

Recently, more attention has been paid in Italy to Septoria blotch and to S. nodorum in particular, since this pathogen, already spread in the North (17), has recently been recorded to a limited extent in Central and Southern Italy (20, 21).

Breeding for high-yielding varieties and for resistance to lodging and to rusts resulted in higher susceptibility to these pathogens, since the short and dense canopy typical of dwarf varieties, often increased by high levels of nitrogen fertilisation, is favourable to short-distance spread of inoculum. Durum wheat varieties carrying Norin 10 genes (like 'Creso' and 'Val-group') are particularly susceptible to Septoria, but, under favourable climatic conditions, sporadic heavy attacks were noted on traditional varieties, like 'Appulo', 'Capeiti' and 'Cappelli' (37). Bread wheats are generally less damaged. Nevertheless, it is difficult to indicate completely resistant varieties, but it should be possible, in the future, to select, among the moderately susceptible varieties, those tolerant both to Septoria tritici and S. nodorum. In S. nodorum, seed is often the most important source of inoculum, even if soil-borne inoculum may also be important. As a consequence, stubble and crop debris must be accurately incorporated in the soil and rotation recommended. Both Septoria species can also attack a wide range of grasses, which may represent an additional source of inoculum.

Foot rot is a complex disease, caused in Italy mainly by Fusarium spp.; Fusarium culmorum (W.G. Smith) Sacc. and F. nivale (Fr.) Ces. are frequently associated with foot rot in the South. This has been indicated as an important consequence of monoculture, which has probably also favoured, although to a lesser extent, Gaeumannomyces graminis (Sacc.) V.Arx.et Olivier var.tritici Walkner and Pseudocercosporella herpotrichoides (Fron.) Deighton. F. nivale has been isolated both from durum and bread wheats, while F. culmorum was more aggressive and frequent on durum wheat (33).

The importance of Fusarium spp. as causal agents of foot rot has recently been evidenced also in the North of Italy on some bread wheat cultivars (among which 'Gemini' was found to be particularly affected) and on durum wheat 'Creso'. Also in the North, sharp eye-spot (Rhizoctonia cerealis Van der Hoeven) and eye-spot (Pseudocercosporella herpotrichoides) are less widespread and cause less damage. Regarding effects of rotation, wheat after wheat has been slightly more favourable to the disease than beet and corn, whereas wheat was less susceptible when grown after potatoes (25). Gaeumannomyces graminis seems to be more frequent in continuous wheat culture and more damaging under particular environmental conditions (low temperatures and high humidity) (32). Sharp eye-spot has significant negative effects on grain yield only when the intensity of infection is high. Dry conditions and sandy and acid soils are predisposing factors for disease development (19). Eye-spot and Fusarium spp. incidence decreased with higher nitrogen rates given directly to the wheat. This is probably in connection with antagonism among soil microbiota components (2, 34).

Since control of this complex disease seems to be difficult, because the causal agents are able to survive many years in the soil and have many alternate hosts, it seems imperative to avoid cultural practices known to enhance the incidence of the foot rot complex, namely:

- stress due to water-logging;
- dense sowing.

A rational management of crop sequence and crop debris is recommended.

The variability of the response of different cultivars needs further investigation.

Tilletia laevis Kühn (= T. foetida (Wallr.) Liro), T. caries (DC.) Tul. (= T. tritici (Bjerk.) Wolff) and T. brevifaciens (= T. intermedia (Gassner) Savul.) represent a potentially damaging complex of pathogens. Bread wheats generally show a higher level of infection (24), while tetraploid wheats have, on the whole, a higher level of bunt resistance. It would seem that some Tilletia strains have a higher degree of virulence, since they are able to attack a wide spectrum of durum wheat varieties (3).

All conditions (both climatic and agrotechnical) which are favourable to a quick germination and growth of the plant have a negative effect on seed contamination; seed treatment is strongly recommended. In fact, infection is frequent on crops from undressed seeds, like emmer (Triticum dicoccum), which is still grown as fodder in restricted mountainous areas in Southern Italy (1). Seed dressing is also recommended for Ustilago tritici, which causes sporadic damage to wheat crops.

2.1 Main Virus Diseases

Barley Yellow Dwarf Virus (BYDV) was recently recorded in several fields of bread wheat ('Adria', 'Argelato', 'Libellula', 'Marzotto') and of durum wheat 'Creso' in Northern Italy (22).

More frequently Soil-Borne Wheat Mosaic Virus (SBWMV) has been recorded both in Northern and in Central Italy (4, 18, 36). However, the most widely-grown durum wheat cultivars ('Appulo', 'Capeiti', 'Creso', 'Trinakria'), as well as the formerly popular 'Cappelli' are all highly resistant (40).

2.2 Conclusions

Integrated control of obligate wheat pathogens (rusts and powdery mildew) is mainly based on the utilisation of genetic resistance and on knowledge of variability of pathogens.

Sources of resistance from related and wild species have been widely exploited in Italy (16, 23, 29,39).

The possibility of introduction of multilines to control rusts has also been investigated (9). Concerning Septoria spp., the difficulty of finding, up to now, sources of genetic resistance, has forced the attention on agrotechnical practices.

Foot rot is largely influenced by cultural practices, but the difficulty of prevention is largely due to the complexity of the disease and the variable frequency in time and place of the pathogens, with different requirements for optimal development.

It should be mentioned, however, that two factors are decreasing the potential damage of the complex of pathogens affecting wheat in Italy, namely:

- the utilisation of a large number of varieties, particularly in the mosaic of small farms still operating;
- the large variability of the environment which prevents the development of epidemics over large areas.

Obviously the increase of monoculture and use of a smaller number of high-yielding varieties should be carefully monitored for their possible implications in the incidence of damage caused by these pathogens.

3. BARLEY

More attention has been paid in recent years to barley diseases as a consequence both of the increase of the area grown, especially of autumn-sown barley, and of the introduction of new varieties.

3.1 Main Barley Diseases in Italy

Puccinia hordei Otth. is a common disease and can cause epidemics, particularly in Central and Southern Italy. The potential damage due to this rust is represented mainly by two factors:

- susceptibility of the most widely cultivated varieties (8) (Tables 3 and 4);
- possibility of rapid spread in the field from initial inoculum as soon as the optimal conditions required (essentially temperature) are satisfied.

TABLE 3. Field Reaction of Several Two-Rowed Barley
Varieties to Main Fungal Pathogens

Variety	P.hordei	E.graminis	R.secalis	U.nuda
ALPHA	S	MS	MS	MS
ARAMIR	MS	MR	S	R
ARDA	S	MS	MR	MS
CORNEL	S	MS	S	R
HAVILA	S	MS	S	MS
GEORGIE	MS	MR	S	R
IGRI	S	MS	R	S
PORTHOS	S	MS	S	S
TIPPER	MS	MS	S	MS

TABLE 4. Field Reaction of Several Six-Rowed Barley
Varieties to Main Fungal Pathogens

Variety	P.hordei	E.graminis	R.secalis	U.nuda
ARMA	S	MS	R	MS
BARBEROUSSE	S	MS	MR	S
ETRUSCO	S	MR	MS	S
GERBEL	S	MR	R	S
MIRCO	S	MR	R	S
PLAISANT	S	MS	MS	R
ROBUR	S	S	S	S
SELVAGGIO	S	S	S	MS
THIBAUT	S	MS	MS	MS

The most common race isolated in Italy, designated as R-1, is
avirulent on Estate (Pa 3) and Cebada Capa (Pa 7); these two genes are
effective also towards the other variants of the rust isolated, that are
less virulent than R-1 (5).

Erysiphe graminis DC. f.sp. hordei Marchal is widespread in Italy,
especially in some areas in the South, where favourable climatic conditions
allow for its quick development (10).

Factors of virulence were analysed from mildew samples collected in
several fields; only the genes transferred from Emir and Lyallpur 3645
derivatives were able to provide a complete resistance to variants isolated
(7). In fact, in 'Aramir' and 'Porthos' the resistance gene from Emir seems
to confer a certain degree of resistance to the natural populations of
powdery mildew (Table 3).

Rhynchosporium secalis (Oud.) Davis, reported from Italy for the first
time in 1970 (35), is particularly damaging in cool and humid environments
both in the North and in the South.

Two-rowed varieties ('Cornel', 'Georgie', 'Gitane', 'Havila',
'Porthos') in autumn sowing, are largely susceptible.

Isolates derived from heavily infected fields fit into 17 virulence
groups; Rh 4 and Rh seem to be the most promising genes for resistance (6).

Barley stripe caused by Drechslera graminea (Rabenh. ex Schlecht.) Shoemaker is potentially damaging to barley crops; for its control the need for seed dressing has been stressed. 'Etrusco', 'Perga', 'Novoperga' and 'Vertulio' were severely affected in recent years.

Ustilago nuda Rostr. has recently been affecting a large number of varieties, spreading both in the North and in the South. Most likely, epidemics are caused by the progressive substitution of local populations (generally more tolerant to this disease) with new high-yielding varieties coming from Northern Europe. Some varieties, like 'Igri', 'Panda', 'Arma', 'Barberousse', 'Jaidor', 'Robur', 'Selvaggio' have been heavily affected during recent years (14). To control Ustilago spp. seed dressing is strongly recommended.

Barley Yellow Dwarf Virus (BYDV) has been reported in the North of Italy strongly affecting barley crops. At present no genetic tolerance has been identified in widely-grown varieties, except for 'Alpha' (38). Indirect control is possible by delaying sowing, most likely because of a lower incidence of aphid vectors of the virus (31).

3.2 Conclusions

Most of the barley varieties now cultivated in Italy come from foreign countries, in particular from Northern and Eastern Europe, and most of them are susceptible to the main pathogens present in Italy. Severe damage is especially due to attacks of the pathogens (like BYDV and powdery mildew) during the early growth stages.

Systemic infection by Drechslera graminea and Ustilago nuda should be prevented by a complex of measures:

- seed dressing (slurry is the most suitable method);
- isolation of fields for certification;
- inspection of imported seeds.

Further efforts for breeding for resistance to other pathogens mentioned, as well as breeding of barley cultivars for resistance to stress (cold, drought), is needed, due to the large variability of climatic conditions and as a necessary precursor to a wider cultivation of this cereal in Italy.

REFERENCES

(1) ANTONICELLI, M., PADALINO, O., GRASSO, V., 1970. Il Farro (Triticum dicoccum Schrank), nuovo ospite di Tilletia caries Tul. in Italia. Phytopathologia Mediterranea, 9,1, 57-58.
(2) BIANCHI, A.A., 1975. Mal del piede del frumento in relazione alla precessione colturale ed alla concimazione azotata. Atti Giornate Fitopatologiche, Torino 11-14 November, 537-545.
(3) BOZZINI, A., 1971. First results of bunt resistance analysis in mutants of durum wheat. Proceedings of 'Mutation Breeding for disease resistance', FAO/IAEA, Vienna, 12-16 October 1970, 131-138.
(4) CANOVA, A. and QUAGLIA, A., 1960. Il mosaico del frumento. Informatore fitopatologico, 10, 206-208.
(5) CEOLONI, C., 1979. Puccinia hordei in Italy: a preliminary survey on the virulence characteristics of the fungus in our country. Cereal Rusts Bulletin, 6,2, 11-16.
(6) CEOLONI, C., 1979. Pathogenic differentiation of barley powdery mildew, leaf rust and scald during 1978. Barley Newsletter, 1978, 132-134.

(7)   CEOLONI, C. and VALLEGA, J., 1979. Reaction of barley varieties and
        lines to virulence genes carried by two Italian mildew races. Genetica
        Agraria, 33, 53-74.

(8)   CORAZZA, L., 1984. Varieta' di orzo a confronto per la resistenza alle
        principali crittogame parassite. L'Informatore Agrario, 33,63-66.

(9)   CORAZZA, L., 1984. Prospettive sull'uso di varieta' multilinee di
        frumento tenero in Italia. L'Informatore Agrario, 35, 43-46.

(10)   CORAZZA, L., 1985. Il comportamento in campo di varieta' di orzo nei
        confronti dei parassiti fungini nel 1984-85. L'Informatore Agrario,
        34, 69-71.

(11)   CORAZZA, L., 1986. Virulence factors of Puccinia graminis f.sp.
        tritici identified in Italy in 1984. Cereal Rusts Bulletin, 14,1,
        30-38.

(12)   CORAZZA, L. and BASILE, R., 1983. Le ruggini del frumento in Italia
        nel 1982 ed identificazione di alcune razze fisiologiche di Puccinia
        recondita Rob. ex Desm. f.sp. tritici Eriks. et Henn. e di Puccinia
        graminis Pers. f.sp. tritici Eriks. et Henn. Annali Istituto
        Sperimentale per la Patologia Vegetale, 8, 39-46.

(13)   CORAZZA, L. and BASILE, R., 1984. Evoluzione razziale della Puccinia
        graminis Pers. f.sp. tritici Eriks. et Henn. su alcune varieta' di
        frumento tenero e duro coltivate in Italia (1953-1983). Annali
        Istituto Sperimentale per la Patologia Vegetale, 9, 141-152.

(14)   CORAZZA, L. and CHILOSI, G., 1986. Comportamento in campo di varieta'
        di orzo nei confronti dei principali parassiti fungini nel 1985-86.
        L'Informatore Agrario, 34, 51-55.

(15)   CORAZZA, L., DI BARI, V., CHILOSI, G., 1986. Alcune considerazioni
        sullo sviluppo epidemico di ruggine nera del frumento duro in
        relazione ad interventi irrigui e diversi livelli di concimazione
        azotata. In press.

(16)   CORAZZA, L., PASQUINI, M., PERRINO, P., 1986. Resistance to rusts and
        powdery mildew in some strains of Triticum monococcum L. and Triticum
        dicoccum Schubler cultivated in Italy. Genetica Agraria, 40, 243-254.

(17)   CORINO, L., 1978. Effects of Septoria nodorum on some durum and bread
        wheat varieties. Genetica Agraria, 32, 359-361.

(18)   CORINO, L. and GRANCINI, P., 1975. Observation on the SBWMV in Italy.
        EPPO Bull., 5(4), 449-453.

(19)   CORTE, A., 1969. La rizottoniosi del culmo del frumento ('sharp
        eyespot') in Italia. Rivista di Patologia Vegetale, 2, 37-67.

(20)   FRISULLO, S. and NALLI, R., 1980. Segnalazione di Septoria nodorum
        (Berk.) Berk. nei pressi di Roma. Annali Istituto Sperimentale per la
        Patologia Vegetale, 6, 75-80.

(21)   FRISULLO, S. and NALLI, R., 1982. Presenza di Septoria nodorum (Berk.)
        Berk. nell'Italia centrale e meridionale. 'Giornate internazionali del
        grano duro. Industra e ricerca scientifica', Foggia, 3-4 May, 305-307.

(22)   GIUNCHEDI, L. and CREDI, R., 1978. Il virus del nanismo giallo
        dell'orzo, riscontrato anche in Italia su coltivazioni di frumento.
        Informatore Fitopatologico, 6, 3-7.

(23)   GRAS, M.A., 1980. Disease resistance in wheat: I, Triticum dicoccum as
        a source of genetic factors against rust and mildew. Genetica Agraria,
        34, 123-132.

(24)   GRASSO, V., 1948. Le specie di Tilletia del frumento esistenti in
        Italia e loro distribuzione geografica. Annali Sperimentazione
        Agraria, 3, 525-547.

(25)   INNOCENTI, G. and BRANZANTI, B., 1986. Indagine sul mal del piede del
        frumento in Emilia Romagna. Informatore Fitopatologico, 10, 32-34.

(26)   LEVINE, M. and BASILE, R., 1960. A review and appraisal of thirty
        years research on cereal uredinology in Italy. Bollettino Stazione di
        Patologia Vegetale, 17, 1-36.

(27) MONOTTI, M. and RAGGI, V., 1978. Produttivita' e resistenza alla ruggine gialla di varieta' di frumento tenero. Sementi Elette, 5, 21-27.

(28) MONTALBINI, P., CAPPELLI, C. and RAGGI, V., 1977. Epidemia di Puccinia striiformis Westend in Umbria. Informatore Fitopatologico, 8, 3-5.

(29) PASQUINI, M., GRAS, M.A. and VALLEGA, J., 1978. Haynaldia villosa (L.) Schur. come fonte di resistenza alle ruggini e all'oidio, da incorporare nelle specie di frumento coltivate. Atti Giornate Fitopatologiche, 349-351.

(30) PASQUINI, M., BIANCOLATTE, E., CECCHI, V., VALLI, M., CALCAGNO, F., IUDICELLO, P., LENDINI, M., CORAZZA, L., ROMANI, M. and ARDUINI, F., 1986. Prove epidemiologiche 1985-86: comportamento in campo rispetto alle malattie di frumenti duri e teneri.

(31) PERESSINI, S. and COCEANO, P.G., 1986. Incidenza delle infezioni di virus del nanismo giallo dell'orzo (BYDV) su orzo e frumento in rapporto a epoca di semina e localita'. Informatore Fitopatologico, 1, 29-32.

(32) PICCO, A.M. 1985. Segnalazione di 'mal del piede' del frumento tenero da Gaeumannomyces graminis (Sacc.) V.Arx et Olivier var. tritici Walkner, nel nord Italia. Rivista di Patologia Vegetale, 21, 67-78.

(33) PIGLIONICA, V., 1975. Il mal del piede del frumento. Italia Agricola, 112, 114-120.

(34) RAGGI, V. and MONOTTI, M., 1978. Indagine sull'influenza dell'azoto e sul comportamento varietale del frumento tenero agli attacchi di mal del piede prodotto da alcune specie di Fusarium. Annali Facolta' di Agraria Universita' di Perugia, 33, 21-31.

(35) RIBALDI, M., LORENZETTI, F. and CECCARELLI, S., 1974. Prime osservazioni in Italia sulla maculatura fogliare dell'orzo causata da Rhynchosporium secalis (Oud.) Davis e miglioramento genetico per la resistenza. Rivista di Patologia Vegetale, 10, 60-89.

(36) RUBIES-AUTONELL, C. and VALLEGA, V., 1985. Il mosaico comune del frumento riscontrato anche nel Lazio. Informatore Fitopatologico, 7-8, 39-42.

(37) SINISCALCO, A., CASULLI, F., CARIDDI, C., TOMMASI, F., PARADIES, M., PIGLIONICA, V., SISTO, D., CICCARONE, A., BASILE, R., CORRAZZA, L. and NALLI, R., 1981. Risultati delle prove di campo eseguite nel 1979-80 e 1980-81 sul comportamento di frumenti verso ruggini ed altri parassiti. Bollettino Informativo n.7, 1-105.

(38) SNIDARO, M., 1980. Virosi del nanismo giallo dell'orzo (BYDW). L'Informatore Agrario, 35, 11937-119341.

(39) VALLEGA, V., 1978. Search for useful genetic characters in diploid Triticum spp. proceedings 5th International Wheat Genetics Symposium, New Delhi, 156-162.

(40) VALLEGA, V. and RUBIES-AUTONELL, C., 1985. Reactions of Italian Triticum durum cultivars to soil borne wheat mosaic. Plant Disease, 69,1, 64-66.

(41) VALLEGA, V. and ZITELLI, G., 1979. Epidemics of yellow rust on wheat in Italy. Cereal Rusts Bulletin, 6,2, 17-22.

(42) ZITELLI, G., PASQUINI, M., VALLEGA, V., BIANCOLATTE, E., CECCHI, V., VALLI, M., LENDINI, M., BASILE, R., CORAZZA, L., NALLI, R., CONCA, G., CALCAGNO, F., GALLO, G., VENORA, G., RAIMONDO, I., PEZZALI, M. and ARDUINI, F., 1984. Il comportamento in campo di frumenti duri e teneri nei confronti delle malattie fungine durante il 1983-84. L'Informatore Agrario, 38.

(43) ZITELLI, G., PASQUINI, M., BIANCOLATTE, E., CECCHI, V. and VALLI, M., 1985. Risultati delle prove eseguite nel 1984-85 sul comportamento in campo di frumenti duri e teneri nei confronti delle malattie fungine. L'Informatore Agrario, 37, 69-78.

Integrating varietal mixtures and fungicide treatments: Preliminary studies of a strategy for controlling yellow rust of wheat

C.De Vallavieille-Pope, H.Goyeau, F.Pinard, C.Vergnet & B.Mille

Laboratoire de Pathologie Végétale INRA, Thiverval-Grignon, France, and SRIV CNRA, Versailles, France

Summary

Deployment of strategies for integrated crop protection against cereal rusts in France is prevented by a lack of information on wheat resistance genes and rust agent virulence factors. Accordingly, a study consisting of a survey of yellow rust virulences (Puccinia striiformis Westend) on wheat and a comparison of mixtures and pure cultivars having different resistance factors was initiated. The comparison concerned the increase of disease over time and the efficiency of fungicide treatments. In a first experiment a varietal mixture (Clement, Joss Cambier, Talent) inoculated with a yellow rust race (232 E 137) had lower disease intensity (area under the disease progress curve) and slower disease progress (apparent infection rate) than the means of the three components cultivated in pure stands. However, the apparent increase in yield of the mixture compared with pure cultivars could be explained by the evolution of the cultivar tiller proportion during the season because the resistant variety was dominant. In a second experiment, where initial inoculum was higher and epidemic duration shorter, no advantage of four three-component mixtures over the means of pure cultivars was encountered. Nevertheless, at the end of the season the proportion of cultivars in the mixtures had evolved significantly from the initial 1:1:1 ratio. Fungicide treatment efficiency was different between varietal mixtures and pure varieties. The same yield was obtained with two of the three mixtures (1 susceptible, 2 resistant) when sprayed once as was obtained with the susceptible cultivar sprayed three times.

1. INTRODUCTION

Genetic and chemical control are the two principal strategies for the management of yellow rust epidemics on wheat caused by Puccinia striiformis Westend. In addition, the chemical control is more efficient when the host population is fairly resistant (4). Therefore, our objective was to integrate these two strategies. In order to manage the genetic components it was hypothesised that the durability of resistance genes is increased by reintroducing genetic diversity among host populations by using varietal mixtures (7, 11).

Little is known about host diversification of yellow rust on wheat. Two multilines were described as having improved yields (2, 5), but no epidemiological information was given for varietal mixtures. In addition, the use of such mixtures is new in France. This study can, therefore, be considered as a model. The expected effects of using mixtures of cultivars with different resistance factors are more dispersed susceptible host populations, a physical barrier of resistant varieties to inoculum, and induced resistance if the pathogen population is composed of several races.

Lack of information about resistance genes and virulence factors in France prevents the management of resistant varieties. This study concerns the start of a virulence survey and an evaluation of three-component mixtures inoculated by a yellow rust race. The evaluation is based on epidemiological criteria, yields and fungicide treatment effects.

2. MATERIALS AND METHODS

Yellow rust race identification was carried out by inoculating differential-host seedlings in controlled growth chambers using the technique developed by Johnson et al. (6).

The choice of cultivars for mixtures (Table 1) was imposed by the limited amount of data on resistance factors in French varieties: Clement, a variety susceptible to the race 232 E 137; Joss Cambier and Top, susceptible at the seedling stage to 232 E 137 and resistant at the adult stage under specific conditions; and Talent, overall resistant to 232 E 137. In the 1985 experiment, one three-component mixture was compared with the three pure stands (Clement, Joss Cambier, and Talent). In the 1986 experiment, four mixtures of three components were analysed.

TABLE 1. Resistance Characteristics of Varieties Used in Experimental Mixtures

Variety	Resistance factor	Susceptibility to race 232 E 137
Clement	Yr 9	Susceptible
Top	Yr 3 + ?	Adult resistance
Joss Cambier	Yr 2 + Yr 3a + ?	Adult resistance
Talent	Yr 7	Resistant

Field experimental design for epidemiological studies: a randomised block design with four replicates was used. Each plot (9 m x 9 m) was surrounded by a 9 m barrier of the overall resistant variety, Talent (according to the EOPP recommendations, (1)). These experiments took place at the INRA Experimental Station, La Verrière, France.

Inoculation. The race chosen for inoculation in the field experiments, is 232 E 137, which is fairly common in France according to previous surveys conducted by Dr. Stubbs, IPO Wageningen, The Netherlands. In the 1985 experiment, plot centres were inoculated by planting 10 sporulating seedlings twice with a 10 day interval. In the 1986 experiment, five foci of three sporulating seedlings were planted three times at two week intervals in each plot.

Disease assessment. Disease intensity was evaluated weekly from inoculation time to the end of the epidemic on 12 tillers with 40 replicates per plot. In 1985, the sporulating area was measured in square centimetres. The total foliar area of each cultivar was estimated with four samples of 12 tillers using Cunow LI 300 equipment. These two parameters were used to calculate the percentage of sporulating area. The 1985 results and the more extensive 1986 experimental design led us to use the international scale (0-10) for percent disease (12), although a direct correlation between percent sporulating area and percent disease could not be established.

Host population evolution in the mixtures was evaluated at the end of the season by their leaf and head morphological characteristics. Twenty replicates of one-linear-metre samples were collected from each mixture in order to estimate the tiller proportion of each component.

Experimental design for evaluating yields and fungicide effects. In 1986 a randomised block design with four replicates and two factors (varieties and fungicide sprays) was used. Three fungicide treatments were made: (i) unsprayed control; (ii) one spray at the stem elongation stage (triadimenol, 12.5 g/ha); and (iii) a seed treatment in addition to three sprays at the early tillering, stem elongation, and heading stages. Each plot (1.25 m x 7 m) was inoculated with a set of three sporulating seedlings planted three times at two week intervals.

3. RESULTS

3.1 Virulence Survey

Samples collected from breeders' nursery traps were tested because no natural epidemic had occurred during the previous two years. The races analysed were complex, combining five or six virulences (Table 2).

TABLE 2. Race Identification in 1984 and 1985

Race	Number of isolates 1984	1985	Virulence factors
41 E 136		1	SD*, 1, 2, 3
43 E 138	7	2	SD, 1, 2, 3, 7
171 E 138	1		SD, 1, 2, 3, 7, 9
43 E 170	4	7	SD, 1, 2, 3, 7, 11
232 E 137		1	SD, 2, 3, 4b, 9, 12

SD = Strubbes Dickkopf

3.2 Comparison of Disease Progress Between Varietal Mixtures and Means of Pure Cultivars

Comparisons between the varietal mixtures and the means of the three components cultivated in pure stands (expected mean value) were carried out for three criteria: (i) the area under the disease progress curve, (ii) the apparent infection rate (r), and (iii) the yield. The apparent infection rate is the linear regression coefficient of ln (x/1-x) over time, where x is the percentage of diseased area (10).

In the 1985 experiment, the mixture (Clement-Joss-Cambier-Talent) had three notable characteristics:

(i) about half the expected degree of disease (Table 3, area under the disease progress curve, differences at P=0.05 with the Mann-Whitney test);

(ii) an apparent infection rate significantly lower than the expected value (Table 4, differences at P=0.05 with the Mann-Whitney test);

(iii) a yield 7% higher than expected (Table 5, significant differences at P=0.05 by variance analysis and Newman-Keuls ranking).

TABLE 3. Area Under the Disease Progress Curve (1985 data = per cent sporulated area, 1986 data = per cent diseased area).

Year	Mixture components	Mixture	Mean value
1985	Joss Clement Talent (2)	0.251	0.527 *
1986	Top Clement Talent (1)	8.48	6.57 ns
1986	Joss Clement Talent (2)	6.43	7.29 ns
1986	Top Joss Clement (3)	1.89	1.42 ns
1986	Top Joss Clement (4)	7.20	6.01 ns

* = significant difference at P=0.01 between mixtures and their respective expected mean values, Mann-Whitney test.
ns = non-significant difference between mixtures and their respective expected mean values.

TABLE 4. Apparent Infection Rate (r) in Mixtures and their Expected Mean Values

Year	Mixture component	r in mixtures	r expected values
1985	Joss Clement Talent (2)	0.526	0.767 *
1986	Top Clement Talent (1)	0.469	0.411 ns
1986	Joss Clement Talent (2)	0.394	0.422 ns
1986	Top Joss Talent (3)	0.354	0.330 ns
1986	Top Joss Clement (4)	0.443	0.401 ns

* = significant difference at P=0.01 between mixtures and their respective expected mean values, Mann-Whitney test.
ns = non-significant difference between mixtures and their respective expected mean values.

TABLE 5. Comparison of Yields (qx/ha) of Varietal Mixtures and their Expected Mean Values

Year	Mixture	Yield: Observed values	Yield: Expected mean values with 1:1:1 ratio	Yield: Expected mean values with observed tiller ratio
1985	(2)	52.5	49.0 (*)	51.7 (=)
1986	(1)	52.1	54.4 (ns)	53.9 (=)
1986	(2)	51.4	54.3 (ns)	53.9 (=)
1986	(3)	60.6	57.7 (ns)	57.5 (=)
1986	(4)	53.2	53.2 (ns)	53.9 (=)

* = significant difference at P=0.05 between mixture yields and their expected mean values with 1:1:1 ratio, variance analysis and Newman-Keuls ranking.
(ns) = non-significant difference between mixture yield and their expected mean values with 1:1:1 ratio.
(=) = non-significant difference between mixture yield and their expected mean values with observed tiller ratio.

In the 1986 experiment, no significant differences were observed between the four mixtures and the means of their respective three components grown in pure stands for the same three criteria (Tables 3, 4 and 5).

3.3 Effects of Fungicide Applications on Mixtures and Pure Stands

Firstly, no differences were detected between the mixtures sprayed by fungicides and the expected mean values of cultivars grown in pure stands.

No yield improvement was obtained by any of the fungicide treatments on the three resistant varieties and the mixture of these three varieties (Figure 1, variance analysis). In the case of the mixtures including the susceptible variety, Clement, the first treatment improved the yield, and the second treatment was efficient only for the mixtures that lacked the overall resistant variety Talent (at P=0.05, Newman-Keuls test). In the case of Clement in pure stand, each application improved the yield. Furthermore, the same yield was obtained by the susceptible variety treated twice as was obtained by two of the mixtures (1 susceptible, 2 adult resistant) treated once.

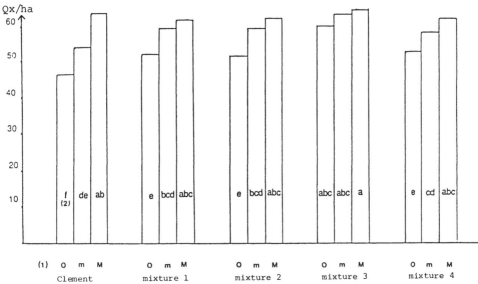

Fig. 1. Effect of fungicide applications on the yield of the susceptible variety (Clement) and four mixtures.

    (1) O: no application

        m: one triadimenol spray at stem elongation stage

        M: one triadimenol seed treatment, and three foliar sprays (tillering, stem elongation, heading stages)

    (2) Figures with the same letters are not significantly different by Newman-Keuls multiple range test, P=0.05.

3.4 Pathosystem Evolution During the Epidemic

3.4.1 Host population evolution

The three-component mixtures were sown in equal proportions. At the end of the season the cultivar proportions were observed in each mixture

(Table 6). The host population was significantly different from the expected proportion (Student test P=0.05).

TABLE 6. Percentage of Each Cultivar Tiller on Mixtures Evaluated at the End of the Season

Year	Mixture	Top	Joss	Clement	Talent
1985	(2)	-	30.8	24.1	45 *
1986	(1)	29	-	37.4	33.5*
1986	(2)	-	36.8	35.3	27.8*
1986	(3)	26.5	43	-	30.4*
1986	(4)	24.4	50.3	25.1	- *

* = The proportions of the mixture components are significantly different at P=0.05 Student test.

3.4.2 Pathogen population

Of the 12 leaf samples collected from the mixture by the end of the epidemic, six had an extra virulence, the virulence 1, although the matching resistance factor was not carried by any of the three cultivars used in the experiment.

In 1986 the virulence 7 pathogenic on the resistant cultivar, Talent, was selected. A focus was observed in a pure stand plot of Talent six to seven weeks after inoculation.

4. DISCUSSION/CONCLUSION

The surveys showed races accumulating five to six virulences, a fairly high number. The most complex race described elsewhere (9) had nine virulences. These findings suggest that some virulences are unnecessary sensu Van Der Plank, i.e., are present even if the matching resistance genes are not in the host population. However, the state of current knowledge about resistance factors does not permit further discussion. The unnecessary virulence 1 found in the field mixture was already shown to be easily selected in field conditions (9). However, the virulence 7 was selected after two epidemiological seasons, reflecting the rapidity with which a resistance gene can be broken down.

In the 1985 experiment, the varietal mixture was more successful in decreasing the amount of disease and slowing disease progress than was the mean of the three components of the mixture grown in pure stands. This improvement appeared as a 7% increase in yield over the expected mean value. When the cultivar tiller proportions evaluated at the end of the season were taken into account in calculating the expected mixture yields, no significant differences remained. The yield advantage of the mixture could be explained by the dominance of Talent, the overall resistant variety. Nevertheless, the mixture was able to reduce the disease, but in proportions that had no effect on yield. It should be noted that competition between cultivars is a characteristic often neglected when evaluating varietal mixtures. Furthermore, it appears, that this phenomenon could determine the 'mixture effect' at least in specific environmental conditions.

In the 1986 experiment, no difference was detected between the mixtures and their expected values for disease intensity and yield. Two hypotheses are possible: (i) the quantity of inoculum was equivalent for both years but was more dispersed in the second, and (ii) the number of infectious cycles in 1986 was smaller than in 1985 (seven in 1985 and only four in 1986 using Zadoks' model (12) modified by Rapilly (8)). According to Barrett (3) the expected dilution effect of the mixtures has negligible epidemiological consequences when a large amount of inoculum comes from outside the plot, and this phenomenon is heightened if the epidemic duration is short.

Despite the fact that no improvement in yield or reduction in disease intensity was observed in the 1986 mixtures, an advantageous effect of the mixtures, including both the susceptible and the overall resistant cultivars, was observed if the chemical applications were taken into account. These mixtures gave the same yield when sprayed once as did the susceptible cultivar when sprayed three times. When epidemic conditions do not favour the dilution of inoculum in these mixtures, the latter act as moderately resistant cultivars vis-à-vis chemical treatments. However when epidemic conditions are favourable, these mixtures may provide better disease resistance and yields in the absence of chemical treatment than would, similarly untreated, moderately resistant cultivars in pure stands.

This epidemiological criterion, in addition to the variety competition phenomenon mentioned above, points out some of the conditions needed to obtain functional diversity. Among these conditions is the ability of cultivars to respond to inoculum dilution. This characteristic can be tested in order to select cultivars that can be grown in mixtures.

REFERENCES

(1)   ANONYMOUS, 1984. Rapport de la 1ère réunion d'experts OEPP sur les stratégies d'utilisation des gènes de résistance. Paris, 21-22 February, 1984.
(2)   ALLAN, R.E., LINE, R.F., PETERSON, C.J.Jr., RUBENTHALER, G.L., MORRISON, K.J. and ROHDE, C.R., 1983. Crew, a multiline wheat cultivar. Crop. Sci., 23, 1015-1016.
(3)   BARRETT, J.A., 1981. Disease progress curves and dispersal gradients in multilines. Phytopathol. Z., 100, 361-365.
(4)   BINGHAM, J., 1981. Breeding wheat for disease resistance. In, Strategies for the control of cereal disease, edited by JENKYN, J.F. and PLUMB, R.T., Blackwell Scientific Publications, Oxford, PP. 3-14.
(5)   GROENEWEGEN, L.J.M., 1977. Multilines as a tool in breeding for reliable yields. Cereal Res. Commun., 5, 125-132.
(6)   JOHNSON, R, STUBBS, R.W., FUCHS, E. and CHAMBERLAIN, N.H., 1972. Nomenclature for physiologic races of Puccinia striiformis infecting wheat. Trans. Br. Mycol. Soc., 58, 475-480.
(7)   MUNDT, C.C. and BROWNING, J.A., 1985. Genetic diversity and cereal rust management. In, The Cereal Rusts, edited by ROELFS, A.P. and BUSHNELL, W.R. Academic Press, vol. II, pp.527-560.
(8)   RAPILLY, F., 1976. Essai d'explication de l'épidémie de rouille jaune sur blé en 1975. Le Sélectionneur Francais, 22, 47-52.
(9)   STUBBS, R.W., 1985. Stripe rust. In, The Cereal Rusts, edited by ROELFS, A.P. and BUSHNELL, W.R. Academic Press, vol. II, pp.61-101.
(10)   VANDERPLANK, J.E., 1963. Plant diseases: Epidemics and control. Academic Press, New York.
(11)   WOLFE, M.S., 1985. The current status and prospects of multiline cultivars and variety mixtures for disease resistance. Ann. Rev. Phytopathol., 23, 251-273.
(12)   ZADOKS, J.C., 1961. Yellow rust on wheat. Studies in epidemiology and physiologic specialisation. Tijdschr. over Plantenz., 67, 69-125.

Recent advances towards reliable forecasting of *Septoria* diseases in winter wheat

D.J.Royle & M.W.Shaw

Department of Agricultural Sciences, Long Ashton Research Station, University of Bristol, UK

Summary

A cooperative survey between member participants of an International Organisation for Biological Control (IOBC) Working Group was carried out in 1981-83. Data on the development of Septoria spp. in winter wheat crops across Western Europe were gathered and related to crop and weather factors. Hypotheses have been constructed concerning the ways in which S. tritici and S. nodorum cause attacks which could be distinguished as explosive or gradual. An epidemiological framework is described for S. tritici, based on these hypotheses and on the results of supporting experiments; a forecasting system is being developed and tested from this framework. In order to indicate the need to apply fungicide sprays, the system relies on monitoring (i) inoculum levels in crops before stem extension, and (ii) the degree by which rain splashes upwards during the period from stem extension to anthesis. Good disease control, with substantial increases in yield, was obtained from sprays guided by the scheme at Long Ashton in 1986, a year in which there was severe disease. However, a cool, wet spring triggered the system frequently and prevented its value for saving sprays from being tested. Development of the forecast system is continuing and, although results so far are very encouraging, its implementation will ultimately rely on advance knowledge of which Septoria sp. will predominate in a given year and locality.

1. Introduction

The use of fungicides is now a well-established and crucial component of modern cereal production techniques. The annual minimum net benefit from using fungicides on cereals in the UK is estimated at £64M, at 1985 prices (3). Since fungicides were first used, in the early 1970s, there has been an increasing dependency on routine, prophylactic treatments due mainly to the attractive profit margins to be had from rising grain prices. The frequent demonstrations in field trials by the Agricultural Development and Advisory Service (ADAS), agrochemical firms and others, of economic yield responses from 2 or 3 sprays routinely applied according to growth stage of the crop, has undoubtedly helped to establish the widespread adoption of this approach. However, yield responses have failed to exceed the costs of treatment in an average of 30% of the trials done by ADAS between 1979-85; this represents about 25% of the crop area of England and Wales which is unnecessarily treated each year. Of course, in severe disease situations, it is necessary to use fungicides several times in a season to maintain production. But, as Cook (3) points out, we might better use these figures to indicate that severe losses are the exception rather than the rule and there is a need to identify those crops which are likely to sustain disease loss and therefore require fungicide treatment.

Current wisdom suggests that the most threatening stem and leaf diseases of winter wheat in the UK are eyespot, mildew, those caused by Septoria spp., and brown rust. Using survey data of disease severity over the last 15 years in England and Wales, Shaw and Royle (13) have calculated that it should be possible to avoid sprays against Septoria nodorum and S. tritici in at least one out of three years, probably more on individual sites. As an alternative to routine spraying, a decision-based approach to disease control in cereals is offered in the ADAS Managed Disease Control (MDC) scheme (1). This was developed from an analysis of yield responses to fungicide treatment in different situations together with data on Septoria weather (15) and uses the occurrence of a 'wet period' (4 or more days with 1 mm or more of rain, or 1 day with more than 5 mm rain, in the previous 14 days) to guide sprays against Septoria between GS 39-55 (decimal growth stage scale (16)). When applied to data from a long run of years at two sites in England it can be shown that the MDC system will only provide for a saving of sprays in less than 5-10% of years. Thus, there is still opportunity to save sprays in at least a further 20-25% of years if an efficient forecast-guided Septoria management system could be produced, and this is regarded as a minimum figure (13).

Such improvements depend on a reliable spray guidance scheme which in turn depends on reliable forecasts of disease. These can be achieved only from better understanding of disease epidemiology and the integration of the various factors involved into a coherent framework. Over the last few years we at Long Ashton have been studying in detail the development of S. tritici and S. nodorum in winter wheat crops together with selected features of the epidemiology of these pathogens. Our aim has been to produce a reliable, robust and biologically based forecast system which can indicate the need to apply control measures. The hypotheses on which the research has been based emerged from a study of Septoria epidemics in crops across southern Britain and continental Europe (11). Most of our detailed work has subsequently been on S. tritici since this species has been predominant in wheat crops over the last few years. In the following account we review the results of these investigations in order to describe the epidemiological framework now emerging from our work. We also consider some promising early experiments in forecasting using an experimental scheme for S. tritici based on it.

2. The IOBC survey

This collaborative exercise was done between 1981-83 under the auspices of a Working Group (The Role of Models in Integrated Crop Protection) of the West Palaearctic Region of the IOBC. The aim was to improve understanding of the relationships between Septoria development, crop and weather by analysing comparable data from experimental plots and commercial fields across W. Europe. According to a carefully structured protocol, data were collected from 30 site/years by up to 10 participants in England, Wales, France, the Netherlands and the Federal Republic of Germany. Despite some problems in obtaining the desired degree of data standardisation, it was possible to make some quantitative assessments of disease distribution within a crop and of the errors associated with pathogen measurements, which are reported in detail elsewhere (14). Observed temporal patterns in the development of Septoria spp. on wheat leaves could also be interpreted descriptively in terms of weather, inoculum and crop growth factors (11). These patterns (Fig. 1) are of two types. (i) Sudden disease outbreaks, in which lesions appear simultaneously on the upper leaf layers of crops, usually after the end of stem extension (GS 39). These can be ascribed to short, heavy rain storms in which pycnidiospore inoculum in basal diseased leaves are elevated up to 60 cm through the crop canopy. (ii) Gradual epidemics, in which disease arises on successive leaves as they appear during stem extension, when there are sustained periods of weather suitable for inoculum transport and infection.

From the long incubation periods that were deduced from the data, (2–4 weeks for S. nodorum and 3–5 weeks for S. tritici), it was concluded that infection of a given leaf layer arises from inoculum borne on leaves well below it, and that the potential amount of disease in a crop in summer may relate to the amount of inoculum present in the spring.

The hypotheses generated from this work have provided the foundation for our approach to forecasting with S. tritici.

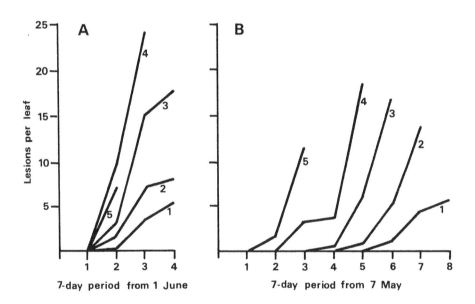

Fig. 1. Typical patterns of Septoria development on leaves 1 (flag) – 5 in crops of susceptible winter wheat. A: a sudden epidemic in response to a salient rain event. B: a gradual epidemic in a season with much rain.

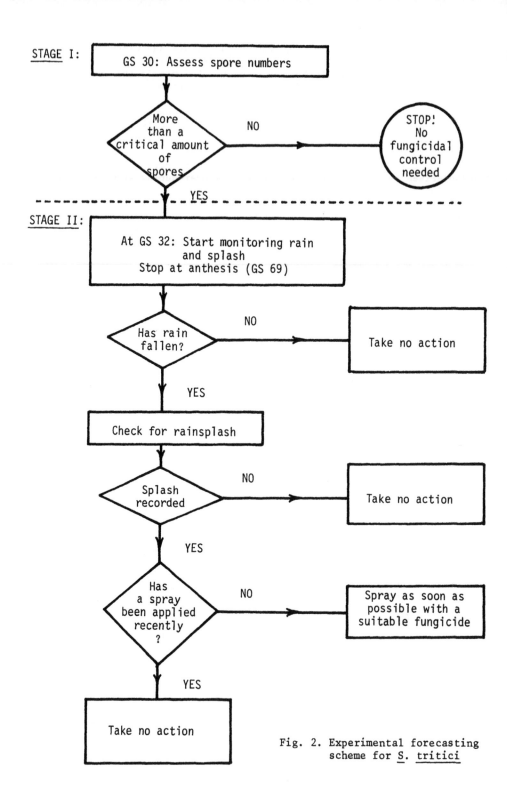

STAGE I: GS 30: Assess spore numbers

More than a critical amount of spores

NO → STOP! No fungicidal control needed

YES

STAGE II: At GS 32: Start monitoring rain and splash
Stop at anthesis (GS 69)

Has rain fallen? → NO → Take no action

YES

Check for rainsplash

Splash recorded → NO → Take no action

YES

Has a spray been applied recently ? → NO → Spray as soon as possible with a suitable fungicide

YES

Take no action

Fig. 2. Experimental forecasting scheme for S. tritici

3. Forecasting framework for S. tritici

3.1 Early-season disease

Field studies using polythene tents (to exclude an airborne source), soil sterilisation and a seed supply from a region in Europe without S. tritici, have established that primary infections of winter wheat crops arise in a time 'window' during autumn and early winter from an airborne source (Shaw and Royle, unpublished). Circumstantial evidence suggests that this source is ascospores of the perfect stage, Mycosphaerella graminicola, although we have not so far found significant numbers of perithecia on residues of previous crops nor identified ascospores in suction traps. When there is severe autumn infection, as in 1985, there may be 50 lesions m^{-2} on primary leaves of a crop. Lesions appear 5-12 weeks after infection, depending on temperature, and contain pycnidia whose splash-dispersed pycnidiospores initiate secondary infections. This means that almost all sowings suffer at least one generation of secondary infections before the spring when stem extension (GS 30-31) starts.

During winter there are long and frequent wetness periods so that, if autumn infection occurred, most crops will have large numbers of active pycnidia on dead leaf tissue in the spring. In our experience, there appear to be very few crops, irrespective of variety, in which some disease cannot be detected in the autumn and, unless sown very late, most crops of current wheat varieties sustain a lot of disease in the early part of the season. We do not yet know how variable are autumn disease levels from year to year. Inoculum amounts at the start of stem extension, however, integrate the effects of primary inoculum and secondary development in winter and, since there appears to be no further ingress of inoculum into a crop from external sources, they represent the total potential risk to a crop later in the year.

In a field, infection in the autumn occurs spatially very uniformly (14). This uniformity is carried through to the spring and facilitates sampling for an assessment of inoculum levels for the initial stage of our experimental forecast scheme (Stage I, Fig. 2). Results from the IOBC study suggested, empirically, that 10^4 spores/tiller might be a threshold level above which crops are at risk from disease. However, data in 1986 now indicate this value to be too high and that even crops with very low numbers of spores in spring may, if environmental conditions allow, sustain relatively severe attacks.

3.2 Summer disease development

Stage II of the forecast scheme allows decisions to be made on the timing of fungicide sprays according to the occurrence of specific rain events between GS 32-69. This section considers the rationale of these steps.

As the stem of a tiller extends during the spring, a leaf is produced typically about every 10 days. In the currently popular cultivar Longbow, which is highly susceptible to S. tritici, the stem grows about 20 cm each week during the time when the last three leaves are being produced. We have already noted from the IOBC results the long latent period of S. tritici. At typical average temperatures for May, (11^{0}C = 35-year mean at Long Ashton), the latent period is about 4 weeks. In a crop sown on an average sowing date five stem leaves are produced on each strong tiller. Then, by the time infectious lesions are present in any numbers on leaf 5 (where flag = 1), even if it became infected as soon as it appeared,

211

leaf 2 is emerging probably 40 cm above the ground, 30 cm above leaf 5. This means that infection of leaf 2 cannot occur unless spores from pycnidia in basal leaves are transported upwards in sufficiently large numbers to heights of 30-50 cm. A similar argument applies to the flag leaf and is relevant for both the sudden and gradual patterns of disease development recognised in the IOBC survey data. This transport can only occur by rain splash, since pycnidiospores of S. tritici are not dispersed in dry conditions (4). Thus, a potential limiting factor in S. tritici epidemics is availability of the type of rain whose drops will penetrate gaps in the crop canopy, impact on to pycnidia near the base and disperse spores upwards to considerable heights.

Subsequent infection conditions are of course another possible limiting factor. A continuous wet period of 20 h is reported to be needed for infection (5). However, the precise effects of wetness have not been studied, but wetness after infection does not appear to influence greatly the expression of S. tritici. The period of wetness arising from a particular rainy day clearly depends on many factors; a wetness period well in excess of 20 h may occur on parts of leaves, even in summer (2). Despite these considerations, we regard upward splash as the more common limiting factor (11). Preliminary results suggest that interruption of wetness by dry periods when the humidity is at c. 85% for up to 12 h does not greatly reduce infection, making the conditions required much more lax.

The characteristics of rain which make for the upward transport cannot reliably be related to rainfall intensity, as measured with a 0.5 mm capacity tipping bucket rain gauge. All we know at present is that in low total amounts of rain (c. 2 mm) there is little inoculum transport. Even so the 'splashiness' of rain varies considerably. For example, the most splashy storm in the period GS 34-65 in 1965 at Long Ashton transported twenty-five times as much material above a given height as any other in that period (calculated from data in (12)). Thus, 'splashiness', as defined in this manner, is potentially a very useful measure for use in forecasting S. tritici. Accordingly in Stage II of our experimental scheme (Fig. 2) we record it directly with a simple apparatus (a 'splashmeter') which detects the upward splash of an aqueous dye. For this purpose the 'splashmeter' is located on open ground, but near to the crops for which the predictions are required. This avoids the complications of placing the device within a single selected canopy which may or may not be typical of other sowings, varieties or cultivation influences.

4.   Evaluating the forecast scheme

In 1986 we began to examine the scheme for guiding control of S. tritici. As explained above, the scheme is based on inoculum and splash measurements, and assumes that adequate wetness for infection will always be provided by rain events which cause enough spore transport to make an outbreak possible. At Long Ashton, assessments of inoculum levels at GS 30 were made for crops sown at six different dates. All but the last sowing had an inoculum level above the working threshold of 10^4 spores/tiller, i.e. that judged to be the minimum sufficient for serious attacks. Starting at the appearance of leaf 2 (judged by stem height from the IOBC data) until anthesis, fungicide sprays were applied following detection by the splashmeter of rain which was at least as 'splashy' as that which caused damaging attacks at Long Ashton in 1985. At present, we have little knowledge of the period after application during which a fungicide will protect against new infections; we assumed a period of 2

212

weeks for prochloraz (6) which was the fungicide used at Long Ashton.

Full results are to be published later, but all sowings were severely affected by S. tritici, the fungicides giving large increases in yield. Despite there being very few spores at GS 31 in the final sowing, damaging attacks developed in this crop also. This indicates that in future the inoculum threshold may need to be interpreted according to the weather conditions following assessment. The spring of 1986 was cold and very wet and was relatively more favourable to disease progress than to crop growth. There were frequent occasions when splash was recorded by the splashmeter; the potential of the scheme to save sprays was not really tested since the control actions indicated were little different from routine sprays at GS 39 and 59. Whether our approach will save fungicides will depend on observations in drier seasons and on optimisation of such factors as the minimum interval between sprays.

5. Septoria nodorum

The forecasting approach described above is appropriate only for S. tritici even though this has several epidemiological features in common with S. nodorum, as the IOBC survey showed. In many ways, the biology of this pathogen is harder to study than S. tritici. It is not clear how it enters the average crop, nor why it seems recently to have declined in importance in the UK. Seed infection remains a possible primary infection route; a very low level of infection could still multiply to produce substantial disease levels by the summer. O'Reilly & Downes (7) have shown that trash may produce high levels of infection which only become apparent in the spring, and Rapilly et al. (9) have shown that spring infection by ascospores is possible. The recent demonstration of persistent and high host specificity (8) makes the role of alternative hosts unclear, but they cannot yet be ruled out as a source. However, if entry into a crop is primarily via seed or trash then early spring (GS 30-31) inoculum levels should still serve to set a risk threshold, as suggested by the results in the IOBC survey (11). Certainly at Long Ashton in 1985 and 1986, action on the basis of spring inoculum levels would have been correct; although in 1986 the disease did eventually reach detectable levels, it never covered more than 0.5% of a leaf layer before the leaves died.

Although the latent period of S. nodorum can be much shorter than that of S. tritici at high temperatures and if wetness is also present, it is at least as long as that of S. tritici at average temperatures below $12^{\circ}C$. This means that at typical average May temperatures for the UK ($11^{\circ}C$) inoculum transport is still likely to be a limiting factor, and the scheme proposed for S. tritici should be applicable. However, since late season multiplication could be rapid, an extra monitoring stage in the scheme around anthesis may be necessary, with control applied if disease severity, though not yet damaging, could become so in unfavourable conditions.

6. Conclusions

There is an urgent need for new forecast methods which avoid the many shortcomings of past empirical rules (10) and which are based on a sound understanding of those features of pathogen life-cycles, and the factors which influence them, which critically determine disease progress in crops. We have outlined in this paper some of the work at Long Ashton which is focusing on this problem as applied to the winter wheat system, in particular to Septoria which offers an excellent model for such studies. Our results so far to forecast S. tritici in the short-term are encouraging and indicate that a soundly-based forecast approach could be very cost-effective. However, until we can blend the forecasting of S. nodorum into a common control scheme, or else establish what determines which of the two species will be dominant in a given year and locality, then any formally applied management scheme may run the risk of foundering. We are now addressing these problems and are also beginning to examine the options for control of Septoria when other diseases, particularly eyespot and mildew, also require control. Meanwhile, the substantially better understanding of the factors governing Septoria progress in crops has a direct bearing on improving the broad intelligence upon which advice can be given to farmers.

Acknowledgement

Long Ashton Research Station is financed through the Agricultural and Food Research Council.

REFERENCES

1. ANON. (1986). Use of Fungicides and Insecticides on Cereals. Booklet 2257(86). M.A.F.F. (Publications), Alnwick, Northumberland.
2. BRAIN, P. and BUTLER, D.R. (1985). A model of drop size distribution for a system with evaporation. Plant, Cell and Environment, 8, 247-252.
3. COOK, R.J. (1985). Thresholds and economics of disease control. In: Marketable Yield of Cereals Course Papers. National Agricultural Centre Arable Unit, December 1985, pp. 1-6.
4. EYAL, Z. (1971). The kinetics of pycnospore liberation in Septoria tritici. Canadian Journal of Botany, 49, 1095-1099.
5. HOLMES, S.J.I. and COLHOUN, J. (1974). Infection of wheat by Septoria nodorum and S. tritici in relation to plant age, air temperature and relative humidity. Transactions of the British Mycological Society, 63, 329-338.
6. JORDAN, V.W.L., HUNTER, T. and FIELDING, E.C. (1986). Biological properties of fungicides for control of Septoria tritici. Proceedings of the British Crop Protection Council Conference: Pests and Diseases, pp. 1063-1069.
7. O'REILLY, P. and DOWNES, M.J. (1986). Forms of survival of Septoria nodorum on symptomless winter wheat. Transactions of the British Mycological Society, 86, 381-385.
8. OSBOURN, A.E., SCOTT, P.R. and CATEN, C.E. (1986). The effects of host passaging on the adaptation of Septoria nodorum to wheat or barley. Plant Pathology, 35, 135-145.
9. RAPILLY, F., FOUCAULT, B. and LACAZEDIEUX, J. (1973). Etudes sur l'inoculum de Septoria nodorum Berk. (Leptosphaeria nodorum Muller) agent de la septoriose du ble. I. Les ascospores. Annales de Phytopathologie, 5, 131-141.
10. ROYLE, D.J. and BUTLER, D.R. (1986). Epidemiological significance of liquid water in crop canopies and its role in disease forecasting. In: Water, Fungi and Plants (eds P.G. Ayres & L. Boddy). Cambridge University Press, pp. 139-156.
11. ROYLE, D.J., SHAW, M.W. and COOK, R.J. (1986). The natural development of Septoria nodorum and S. tritici in some winter wheat crops in Western Europe, 1981-83. Plant Pathology, 35, (in press).

12. SHAW, M.W. (1987). Assessment of upward movement of rain splash using a fluorescent tracer method and its application to the epidemiology of cereal pathogens. Plant Pathology, 36 (in press).
13. SHAW, M.W. and ROYLE, D.J. (1986). Saving Septoria fungicide sprays: the use of disease forecasts. Proceedings of the British Crop Protection Council Conference: Pests and Diseases, pp. 1193-1200.
14. SHAW, M.W. and ROYLE, D.J. (1987). The spatial distributions of Septoria nodorum and S. tritici within crops of winter wheat. Plant Pathology, 36 (in press).
15. TYLDESLEY, J.B. and THOMPSON, N. (1980). Forecasting Septoria nodorum on winter wheat in England and Wales. Plant Pathology, 29, 9-20.
16. ZADOKS, J.C., CHANG, T.T. and KONZAK, C.F. (1974). A decimal code for the growth stage of cereals. Weed Research, 14, 415-421.

Barley mildew in Europe: Analysis of virulence and of fungicide resistance for improvement of disease control

E.Limpert
Lehrstuhl für Pflanzenbau und Pflanzenzüchtung, Technische Universität München, Freising-Weihenstephan, FR Germany

Summary

The development of improved strategies for deployment of host resistance and of fungicides for disease control requires knowledge of the composition of the pathogen population. This is of particular importance for wind-dispersed diseases like Erysiphe graminis f. sp. hordei, causing mildew of barley. By means of a Schwarzbach jet spore sampler representative samples of spores have been taken from the atmosphere along transects through important barley growing areas from France and Italy to Denmark and from Scotland to Austria. Progenies of single spores were analysed both for virulence against the main resistance genes present in European commercial varieties and for resistance against the most important groups of systemic fungicides. Distinct patterns of frequencies of virulence and of different levels of fungicide resistance were obtained. The results can be used as a basis for cultivar diversification and for the choice of appropriate fungicides. General rules of evolution could be recognised during the recent past and might be used to direct the future evolution. Besides the influence of the regional barley cultivation, basic features of the drift of mildew populations throughout Europe were detected, demonstrating large parts of Europe to be an epidemiologic unit. A system for the Europe-wide use of host resistance and of fungicides is described to provide a demonstration of the basic principles involved.

1. INTRODUCTION

1.1 Background and Objectives.

Barley mildew is one of the most important diseases of barley in Europe, causing losses estimated to surpass one thousand million DM on an average per year in the countries of the European Community (1).

To control the disease and to reduce damage and production costs, resistant varieties have been bred and grown for more than 50 years. For the same purpose systemic fungicides with specific action have been developed and used during the last decades.

However, previously successful methods of control have been regularly overcome after extensive use due to changes in the pathogen population caused by physiologic races able to overcome the means of control (2). Therefore the efficiency of the different means depends on the composition of the population. During the recent past the actual composition was determined. Because of the spread of the parasite by wind the analysis of the pathogen population in large European areas improved the knowledge of the epidemiology of barley mildew in Europe, with which strategies for the Europe-wide use of integrated means of control can be developed.

1.2 Basic Features of the Method

The investigation of the mildew population in large areas has become feasible through the development of a method of sampling and analysing spores, described in more detail elsewhere (1, 3). By means of an apparatus mounted on the roof of a car and containing a jet spore sampler (4) representative random samples of the spores were taken from the atmosphere along transects through regions of interest for barley cultivation and for the epidemiology of barley mildew. The progenies of single colonies were subsequently analysed in the laboratory for their ability both to attack barley varieties with defined resistance genes and to overcome treatments of susceptible varieties with certain fungicides. The ability of the pathogen to overcome host specific resistance genes is termed 'virulence' and pathotypes able to overcome fungicide treatment are 'fungicide-resistant'.

2. DISTRIBUTION OF VIRULENCE CHARACTERS IN EUROPE AND THEIR CAUSES

The frequencies of virulence against ten important resistance genes present in commercial varieties on the European market have been determined during the last years. The distribution of two of them will be regarded more closely, and further details and results about the population in Europe in 1985 and 1986 are described elsewhere (5).

The frequencies of VLa, the specific matching virulence allowing attack of host resistance MlLa, showed a distinct regional pattern (Figure 1). Highest frequencies of up to 90% were observed in large parts of central France, intermediate values were present in the UK, in Austria, in Denmark and the neighbouring part of northern Germany, and lowest values in the remaining parts of Germany, in Belgium and in marginal areas of France; in the latter regions adequate control of the disease on the varieties containing MlLa can be expected.

This pattern reflects mostly the actual and previous utilisation of corresponding varieties (1, 5, 6). Thus the selective forces of the composition of the host are the main reason for the composition of the parasite population.

Another pattern was found for the distribution of Va6. This virulence was present almost everywhere close to 100% and the corresponding resistance, Mla6, still had some practical value only in western France (Figure 2).

Mla6 varieties were grown extensively during the last decades, leading to an almost complete change of the composition of the pathogen population. However, as these varieties have been grown mainly in Germany and in the UK and as they were almost absent in France, Denmark and Austria, the frequency of Va6 in these latter countries can only be explained by the dispersal of the pathogen population by wind. Therefore, next to the selection of the host grown in a region, wind dispersal of the pathogen is the most important influence on the composition of a population.

In the latter countries, Va6 was of no selective advantage for the survival of the pathogen. The virulence was thus unnecessary for the respective spores in the regions concerned, and unnecessary virulence provides direct evidence for the introgression of mildew from another region.

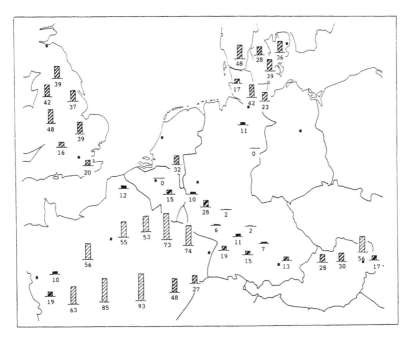

Fig. 1. Distribution of VLa in Europe in 1985. The results are based on 2754 isolates tested individually. Different hachures accentuate small values in order to highlight those regions in which the resistance is most effective.

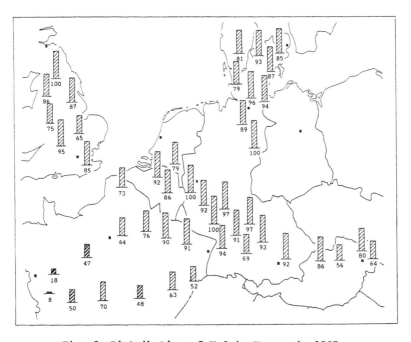

Fig. 2. Distribution of Va6 in Europe in 1985.

New and unusual virulence characters provide an even better tool for investigations of wind dispersal of pathogens. They are selected in regions where the varieties with corresponding resistance are cultivated and occur there for the first time in measurable amounts. The subsequent emigration of spores from such regions can easily be traced, as they are labelled by their unusual character, which is not needed for survival in other regions.

Several examples (1) indicate that the distribution of spores is enhanced in and reduced against the main wind direction. Two of them are depicted in Figure 3. Va1 was selected in England, where it occurred most frequently. In other European regions (besides Denmark) the virulence was

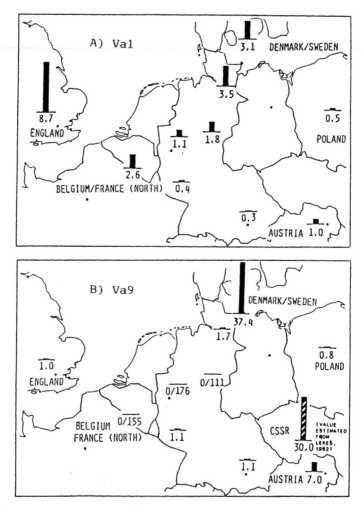

Fig. 3. **Distribution pattern of virulence characters in north western Europe in 1981.** The frequency (%) of Va1 (A) was highest in England and decreased steadily to the east whereas Va9-frequency (B) sharply decreased from its foci to the west. These patterns are probably due to the spread of mildew spores with prevailing winds from the west.

unnecessary. The observed distribution of the virulence indicates a considerable introgression from England to the continent, where the frequency steadily decreased eastward.

On the other hand Va9 (Figure 3B) was found most often in Denmark, Czechoslovakia and Austria. From these centres of origin the frequency of this virulence decreased sharply to the west, thus indicating that the spread of mildew was reduced in this direction. This conclusion was confirmed by similar data of the distribution of Va13 in 1985 and 1986 (5).

A distribution pattern similar to that described in Figure 3A was observed within southern Germany in 1979 (Limpert, unpublished), where a steady decrease of the frequency of Va7+Vk was pronounced, from approximately 80% in the north west to 20% in the south east, pointing to Rheinland-Pfalz in the north west as the region of origin. Since the virulence was unnecessary in Bavaria, the evolution of its frequency observed there in the region of Weihenstephan was striking (Figure 4). In addition, this development was rather similar to that partly observed, partly estimated, in the regions of origin, indicating that not just single spores but rather the entire mildew population from Rheinland-Pfalz followed the drift and passed Weihenstephan. As there was an interval of time of three years between these developments, it is concluded that the mildew population needed this lapse of time to drift from Rheinland-Pfalz to Weihenstephan. This corresponds to an annual drift of approximately 110 km on average.

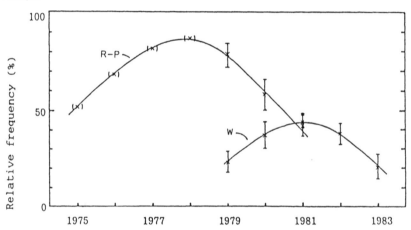

Fig. 4. **Evolution of the frequency of Va7+Vk in Rheinland-Pfalz (R-P), the region of origin in south west Germany, and near Weihenstephan (W), in Bavaria, where the virulence was unnecessary.** The difference in time between the maximum and the subsequent decrease of the virulence frequencies in the two regions indicates that the mildew population needed three years to cover the 330 km distance to Weihenstephan.

These examples show that vast areas of Europe have to be regarded as an epidemiologic unit and even in regions where winter barley is cultivated and mildew is regarded to be endemic a considerable influence from neighbouring regions is present. In the face of this knowledge, the high frequency of VLa in France (Figure 1) may threaten the large number of MlLa varieties on the European market, because the mildew population in France is in a favourable position to spread by wind to most other parts of Europe during the coming years.

4. DISTRIBUTION OF FUNGICIDE RESISTANCE IN EUROPE

Economically important systemic fungicides with specific action are the inhibitors of sterol biosynthesis. For each of the two main groups amongst them one representative was chosen, triadimenol for the inhibitors of demethylation and fenpropimorph for the morpholins, which act differently. The reaction of each mildew isolate was tested on leaf segments of plants of a susceptible variety, treated with different doses of fungicide. In comparison with the reaction on untreated plants inoculated simultaneously, the highest dose of fungicide was determined, which allowed at least 50% sporulation. Finally, levels of resistance were determined, comparing this dose with that obtained from four standard isolates, representing the original population before introduction of these fungicides (1).

Regional differences of resistance to triadimenol were obvious, and due to positive cross resistance these results apply to all fungicides belonging to the group of demethylation inhibitors. To give a survey the distribution of frequencies of resistance $\geqslant 64$ is illustrated in Figure 5.

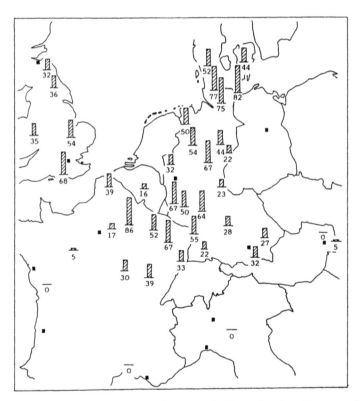

Fig. 5 Survey of resistance against triadimenol in Europe in 1986.
Indicated are frequencies of resistance factors $\geqslant 64$.

The most resistant population (see as well Table 1) was noticed in Schleswig-Holstein in northern Germany (regions D1, 2, 3a, Table 1), followed by that in Champagne in France (F12) and by that in southern UK (GB1). The values decreased, with some exceptions, from north to south in Germany and in France, and in the opposite direction in the UK. Intermediate frequencies were present in Denmark, where the population showed a very broad spectrum of resistance levels (Table 1). Extremely low values occurred in Austria, in western and southern France, and in Italy. In general, the basic features which were revealed for selection and spread of virulence against host resistance have been found to hold for fungicide resistance as well.

The significance of the different levels of resistance and of the different composition of mildew populations (Table 1) can be deduced from the decline of the concentration of seed-applied triadimenol in the growing plant and from the duration of effect of the treatment against the original, sensitive mildew population (7). This population was able to attack the plant from GS 30 or GS 31, whereas a resistance level of 4 should enable the pathogen to attack from GS 21, thus reducing the duration of effect of the fungicide by two to three weeks, and a level of 40 and more must be considered to allow the attack of the pathogen in the field even on the first leaf of seedlings grown from seed treated with the recommended dose (1). Therefore it is supposed that Figure 5 shows the part of the regional population which cannot be controlled any more with the recommended dose.

The population in northern Italy was investigated only in 1986 and differed markedly from that in other areas (Table 1). It was the most sensitive population observed and is supposed to represent a rather original state of composition. Even mildew was present reacting significantly more sensitively than the standard isolates used.

Against the morpholine, fenpropimorph, regional differences were far less pronounced, indicating that the efficiency of the fungicide was similar in the investigated area. The same degrees of sensitivity close to that of the standard isolates, which had a mean position, were established in both years. However, first signs of a change in the composition of the population towards reduced sensitivities were found in some regions in 1986. They should be kept under observation to answer the question, whether there is a risk of resistance in practice for this group of fungicides as well.

5. PATTERNS OF EVOLUTION AND THEIR CONSEQUENCES

From the development of virulence frequencies observed in different regions a characteristic picture of the population dynamics of host and pathogen can be drawn (Figure 6). The introduction of a new resistance is followed by a gradual adaptation of the pathogen population and during the first few years of utilisation the resistance is most effective. After the maximum of the frequency of the resistance is passed over, the frequency of virulence continues to increase and reaches its maximum a few years later. During this period there is an increasing potential of the pathogen to cause severe damage. Afterwards a marked decrease of the virulence frequency is established which is, according to Figure 4, mainly due to emigration of the mildew population in the main wind direction. The varieties with the corresponding resistance will be more and more healthy and, finally, a new successful cycle of utilisation of this race-specific resistance may start in that region. In a similar way resistance Mla12 was used in the UK during two cycles (8).

Starting from the fact that varieties with new resistances can be used with most success during the first years after introduction (Figure 6), and that the parasite population in Europe mainly drifts from west to east (Figures 3 and 4) a system for a Europe-wide use of these race-specific resistances was developed (Figure 7). Even if it is not applicable as a whole it does provide a demonstration of the basic principles involved.

TABLE 1. Frequency and Level of Triadimenol-resistance of Random Samples of Barley Mildew from Different Parts of Europe in 1986

No. Date Distance	n											
		Percentage of isolates per dose of triadimenol (mg a.c./kg seed, I) and resistance levels* (II), respectively										
	I	0.73	1.46	2.9	5.9	11.7	23.4	46.7	93.7	187.5	375	750
	II	0.5	1	2	4	8	16	32	64	128	256	512
Standard isolates	4	0	100	0	0	0	0	0	0	0	0	0
Great Britain												
GB 1 0807 Dover-Reading	36	0	0	0	3	9	11	9	11	34	23	0
GB 3 0807 Birmingham-Preston	23	4	4	9	9	0	22	17	22	4	9	0
GB 7 0907 Chiras.-Kelso-Coldstr.	33	0	0	0	6	12	29	21	9	18	6	0
GB 8 0907 Coldstr.-Newcastle	36	0	3	0	3	6	6	47	25	11	0	0
GB11 0907 Newark-Cambridge	22	0	0	0	0	9	14	23	23	5	18	9
France												
F 2 0305 Beaucaire-Montpellier	24	0	33	50	13	0	4	0	0	0	0	0
F 4 2906 Mulhouse-Besancon	31	0	0	10	23	17	0	17	30	3	0	0
F 5 2906 Besancon-Beaune	37	0	0	0	3	14	25	19	14	22	3	0
F 7 2906 Beaune-Le Blanc	13	0	0	15	0	0	8	47	15	15	0	0
F 8 2906 Le Bl.-Ancenis-Le Mans	13	0	0	8	8	23	46	15	0	0	0	0
F 9 3006 Le Mans-Illier	20	0	15	10	5	45	20	5	0	0	0	0
F11 3006 Paris-Reims	23	0	0	4	0	0	17	62	17	0	0	0
F12 3006 Reims-St. Menehould	35	0	0	0	0	2	3	9	83	3	0	0
F13 3006 St. Menehould-Metz	48	0	2	0	0	6	11	29	23	29	0	0
F14 3006 Metz-Strasbourg	24	0	4	4	4	5	8	8	50	17	0	0
F15 0707 Calais-fr.Mons	46	0	0	0	9	28	11	13	13	17	6	2
Belgium												
B 1 0707 Mons-fr.Aachen	25	0	4	8	8	12	16	36	12	0	4	0

Denmark												
DK 1 1407 Kolding-fr.Flensburg	48	0	10	8	2	2	10	15	8	10	22	10
DK 3 1407 Halsskov-Riso	48	0	4	12	7	2	17	12	12	15	17	2
Germany (FRG)												
D 1 1407 Flensburg-Kiel	22	0	0	0	0	0	9	14	59	14	4	0
D 2 1407 Kiel-Hamburg	16	0	0	0	0	0	12	13	25	50	0	0
D 3a 1507 Puttgarden-Neustadt	32	0	0	0	0	6	3	9	9	22	31	19
D 4 1307 Dörpen-Oldenbg.-Bre.	12	0	0	0	0	8	0	42	42	8	0	0
D 8a 1706 Osnabrück-Minden	24	0	0	0	0	8	12	25	42	8	5	0
D 9 1307 Venlo-Rheine	24	0	0	0	4	13	13	37	25	8	0	0
D13 1107 Aachen-Meckenheim	27	0	0	22	0	0	0	33	30	15	22	0
D15 1107 Meckenheim-Speyer	12	0	0	0	0	0	0	50	50	0	0	0
D16 1307 Limburg-Aschaffenbg.	22	0	0	0	0	0	18	18	41	18	5	0
D20a 0107 Crailsheim-Nürnberg	35	0	0	3	3	3	23	40	20	3	5	0
D21 0107 Bruchsal-fr.Strasburg	36	0	0	0	0	3	6	36	33	14	8	0
D24 2806 Stuttgart-Donauesch.	36	0	8	22	0	3	11	33	6	11	6	0
D27 2306 Freising-Dingolfing	37	0	5	0	8	8	3	49	13	11	3	0
D29 2206 München-Salzburg	36	0	6	3	0	6	14	42	14	14	2	0
Italy												
I 2 0205 Verona-Piacenza	27	4	48	41	4	4	0	0	0	0	0	0
Austria												
A 4 2206 Krems-Mistelbach	54	0	50	31	9	0	4	6	0	0	0	0
A 5 2206 Marchfeld	21	0	24	43	14	0	5	9	5	0	0	0

* resistance levels: dose test isolate/dose standard; dose standard isolates = 1.46 mg/kg

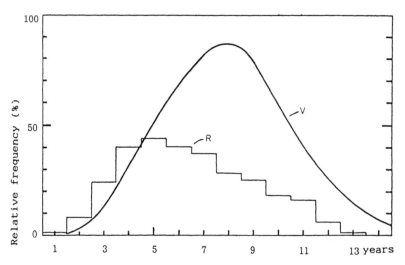

Fig. 6. Characteristic development of the frequency of a race-specific resistance (R) and the specific matching virulence.

Period	Zones			
	A	B	C	D
1	Ly	Ar	Sp	Ru,MC
2	Ar	Sp	Ru,MC	Ly
3	Sp,NR1	Ru,MC	Ly	Ar
4	Ru,MC	Ly	Ar,NR2	Sp,NR1
5 ($\hat{=}$ 1')	Ly	Ar,NR2	Sp,NR1	Ru,MC

Fig. 7. System for a Europe-wide use of race-specific resistance against barley mildew. The whole area of cultivation is divided into several zones (A-D) across the main wind direction (MWD), in each of which certain resistance genes will be used during a certain period. Every new period cultivation moves for one stage against MWD or from A to D, respectively. Varieties with new resistance genes can be integrated at any position. During the fifth period positions of departure are reached again; diversity has increased, however, in the meantime by the addition of new resistance genes (NR1, NR2). Symbols of resistance genes: Ly = Lyallpur (Mla7+Mlk); Ar = 'arabische' (Mla12); Sp = Spontaneum (Mla6); Ru = Rupee (Mla13); MC = Monte Cristo (Mla9).

6. PERSPECTIVES

The significance of these data for integrated crop protection in cereals is evident. First results obtained with a similar approach with wheat mildew (9), another important disease in Europe, demonstrate that further investigations would be advantageous, and there are other wind dispersed diseases like net blotch or cereal rusts worth investigation.

The importance of the dispersal of the barley mildew population throughout Europe indicates that a closer coordination between the European countries would be sensible. Therefore, the Commission of the European Communities supported a meeting of scientists on which a common basis for monitoring cereal mildew populations at a European level was developed (10).

Within the former Europe, composed of single nations acting separately, most of the problems and questions referred to in this paper were not even recognised. It would be a challenge for the European Community to take them up and to solve them by further Europe-wide research and coordinated efforts.

ACKNOWLEDGEMENTS

This work has been supported by the German Research Foundation and by the Commission of the European Communities. Their help is gratefully acknowledged.

REFERENCES

(1)   LIMPERT, E., 1985. Ursachen unterschiedlicher Zusammensetzung des Gerstenmehltaus, Erysiphe graminis DC f.sp. hordei Marchal und deren Bedeutung für Züchtung und Anbau von Gerste in Europa. Diss. München 183 pp.

(2)   WOLFE, M.S. and SCHWARZBACH, E., 1978. The recent history of the evolution of barley powdery mildew in Europe. The Powdery Mildews, D.M. Spencer (ed.) Academic, London, 129-157.

(3)   LIMPERT, E., SCHWARZBACH, E. and FISCHBECK G., 1984. Influence of weather and climate on epidemics of barley mildew, Erysiphe graminis f.sp. hordei. In, Interactions between climate and biosphere, H. Lieth, R. Fantechi and H. Schnitzler (eds.), Progress in Biometeorology Vol.3, Swets and Zeitlinger B.V., Lisse, 146-157.

(4)   SCHWARZBACH, E., 1979. A high throughput jet trap for collecting mildew spores on living leaves. Phythopath. Z. 94, 165-171.

(5)   LIMPERT, E. and FISCHBECK, G., in press. Distribution of virulence and of fungicide resistance in the European barley mildew population. In (10).

(6)   ANDRIVON, D. and POPE-DE VALLAVIEILLE, C., in press. Cereal powdery mildew in France. In (10).

(7)   STEFENS, W., FÜHR, F., KRAUS, P. and SCHEINPFLUG, H., 1981. Aufnahme und Verlagerung von (^{14}C) Triadimenol in Sommergerste und Sommerweizen nach Saatgutbeizung. Mitt. Biol. Bundesanst. Land-Forstwirtsch. 203, 234-235.

(8)   WOLFE, M.S., 1984. Trying to understand and control powdery mildew. Plant Pathology 33, 451-466.

(9)   FELSENSTEIN, F.G., LIMPERT, E. and FISCHBECK, G., in press. Analysis of virulence characters in populations of wheat powdery mildew (Erysiphe graminis f.sp. tritici) in Europe. In (10).

(10) WOLFE, M.S. and LIMPERT, E., eds., in press. Integrated control to reduce the damage caused by cereal mildews: Monitoring pathogen populations. Proceedings of the CEC-workshop, Weihenstephan, 1986. Commission of the European Communities.

Control of barley mildew by integrating the use of varietal resistance and seed-applied fungicides

M.S.Wolfe, P.N.Minchin & S.E.Slater
Plant Breeding Institute, Trumpington, Cambridge, UK

Summary

The barley mildew pathogen, Erysiphe graminis f. sp. hordei, responds rapidly to almost all host resistance genes and fungicides, particularly if their commercial introduction is rapid and on a large scale. However, the pathogen population on each variety or fungicide treatment tends to be adapted to that particular control measure and is usually less well-adapted to other hosts or fungicides.
The inability of the pathogen to become highly adapted to many different components of control simultaneously can be exploited in crops in which adjacent plants are either of different varieties or are treated with different fungicides. In such mixtures, the distance between identical plants, or identically treated plants, is greater than in a conventional crop stand, which helps to slow down the spread of the pathogen.
Mixing of varieties or of fungicides can be integrated in various ways. One of the simplest and most advantageous appears to be fungicide treatment of the seed of a single component of a three variety mixture. Yield is not less than that from a conventionally treated mixture, but cost, selection for fungicide insensitivity and environmental contamination are all reduced.

1.1 Introduction

The principal methods available for the control of powdery mildew of barley, caused by Erysiphe graminis f. sp. hordei, are the use of resistant varieties and of fungicides. Under the present forms of deployment, largely by widespread and continuous exposure of single varieties or fungicides, both methods are prone to the ability of the pathogen to respond rapidly to their use. The pathologist therefore needs to provide varieties and fungicides that are inherently more durable (1), or to recommend methods of deployment that reduce the risk of pathogen response. At the same time, it is essential that improved methods should reduce the costs of grain production and limit the release of fungicides into the environment.

For a number of years, we have developed the idea of using mixtures of resistant varieties to reduce the rate of epidemic spread and increase yield (6,7), and the use of this technique is spreading in Denmark, the FRG and GDR, Poland, Switzerland and the UK. An essential element of the technique in its practical application is to vary the composition of the mixtures by as much as is practicable in order to limit the pathogen response to any one mixture.

In principle, there is no reason why fungicides should not be used in an analogous way (4), provided that the mixtures are heterogeneous in the sense that the different chemical components are applied to different plants. Conventional, or homogeneous, fungicide mixtures are applied uniformly to all plants in a crop stand, which increases selection in the pathogen population for combined insensitivity (resistance) to the components of the mixture.

Ideally, from the pathological point of view, the use of mixtures of varieties and of fungicides should be integrated since this would improve disease control whilst offering an opportunity to further decrease the risk of pathogen response.

1.2 Pathogen response to selection

Mixtures for mildew control operate on two major principles. First, the likelihood of a spore from one infected host plant reaching another similarly susceptible plant diminishes inversely with a power of the distance between the plants (3). Second, genes for pathogen adaptation to a particular host or fungicide are selected in the presence of the host or fungicide, but tend to decline in its absence. The consequence of these principles operating together is that the population of the pathogen selected on a single host plant in a mixture is restricted in its spread through the host population.

An example of changing frequencies of virulence genes in the pathogen population, caused by selection, is given in Table I. The data were obtained from a pathogen population on a field crop of the spring barley variety, Natasha, which has resistance genes Mla12 and Ml(Ab).

Table I. Changes in frequencies of virulence genes in the powdery mildew population on the barley variety Natasha, 1985.

Date	Selected genes		Unselected genes		
	Va12	Va12,(Ab)	Vg	Va6	V(La)
13/6	26	79	14	25	1
27/6	30	84	33	18	7
23/7	30	100	9	3	0

From Table I, during the earlier part of the season, there was a tendency for the frequencies of unselected genes to drift, but in late July, when all isolates from the crop were adapted to Natasha, the frequencies of the unselected genes were generally low.

We have also found consistently in barley mildew populations, that particular combinations of virulence genes occur at a lower frequency than would be expected by chance association. In particular, both Va12 and V(La) tend to occur infrequently in combination with Vk and Va7 (5); this phenomenon has also been observed in Denmark (Houmoller, pers. comm.). For example, the data in Table II were obtained from a large field plot of the susceptible barley variety, Golden Promise, grown at the Plant Breeding Institute in 1985.

Knowledge of this consistent dissociation between virulence genes has proved valuable in choosing components of variety mixtures; clearly, mixtures that contain varieties with Ml(La) or Mla12 together with those with Mlk,a7 will be more effective than other combinations, such as Ml(La) with Mla12 (see 1.6).

Table II. Absolute frequency of isolates with particular combinations of virulence genes in a pathogen population on the barley variety Golden Promise in 1985 (N = non-virulent, V = virulent).

Resistance genes: R1R2		Virulence gene combination: N1N2	V1N2	N1V2	V1V2	Prob. of deviation
Ml(La);Mla12	obs.	16	5	13	16	n.s.
	exp.	12	9	17	12	
Ml(La);Mlk,a7	obs.	15	20	14	1	<0.05
	exp.	20	15	9	6	
Mla12;Mlk,a7	obs.	7	28	14	1	<0.001
	exp.	15	20	6	9	

In the absence of selection, fungicide insensitivity genes may also decline in frequency. An example is shown in Table III from pathogen populations taken from three fields of mildew susceptible winter barleys, the seed of which had been treated with triadimenol prior to sowing.

Table III. Changing frequencies of triadimenol-insensitivity characters in populations of the mildew pathogen on autumn-sown winter barley, 1983-4.

Date	No. of isolates	Proportion of isolates Sensitive	Intermed.	Insens.
26/10	21	40	40	20
7/11	136	46	43	12
29/11	151	49	45	6
4/1	232	33	61	6
29/3	21	5	95	0

Through the autumn and winter, as the fungicide concentration in the plants decreased, isolates with a high degree of insensitivity decreased in frequency. The apparent increase in the frequency of isolates with intermediate insensitivity, relative to sensitive isolates, may have been due to a large influence of infected but untreated, susceptible volunteer plants. This influence would have declined as the crop grew and became more infected.

Unlike the response to selection for combinations of virulence genes, we have no evidence for dissociation between fungicide insensitivity characters. However, it is theoretically possible to exploit the effect with virulence characters to delay development of combinations of insensitivity characters. For example, if varieties with Mla12 were treated only with triazole fungicides, while those with Mlk,a7 were treated only with ethirimol, the poor performance of the corresponding virulence combination would help to delay selection for combined insensitivity to triazoles and ethirimol.

1.3 Heterogeneous fungicide mixtures

Field trials to mimic variety mixtures by fungicide treatments were carried out in 1985 and 1986 on the barley varieties Patty and Triumph, using a proprietary fungicide which is effectively a homogeneous mixture of a triazole with ethirimol. Performance with this single treatment was compared with plots grown from seed of which one half had been treated with a triazole alone and the other half with ethirimol alone, followed by thorough mixing of the seed before sowing (Table IV).

Table IV. Yield (t/ha) of field plots of Patty or Triumph spring barley, untreated, treated with a homogeneous mixture of a triazole and ethirimol, or treated heterogeneously, that is grown from mixed seed treated either with a triazole or with ethirimol.

| Variety | Treatment | 1985 | | 1986 |
		Throws Fm.	PBI	PBI
Patty	untreated	7.23 (100)	6.63 (100)	5.65 (100)
	heterog. mix	7.70 (107)	6.82 (103)	6.05 (107)
	homog. mix	7.68 (106)	7.33 (111)	6.10 (108)
Triumph	untreated	6.24 (100)	5.14 (100)	5.35 (100)
	heterog. mix	7.23 (116)	5.41 (105)	–
	homog. mix	7.41 (119)	5.61 (109)	–
lsd (P = 0.05)		.56	.40	.40

From Table 4, at Throws Farm in 1985 and at PBI in 1986, the heterogeneous mixture performed as well as the homogeneous mixture; only under the very severe infection at PBI in 1985, particularly on Patty, did it not perform as well. Since the heterogeneous mixture provides the added benefit of reduced selection for combined insensitivity, one may therefore argue that this stratagem should be preferred to the conventional homogeneous mixture.

This approach may be useful for application to malting barley varieties, where the malting industry, at least in the UK, is averse to variety mixtures. However, it does not take advantage of a fully integrated approach to the use of varieties and fungicides.

1.4. Choosing the best option for integrating variety resistance and fungicide use

Assuming that variety mixtures present the best option for exploiting variety resistance, there are several ways in which the fungicide may be applied to obtain integrated control. These were compared in a field experiment in 1982 (Table V).

In this experiment, the only treatment of the pure stands that was significantly better than the control, was the normal application of fungicide to all varieties. Among the mixtures, however, all three ways of applying the normal rate yielded significantly more than the untreated control, but these treatments were not significantly different from each other. For practical purposes, cheapness and biological flexibility, it was felt that the treatment in which one component only of the variety mixture was treated, was the best integrated stratagem for further experimentation.

Table V. Mean yield (t/ha) of the spring barley varieties Atem, Carnival and Triumph, and of their mixture, untreated or treated in various ways with triadimenol.

Seed treatment	Mean of pure stands	Mixture
untreated	5.63 (100)	5.88 (104)
1/3 field rate on all seed	5.86 (104)	5.98 (106)
normal rate on 1/3 seed	5.84 (104)	6.00 (107)
normal rate on 1 var. only	5.75 (102)	6.11 (109)
normal rate on all seed	5.98 (106)	6.15 (109)
s.e. diff. (P = 0.05)	.13	.22
s.e. diff. for comparisons between columns (P = 0.05)	.18	

1.5. Treating only one component of a three-variety mixture with fungicide

Several field experiments were carried out between 1982 and 1986 to compare pure stands of barley varieties and their mixtures, untreated, with plots grown from seed which had been treated either conventionally with mildew fungicides, or in which one component only of the variety mixture had been treated. For statistical purposes, plots from the latter treatment were compared with the mean of the same varieties grown as pure stands including the conventional fungicide treatment where appropriate.

In the trials from 1982 to 1984 (Table VI), the data for the 1/3 treatments are means of the three mixtures in which each component alone was treated. In the 1986 trial, the treatments were arranged so that a susceptible treated component (Patty or Triumph) was mixed with two different resistant components. Untreated controls were not available in the 1986 trial.

In all of the trials, it is evident that the yields for the 1/3 mixture treatments are closely similar to those from mixtures in which all of the seed was treated at the field rate. Indeed, the average yield of the seven 1/3 treatments was 6.24 t/ha, compared with 6.21 t/ha for the conventionally treated mixtures.

The results so far indicate, therefore, that a mixture of three varieties in which one component, preferably the most susceptible, is treated with a fungicide, provides the following advantages:
a) high yield, resulting from a high level of disease control; no disadvantage relative to conventional treatment,
b) low cost due to the reduction of fungicide application by two-thirds,
c) reduced release of fungicide into the environment relative to conventional treatment,
d) reduced selection for fungicide insensitivity since only one-third of the host population is treated relative to conventional treatment,

Table VI. Yields (t/ha) of pure stands and mixtures of barley varieties, untreated, or treated with a triazole, ethirimol or the combination, compared with mixtures in which only one component variety was treated.

Culture	untrt.	triazole 1/3	triazole N	Fungicide ethirimol 1/3	Fungicide ethirimol N	triaz/ethir 1/3	triaz/ethir N
1. Pure	5.63	5.84	5.98				
Mixed	5.88	6.11	6.15				
2. Pure	5.19	5.36	5.67	5.25	5.37		
Mixed	5.61	5.68	5.65	5.62	5.44		
3. Mixed (1)						6.68	6.53
(2)						6.94	7.03
(3)						6.31	6.46
(4)						6.31	6.22

N.B. the 1/3 treatments are the means of the three mixtures in which each single component in turn was treated.
1. 1982 trial, Atem, Carnival, Triumph
2. 1983-4 trials, Patty, Carnival, Tasman
3. 1986 trial (1) Sherpa, Everest, Patty-trt.
            (2) CSB 183, Sherpa, Patty-trt.
            (3) CSBN 769, Sherpa, Triumph-trt.
            (4) CSBN 746, CSB 699, Triumph-trt.

e) reduced probability of response to selection for virulence, because of the reduction in absolute population size of the pathogen on the mixture and the increase in the size of the gene combination required by the pathogen to overcome the components of the mixture,
f) increased potential for diversification, by using the same or different fungicides on the same or different components in the same or different mixtures. Exploitation of this advantage may increase the value of e) above.
    The main disadvantages appear to be:
a) novelty,
b) non-acceptance by the malting industry, at least in the UK,
c) reluctance of the seed trade, with some notable exceptions, to provide a mixing service for farmers.

1.6. Yield and yield variation in variety mixtures

    Because of the limited number of trials and the inconsistency among the mixture components in the series of trials on integrated control, it was not possible to indicate a further advantage of the use of variety mixtures for the farmer, namely the association of improved yield with reduced variability. This aspect is summarised briefly for mixtures without fungicide treatment.
    Wolfe et al., (in preparation) followed the performance of four spring barley varieties, Claret, Egmont, Goldmarker and Triumph, and the four three-component mixtures that can be made from them, over the years 1980 to 1986. The yield data only are summarised in Table VII.

Table VII. Yields (t/ha) of four spring barley varieties, Claret, Egmont, Goldmarker and Triumph grown as pure stands or as their four possible three-component mixtures. Diversification yields are the means of the appropriate pure stands.

Variety	Pure culture	Diversification		Mixtures	
Claret (C)	5.45 (107)	CET	5.30 (104)	CET	5.79 (114)
Egmont (E)	5.24 (103)	CEG	5.06 (99)	EGT	5.61 (110)
Triumph (T)	5.20 (102)	CGT	5.05 (99)	CGT	5.46 (107)
Goldmarker (G)	4.49 (88)	EGT	4.98 (98)	CEG	5.43 (106)
means	5.10 (100)		5.10 (100)		5.57 (109)
ranges	6.82 (134)		6.12 (120)		6.56 (129)
	3.55 (70)		4.33 (85)		4.57 (90)

overall s.e. diff. between pure stands and mixtures: 0.07, $P<0.001$

The yields of the pure varieties varied considerably from year to year, and the rank order changed within years. These changes were largely unpredictable, so that a farmer would not have been able to choose the best single variety for the following year. Variability was reduced, but yield was unchanged, by selecting sets of three varieties to grow each year, that is, by diversifying among the available pure cultures. By growing mixtures, yield variability was reduced almost to the same level as with diversification, but yields were considerably increased, giving an overall advantage of nine per cent. The best two mixtures performed outstandingly well, and the worst two performed as well as the best pure variety. The particularly high yields of CET and EGT may have been because both mixtures contained Egmont (Mla12, Ml(La)) and Triumph (Mla7, Ml(Ab)), a variety combination for which the pathogen was not well-adapted (see 1.2).

It is likely that the use of fungicides integrated into these mixtures as in 1.5 above, would have further improved disease control and yields while diminishing, but not eliminating, mixture advantage in terms of mean yield and reduced variability.

1.7. Future needs

Research is needed to confirm the hypothesis that an integrated system, such as fungicide treatment of a single component of a variety mixture, does reduce the risk of selection of fungicide insensitivity. Such analyses need to be carried out in the field, but this presents major difficulties. For example, the initial, natural, inoculum for any field trial in the UK is already at least partly adapted to fungicides and fungicide combinations. Further, on lightly infected crops, such as a treated variety mixture, wind drift of spores into the crop during the growing season may profoudly influence observations on selection in the pathogen population generated within the crop.

In any one year or place, certain variety mixture and fungicide combinations will be more effective than others. However, recent initiatives taken by mildew workers in Europe, may help to provide forecasts of the distribution and movement of virulence and fungicide insensitivity characters over the whole region (see also 2). The first publication of European survey data for virulence and insensitivity in barley and wheat mildew, by the CEC, is scheduled for early in 1987. These forecasts should be tested by comparing the performance of appropriate and inappropriate mixtures in different places, chosen on the basis of the European survey data.

Advances in these two areas, at the field level and at the regional level, would be valuable in providing the information necessary to encourage greater acceptance of these simple techniques more generally in agriculture.

REFERENCES

1. JOHNSON, R. (1984) A critical analysis of durable resistance. Annual Review of Phytopathology 22, 309-30
2. LIMPERT, E. (1985) Ursachen unterschiedlicher Zusammensetzung des Gerstenmehltaus, Eryisphe graminis f. sp. hordei Marchal, und deren Bedeutung für Züchtung und Anbau von Gerste in Europa. Diss. München 183 pp.
3. McCARTNEY, H. A. and FITT, B. D. L. (1987) Spore dispersal gradients and disease development. In Populations of Plant Pathogens; their Dynamics and Genetics. ed. M. S. Wolfe and C. E. Caten: Blackwell Scientific Publications, 109-17
4. WOLFE, M. S. (1981) Integrated use of fungicides and host resistance for stable disease control. Philosophical Transactions of the Royal Society, London, Series B. 295, 175-84
5. WOLFE, M. S. (1984) Trying to understand and control powdery mildew. Plant Pathology 33, 451-66
6. WOLFE, M. S. (1985) The current status and prospects of multiline cultivars and variety mixtures for disease resistance. Annual Review of Phytopathology 23, 251-73
7. WOLFE, M. S. and BARRETT, J. A. (1980) Can we lead the pathogen astray? Plant Disease 64, 148-55

Attempts to combine different methods of control of foot and root diseases of winter wheat

P.Lucas, F.Montfort & N.Cavelier
Station de Pathologie Végétale, INRA, Le Rheu, France
A.Cavelier
SRIV, Le Rheu, France

Summary

The main foot and root diseases which occur in France are take-all and eye-spot. Sharp eye-spot can now be observed more often than in the past but seems to be less injurious than the former two.

Although take-all is a major disease throughout the world and has been studied for many years, no practical method of control has been found in this time. Relationships between cultural practices, nutrients or biological phenomena and the development of the disease have been shown in some cases, but trying to control take-all by techniques based on these observations gives no significant results on a practical level.

Despite this, analytical studies carried out in our laboratories and elsewhere show that some of these techniques may reduce the occurrence of take-all in most cases even if not with a high level of protection. We attempted to combine some of these techniques by carrying out experiments on wheat grown in the field, to achieve a method of cumulative protection (nitrogen fertiliser application, seed treatment, selected variety).

The article discusses results obtained on plant development at different growth stages (using the yield components), on eye-spot, sharp eye-spot and take-all disease levels.

1. INTRODUCTION

In France, the main root and foot diseases on wheat are take-all (Gaeumannomyces graminis (Sacc.) von Arx and Olivier var. tritici Walker) and eye-spot (Pseudocercosporella herpotrichoides (Fron) Deighton). Sharp eye-spot (Rhizoctonia cerealis van der Hoeven) can now be observed more often than in the past but seems to be less injurious than the former two.

If eye-spot is well controlled by fungicides (carbendazimes and benzimidazoles until P. herpotrichoides become resistant, and ergosterol biosynthesis inhibitors now), in practice, only crop rotations give significant results against take-all.

Although take-all is a major disease throughout the world, and has been studied for many years, no practical method of control has been found up to now.

So, take-all remains one of the most difficult cereal diseases to control by resistance breeding, because of no substantial degree of resistance found in wheat, barley or other related species. It is the same with fungicides which have only limited value because of their lack of efficacy or persistence when used as seed treatments or because of their cost if applied to the soil. So, scientists have focussed their attention on studying relations between take-all development and soil characteristics.

The synthesis by Huber (10) shows some results obtained on relations between physico-chemical status of soils and the severity of the disease. The data sometimes look contradictory and the most consistent conclusion is that the ammonium form of nitrogen reduces take-all incidence when compared to nitrate nitrogen. Some more recent works confirm this hypothesis (18, 22, 4).

In the early 1970s studies on the take-all decline (9, 11, 19), observed under wheat monoculture and researches on suppressive soils (20), are the source of several hypotheses on the role of biological agents on take-all limitation: hypovirulent strains of G. graminis tritici (21), Phialophora sp. (6), Pseudomonas sp. (5).

So far, attempts to convert such concepts into biological methods of control have not achieved results because some problems remaining are not yet resolved. In this respect, we obtained irregular results with seed coating by hypovirulent strains of the fungus (15). It seems that beyond problems of quality of seed coating, the method proposed by Lemaire et al. (12) works better in some soils than in others (17). Seed bacterisation with fluorescent pseudomonads seems to give inconsistent results in the field and Weller (24) suggests that more work will have to be done, particularly on the effects of soil physico-chemical factors, on root colonisation.

Considering these results, and with the development of fungicides like triadimenol or nuarimol, which show a good effectiveness in vitro, trials have been made to build up chemical methods of control. So, Bockus (2), in the USA, suppressed infection by take-all for about six weeks with triadimenol seed treatment. In these conditions, early autumn infection by take-all was related to yield loss, and the reduction in take-all with seed treatment resulted in a yield increase. Bateman and Gutteridge (1) observed that under their conditions, where infection developed rapidly in the wheat in autumn and became severe by summer, seed treatment did not significantly suppress take-all. With a special seed coating (= rolling (8)) which allows a greater dose of fungicide (up to 60 or 100 g a.m./ql of seeds) Lucas and Cavelier (14) observed, in the field, a significant decrease of take-all and eye-spot, assessed even in June.

Plant breeding for resistance has never been developed but an original process based on the host capacity for cross protection by hypovirulent strains started up in France in 1973 (13, 7). Hypovirulent isolates of G. graminis tritici inoculated to wheat induce some degree of resistance against the normal aggressive strains. This induced resistance is variable among the wheat cultivars. The method of selection (using the bulk technique) consisted of screening plants, grown from seed coated with hypovirulent isolate, and selecting those with high-weight grains in the presence of a high level of take-all in the field. It appears that some of the cultivars selected in this way show a lower susceptibility to take-all and are able to produce more roots when inoculated with take-all than do some cultivars usually cropped in France (Doussinault and Lucas, unpublished).

Although we cannot propose today an efficient method for the control of take-all, there is evidence that some factors may contribute to minimise the disease incidence. So far, these factors have always been tested one by one; we present in this paper results obtained on a trial in the field where we combine some of the factors known to have consistent effects on take-all.

2. MATERIALS AND METHODS

2.1 Experimental Design

In a field sown with wheat the previous year, three factors were combined in a split plot (source of nitrogen = main plots) factorial (variety x seed treatment = subplots) experimental design (Table 1), with four replications. Each subplot had a surface area of 12 m².

TABLE 1. Factors Combined in the Field Trial.
(Sowing date: 24th October)

Factors	Levels	Codes
Source of nitrogen	Ammonium sulphate	Am
	Ammonium nitrate	Ni
Varieties	Arminda	Arm
	L193	Lig
Fungicide seed treatment	Manebe 160 g a.m./ql	M
	Triadimenol 60 g a.m./ql	T1
	Triadimenol 100 g a.m./ql	T2

2.2 Description of the Combined Factors

2.2.1 Source of nitrogen

The objective was to make the soil less susceptible to take-all by applying ammonium sulphate in place of ammonitrate which is commonly used by farmers. In France ammonitrate is usually applied in two parts, the first one at the beginning of tillering, the other at the beginning of stem extension. This will be our referencee which is compared with fertilisation with ammonium sulphate. In this case, even if the plants did not need it at this stage, we applied 40 units/ha of (NH4)2SO4 just after sowing in plots fertilised with this source of nitrogen. Then 40 units of both fertilisers were applied at the beginning of tillering, and the respective amounts applied at the beginning of stem extension were decided after a soil analysis. At the beginning of tillering, ammonium sulphate was applied 15 days before ammonitrate because of the low temperatures which slow down nitrification (Table 2).

2.2.2 Varieties

Two varieties are compared: 'Arminda' commonly sown in the western area of France and 'L193', a line issued from the plant breeding programme previously described. L193 has also the particularity of being resistant to eye-spot (VPM gene).

2.2.3 Seed treatment with fungicides

Seed pelleting (= rolling) with triadimenol was carried out by Pr. Fraselle*. The technique used (8) consisted of carrying out the

---

* Faculté d'Agronomie de Gembloux B.

TABLE 2. Fertiliser Treatments (Nitrogen kg/ha) and Fungicide Treatments Applied on the Field Experiment (dates and rates)

Date	Growth stage	Nitrogen	
		'Am plots'	'Ni plots'
31.10.85	Seed	40	0
20.01.186	Early	50	0
4.02.86	Tillering	0	50
14.04.86	Beginning of stem extension	40	60

		Fungicides All plots
5.05.86	Stem extension	Prochloraz (450 g a.m./ha)+Carbendazime (120 g a.m./ha)
17.06.86	Flowering	Propiconazole (125 g a.m./ha)+Carbendazime (150 g a.m./ha)

pelleting of seeds with a mixture of inert powders, cements and fungicides which have a systemic action. The purpose of this treatment is to obtain a protection as long as possible with higher rates of fungicide than those allowed by classical seed treatments.

2.3 Field Assessment

Foot and root diseases, and plant development, were assessed on samples taken from the field in March, April and June:

- end of March (fourth tiller emerging): number and range of tillers;
- end of April, beginning of May (1-2 nodes; sampling = 8 x 10 cm on the row per subplot): number of plants, number of stems, number of infected roots per plant, total number of roots per plant, percent of plants with sharp eye-spot, with eye-spot, aerial dry matter produced in each sample (10 cm row).
- end of June (flowering over, same sampling): number of plants, number of ears, aerial dry matter produced on 10 cm row.

For each sample, the root system of each plant was washed and examined for evidence of take-all lesions. Each plant was assigned to one of the four disease severity classes (0, 1, 2, 3) corresponding to nil, 1 to 30, 31 to 60, 61 to 100% of the root system with take-all lesions. For eye-spot and sharp eye-spot, percent of plants and stems with lesions were assessed and each stem with lesions was assigned to one of the five severity classes (0, 1, 2, 3, 4) corresponding to nil, 1 to 10, 11 to 30, 31 to 60, 61 to 100% of the stem section infected.

For each sample, and each disease, the disease intensity was calculated with the formula:

$$D.I. = \sum_{i=0}^{n} \frac{ni \times i}{\sum_{i=0}^{n} ni} \quad .$$

A disease severity may be calculated by taking only the diseased classes:

$$D.S. = \sum_{i=1}^{n} \frac{ni \times i}{\sum_{i=1}^{n} ni} \quad .$$

i = frequency of the class
ni = number of plants or stems assigned to the class i.

2.4 Soil Receptivity to Take-all Depending on Form of Nitrogen

A method has been developed to determine the receptivity to take-all of soils fertilised with ammonium sulphate or with ammonium nitrate. Soils were sampled from the field on 2.12.85, 6.03.86, 22.05.86, about 40 days after fertilisation (Table 2).

The method consists of measuring the ability of a soil to allow expression of pathogenicity by increasing rates of inoculum, in a population of susceptible host plants (16). The bioassay is carried out over 45 days (15°C, photoperiod 16h). Then plants are taken from the pots and a disease index for each soil is calculated and used to draw the curve of receptivity of the soil tested.

3. RESULTS

3.1 Development of Foot Diseases

3.1.1 April assessment

The first recording made at the beginning of stem extension showed a very low level of infection by take-all. We observed a difference between the two varieties, L193 with manebe being less infected than Arminda (Table 7). The two triadimenol seed treatments reduced incidence of the disease (significant at P = 0.05). We observed an effect of the source of nitrogen, significant only at P = 0.09 (Table 3).

Regarding other diseases, L193 appeared to be resistant to eye-spot and we observed an interesting effect of triadeimenol which reduced the percentage of infected plants (cv 'Arminda') from 48 to 10 and 11% (Table 7). Attacks by sharp eye-spot were not important but were more frequent in plots fertilised by ammonium nitrate (Table 3).

3.1.2 June assessment

The most interesting effects on take-all at this stage were those obtained with ammonium sulphate which reduced by half the frequency of diseased plants but seemed to have no effect on the severity of the attacks (Table 4) and with L193 which showed a lower disease severity. We observed no significant effect with triadimenol.

Conversely, it is in the plots fertilised with ammonium nitrate that we observed the fewest attacks by eye-spot. This observation confirmed the resistance of L193 to eye-spot and the effectiveness of the triadimenol against this disease even at this stage of the plants.

TABLE 3. Incidence of the Analysed Factors on the Main Foot Diseases of Wheat. (April assessment, beginning of stem extension). Values followed by the same letters are not significantly different at P = 0.05.

	Number of roots infected per plant take-all	Percent of plants with eye-spot	sharp eye-spot
Am	0.05 *	13	2 a
Ni	0.13 *	13	6 b
Arm	0.09	23 a	4
Lig	0.09	3 b	4
M	0.15 a	26 a	4
T1	0.06 b	7 b	5
T2	0.05 b	6 b	2

* Significant difference at P = 0.10

TABLE 4. Incidence of the Analysed Factors on the Main Foot Diseases of Wheat. (June assessment, end of flowering). The values followed by the same letters are not significantly different (P = 0.05).

	Take-all			eye-spot		sharp eye-spot	
	D.I. (0-3)	D.S. (1-3)	FREQ (%)	D.I. (0-4)	FREQ (%)	D.I. (0-4)	FREQ (%)
Am	0.20*	1.09	15 a	0.41 a	16 a	0.10	4
Ni	0.42*	1.22	32 b	0.20 b	8 b	0.12	5
Arm	0.38	1.29 a	27	0.52 a	20 a	0.06 a	3 a
Lig	0.25	1.02 b	20	0.09 b	4 b	0.16 b	6 b
M	0.35	1.26	23	0.42 a	17 a	0.14	6
T1	0.27	1.16	26	0.27 b	11 b	0.12	5
T2	0.21	1.05	21	0.23 b	9 b	0.07	3

*   Significant difference at P = 0.10
D.I. Disease Index
D.S. Disease Severity
FREQ Percent of diseased plants (take-all) or stems (others)

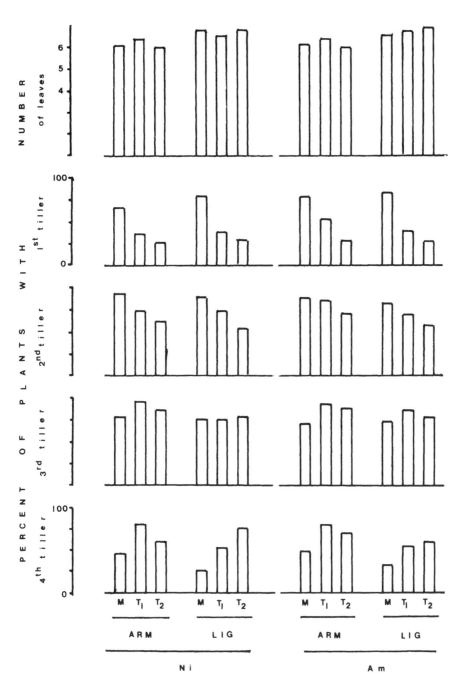

Fig. 1. Tiller sequence appearance (March assessment)

243

TABLE 5. Incidence of the Analysed Factors on Plant Growth and Development Between April and June. (Mean of eight samples of 10 cm row and four replications). Values followed by the same letters are not significantly different at P = 0.05

	Number of plants per 10 cm row		Number of stems per 10 cm row		Number of stems per plant		Aerial dry matter weight/10 cm (cg)		Aerial dry matter weight/plant (cg)	
	April	June	April	June	April	June	April	June	April	June
Am	6.1	5.7	30.5	17.2	5.1	3.0	1475	6938	247	1219
Ni	6.2	5.6	28.7	16.6	4.7	3.0	1439	6640	236	1192
Arm	6.1	5.4 a	35.4 a	17.4 a	5.9 a	3.2 a	1152 a	6490 a	195 a	1215
Lig	6.2	6.0 b	23.7 b	16.3 b	3.9 b	2.7 b	1762 b	7087 b	289 b	1197
M	6.8 a	6.0 a	30.7 a	16.3	4.5 a	2.7 a	1468	6440 a	221 a	1072 a
Tl	5.9 b	5.4 b	29.4 ab	17.2	5.0 b	3.1 b	1451	7067 b	246 ab	1251 b
T2	5.7 b	5.7 ab	28.6 b	17.1	5.1 b	3.2 b	1451	6859 b	258 b	1294 b

Sharp eye-spot remained at a low level; the differences between L193 and Arminda may be understood by a greater susceptibility of L193 or by the relation between eye-spot and sharp eye-spot shown by Cavelier et al. (3).

3.2 Plant Development

As shown in Figure 1, triadimenol inhibits the formation of the first two tillers. The formation of the tiller of the third sheath is not affected and we observed a phenomenon of of compensation, the proportion of plants with a tiller of the fourth sheath being higher in the plots with triadimenol-treated seeds. In fact, when the number of plants per unit length of row was reduced by the triadimenol treatment (April and June assessments), the number of tillers per plant was on the contrary, higher.

Thus, despite disruption of plant development during early stages, we observed, with triadimenol, an increase in the aerial dry matter produced per plant in April and June and per unit length of row in June (Tables 5 and 6).

The study of the interaction 'variety x treatment' (Tables 6 and 7) shows that it is Arminda that is mainly responsible for the increase of dry matter linked to triadimenol treatment. This leads us to think that it may be due in part to the effectiveness of control of eye-spot, L193 being resistant to this disease, and in part to the ability to produce more tillers (Tables 5 and 6) and to get a better compensation. In this way, the thousand grain weight ($P = 0.05$) and the yield ($P = 0.10$) are higher in plots where seeds are treated with triadimenol (Table 8). On the other hand, the reduction of take-all infection by ammonium sulphate does not lead to an increased yield but we saw that this form of nitrogen fertilisation has an opposite effect on eye-spot (Table 4).

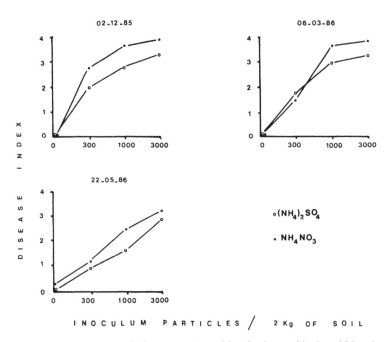

Fig. 2. **Curves of receptivity to take-all of the soil fertilised with ammonium sulphate or with ammonium nitrate on three dates**

245

TABLE 6. Significant Effects (P = 0.05) of Some Interactions Between
Factors (June assessment, end of flowering)

Weight of aerial dry matter produced (cg):

Arm x M	Arm x T2	Arm x Tl	Lig x M	Lig x T2	Lig x Tl
5959 a	6746 b	6767 b	6922 bc	6973 bc	7367 c

Average number of stems per plant:

Lig x M	Lig x Tl	Arm x M	Lig x T2	Arm x Tl	Arm x T2
2.6 a	2.7 a	2.8 a	2.9 a	3.4 b	3.5 b

Disease index of eye spot (scale 0-4):

Lig x Ni	Lig x Am	Arm x Ni	Arm x Am
0.08 a	0.10 a	0.32 b	0.73 c

Percent of plants with eye-spot:

Lig x Ni	Lig x Am	Arm x Ni	Arm x Am
4 a	4 a	13 b	28 c

TABLE 7. Significant Effects (P = 0.05) of Some Interactions Between
Factors. (April assessment, beginning of stem extension)

Average number of stems by 10 cm row:

Lig x Tl	Lig x T2	Lig x M	Arm x T2	Arm x Tl	Arm x M
23.4 a	23.9 a	23.9 a	33.3 b	35.5 c	37.5 c

Number of roots with take-all per plant:

Arm x Tl	Arm x T2	Lig x T2	Lig x M	Lig x Tl	Arm x M
0.02 a	0.04 a	0.07 a	0.08 a	0.11 a	0.22 b

Percent of plants with eye-spot:

Lig x Tl	Lig x T2	Lig x M	Arm x Tl	Arm x T2	Arm x M
2 a	3 a	4 a	10 a	11 a	48 b

TABLE 8. Incidence of the Analysed Factors on Yield and Thousand
Grain Weight. For each factor, the values followed by
the same letters are not significantly different at
P = 0.05

Factors	Nitrogen		Varieties		Fungicides		
Levels	Am	Ni	Arm	Lig	M	Tl	T2
Yield g/m²	642	649	673 a	635 b	635 *	661 *	667 *
Thousand grain weight (g)	41.3	41.1	38.5a	43.9b	40.0a	42.0b	41.5b

* Significant difference at P = 0.10

3.3 Soil Receptivity to Take-all

The curves of receptivity drawn for the three sampling dates show the beneficial effect of ammonium sulphate compared to ammonium nitrate (Figure 2). Thus, there is a good relation between the receptivity as measured by the bioassay and the differences of level of disease in the field. Some complementary studies reveal that soils which received ammonium sulphate have a higher population of fluorescent pseudomonads (16) well known for their antagonistic activity against G. graminis tritici (5, 23).

4. DISCUSSION - CONCLUSION

Generally, operations on a crop are divided into cultural practices, which aim to get an optimal growth and development of the plant population, and crop protection, mainly based on the use of chemicals, which aim to preserve the plant production potential set by the former.

In this experiment, we decided to compare factors from these two kinds of operation and to measure their impact on plant growth and development and on the foot and root diseases.

So, it appears that if the source of nitrogen does not affect the plant growth and development, it has some effect on the sanitary status, the level of take-all being lower in plots with ammonium sulphate, while the level of eye-spot is higher.

Seed treatment with triadimenol seems to delay early infection by take-all and by eye-spot. It is interesting to note that plants from triadimenol-coated seeds did not need the usual spraying carried out at the beginning of stem extension and which is directed only against eye-spot. If this fungicide has a beneficial effect on some diseases, we saw, on the other hand, that it affects plant development, particularly during the early stages. The most injurious effect is on plant density. Later, we observed a phenomenon of compensation, except for L193 which shows a low tillering.

Observations made on plant growth by measuring the aerial dry matter produced do not show any difference at the beginning of stem extension in spite of a lower plant density in plots sown with triadimenol-coated seeds. At the end of flowering, the triadimenol effect appears as an increase of the aerial dry matter produced and later as an increase of the thousand grain weight and, to some extent, as an increase in the yield. Is it an indirect effect through the control of eye-spot, or a direct one on the physiology of the plant?

The main difference between the two varieties lies in the level of resistance of L193 against eye-spot, but we noticed a lower take-all disease severity with this cultivar.

Thus, combination of triadimenol as seed coating and ammonium sulphate as source of nitrogen will reduce take-all incidence while allowing the farmer to avoid the treatment against eye-spot at the beginning of stem extension. In any case, varieties with a high level of resistance against eye-spot would be of great value and we have to check amongst those issued from the plant breeding programme of Doussinault et al. (7) to discover if there are better ones available regarding their performance in the presence of take-all.

REFERENCES

(1) BATEMAN, G.L. and GUTTERIDGE, R.J., 1984. Triadimenol seed treatment: implications for take-all. Abstract paper from 1984 British Crop Protection Conference, Pests and Diseases, 43-45.

(2)   BOCKUS, W.W., 1983. Effects of fall infection by G. graminis var. tritici and triadimenol seed treatment on severity of take-all in winter wheat. Phytopathology 73, 540-543.

(3)   CAVELIER, N., LUCAS, P. and BOULCH, G., 1985. Evolution du complexe parasitaire constitué par Rhizoctonia cerealis Van der Hoeven et Pseudocercosporella herpotrichoides (Fron) Deighton, champignons parasites de la base des tiges de céréales. Agronomie 5 (8), 693-700.

(4)   CHRISTENSEN, N.W. and BRETT, M., 1985. Chloride and liming effects on soil nitrogen form and take-all on wheat. Agron. J., 77, 157-63.

(5)   COOK, R.J. and ROVIRA, A.D., 1976. The role of bacteria in the biological control of Gaeumannomyces graminis by suppressive soils. Soil Biol. and Biochem. 8, 269-273.

(6)   DEACON, J.W., 1973. Control of the take-all fungus by grass leys in intensive cereal cropping. Pl. Path. 22, 88-94.

(7)   DOUSSINAULT, G., LEMAIRE, J.M. and LUCAS, P., 1986. Aspects génétiques de la lutte contre le piétin échaudage p.135-144. In INRA (ed.) Les Rotations Céréalières Intensives. Dix années d'études concertées INRA-ONIC-ITCF- 1973-1983.

(8)   FRASELLE, J. and SCHIFFER, B., 1982. L'enrobage des semences en tant que vecteur phytosanitaire pour une protection à long terme. Meded. Fac. Landbouwwet., Rijksuniv. Gent, 47 665-673.

(9)   GERLAGH, M., 1968. Introduction of Ophiobolus graminis into new polders and its decline. Meded. Lab. Phytopath., No.241.

(10)   HUBER, D.M., 1981. The role of nutrients and chemicals. p.317-342. In M.J.C. Asher and P.J. Shipton (ed.) Biology and Control of Take-all. Academic Press, New York.

(11)   LEMAIRE, J.M. and COPPENET, M., 1968. Influence de la succession céréalière sur les fluctuations du piétin-échaudage Ophiobolus graminis. Ann. Epiphyt, 19 (4), 584-599.

(12)   LEMAIRE, J.M., JOUAN, B., COPPENET, M., PERRATON, B and LECORRE, L., 1976. Lutte biologique contre le piétin échaudage des céréales par utilisation de souches hypoagressives d'Ophiobolus graminis. Sci. Agron. Rennes, 63-65.

(13)   LEMAIRE, J.M., DOUSSINAULT, G., LUCAS, P., PERRATON, B. and MESSAGER, A., 1982. Possibilité de sélection pour l'aptitude à la prémunition dans le cas du piétin échaudage des céréales. Gaeumannomyces graminis. Cryptog. Mycol. 3, 347-359.

(14)   LUCAS, P. and CAVELIER, N., 1985. Résultats préliminaires concernant l'efficacité du triadiménol sur le piétin échaudage et sur d'autres maladies du pied des céréales, p.125-134. In ANPP (ed.) Premières journées d'étude sur les maladies des plantes. Tome 1. 26-27 February 1985. Palais des Congrès, Versailles.

(15)   LUCAS, P., LEMAIRE, J.M., DOUSSINAULT, G., PERRATON, B., TIVOLI, B. and CARPENTIER, F., 1986. Lutte biologique contre Gaeumannomyces graminis (Sacc.) Von Arx and Olivier var. tritici Walker, agent du piétin échaudage, par l'utilisation d'une souche hypoagressive du parasite. Résultats, perspectives, p.113-126. In INRA (ed.) Les rotations céréalières intensives. Dix années d'études concertées INRA-ONIC-ITCF. 1973-1983.

(16)   LUCAS, P. and COLLET, J.M., 1986. Influence de la fertilisation azotée sur la réceptivité d'un sol au piétin échaudage, le développement de la maladie au champ et les populations de Pseudomonas fluorescents. Conférence OEPP sur les Stratégies et Applications de la Lutte Microbiologique contre les Maladies des Plantes. 27-30 October 1986, Dijon. à paraitre.

(17)   LUCAS, P. and NIGNON, M., 1987. Influence du type de sol et de ses composantes physicochimiques sur les relations entre une variété de

blé (<u>Triticum</u> <u>aestivum</u> L. var Rescler) et 2 souches, agressive et hypoagressive, de <u>Gaeumannomyces</u> <u>graminis</u> (Sacc.) Von Arx and Olivier var. <u>tritici</u> Walker, Plant and Soil, 97, 105-117.

(18) MACNISH, G.C. and SPEIJERS, J., 1982. The use of ammonium fertilisers to reduce the severity of take-all (<u>Gaeumannomyces</u> <u>graminis</u> <u>tritici</u>) on wheat, in Western Australia. Ann. Appl. Biol. 100, 83-90.

(19) SHIPTON, P.J., 1972. Take-all in spring sown cereals under continuous cultivation: disease progress and decline in relation to crop succession and nitrogen. Ann. Appl. Biol. 71, 33-46.

(20) SHIPTON, P.J., COOK, R.J. and SITTON, G.W., 1973. Occurrence and transfer of a biological factor in soil that suppresses take-all of wheat in eastern Washington. Phytopathology 63, 511-517.

(21) TIVOLI, B., LEMAIRE, J.M. and JOUAN, B., 1974. Prémunition du blé contre <u>Ophiobolus</u> <u>graminis</u> Sacc. par des souches peu agressives du même parasite. Ann. Phytopathol. 6, 395-406.

(22) TROLLDENIER, G., 1981. Influence of soil moisture, soil acidity and nitrogen source on take-all of wheat. Phytopath. Z. 102, 163-177.

(23) WELLER, D.W. and COOK, A.J., 1983. Suppression of take-all of wheat by seed treatments with fluorescent pseudomonads. Phytopathology, 73 (3), 463-469.

(24) WELLER, D.W., 1985. Proceedings of the First International Workshop on Take-all of Cereals. In C.A. Parker and A.D. Rovira (ed.) Ecology and Management of Soilborn Plant Pathogens. The American Phytopathological Society (St. Paul USA).

Contribution of seed pathology to integrated control of cereal diseases in Italy

A.Porta-Puglia
Istituto Sperimentale per la Patologia Vegetale, Rome, Italy

Summary

Several diseases of cereals can be, or are exclusively, transmitted through the seed. New pathotypes of already existing pathogens can also be spread in this way. The awareness of such problems has grown strongly in Italy in recent years and now several institutions are concerned with the science of seed pathology and related technology. Cooperative investigations are being carried out among such institutions. Surveys conducted on commercial seed samples of barley, wheat, maize, oats and rice, indicated frequent occurrence of Drechslera and Fusarium. Assessments of losses were performed in order to correlate the percentage of seed infection with yield losses. The efficiency of treatments on seeds with diverse levels of infection was also investigated to establish thresholds for appropriate disease management.

Certification schemes have been implemented. Through field inspections, laboratory seed health testing, treatment of pre-basic and basic seeds, the health condition of some cereal seeds is gradually improving, resulting in a reduced need for treatment of crops. A lot remains to be done concerning the choice of more suitable seed multiplication areas. In conclusion, seed pathology is contributing and is likely to contribute more in the future, to integrated control of cereal diseases in Italy.

1. INTRODUCTION

The transmission of cereal diseases through the seed is quite common. Considering only fungi, over 40 pathogenic species were recorded in the world as being seed-transmitted in both maize and rice, nearly 30 on wheat and just a few less in barley (28). Most of these pathogens are present in Italy, causing diseases which reduce either yield or commercial value of the seeds. Some seed-borne fungi can produce mycotoxins which are harmful for man and cattle if present in food or feed. Seed transmission is a powerful means of pathogen dissemination. Diverse pathotypes of widespread species of fungi can easily colonise new areas in this way.

Phytopathological problems related to the seed have been well known in Italy for many years and good pioneering work was performed in the past (1, 7, 9). Nevertheless, the practical impact at farm level is a relatively recent issue. A paper summarising the situation for grain cereals and forage crops was published in 1981 (29). Nowadays several institutions are involved in activities concerning seed-borne pathogens. Seed pathology is taught, and related researches are conducted, in several universities. Experimental institutes of the Ministry of Agriculture are working on some seed-transmitted diseases in the framework of special projects, one of which

251

1) Alpine (2 sub-regions)
2) Continental planes
   (2 sub-regions)
3) Maritime-tyrrhenian
4) Maritime-adriatic
5) Apenninic
6) Insular

Fig. 1. Climatic regions in Italy.

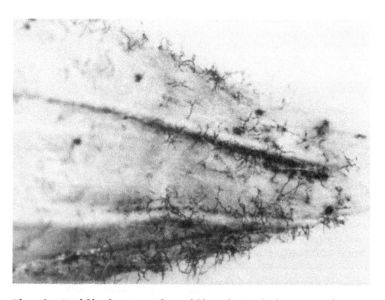

Fig. 2. Conidiophores and conidia of <u>Drechslera graminea</u> on
a barley seed (after incubation on blotting paper).

concerns barley and wheat. The National Research Council (CNR) devoted a specific institute to the study of toxin-producing fungi and the official agency for seed certification (ENSE) is attaching greater importance to seed health testing. These institutions are also involved in cooperative research projects. Moreover, seed-producing firms and farmers show an increased interest in preventing seed-transmitted diseases.

Fungicide treatments are not commonly applied on cereals in Italy but the situation could quickly change unless alternative control measures are proposed in order to reduce losses. The production of healthy commercial seeds could greatly contribute to prevent the massive application of chemicals in the near future.

2. SOME IMPORTANT SEED-BORNE PATHOGENS OF CEREALS IN ITALY

Italy is characterised by largely heterogeneous climatic conditions (Figure 1) which have diverse effects on the epidemiology of plant diseases of cereal crops, including seed-borne diseases. Cereals are grown all over the country (Table 1) and they meet environmental conditions so disparate that important diseases in one area could be negligible elsewhere. Nevertheless, taking into account the estimated economical impact in the field and the observations on the seeds in the laboratory, some outstanding pathogens can be mentioned.

TABLE 1. Areas and Yields of Grain Cereals in Italy (Sources: ISTAT and ENR, 1985)

Crop	Area (ha)	Yield (t) ha^{-1}	Total
Maize	934 770	6.9	6 439 800
Bread wheat	1 293 633	3.6	4 698 900
Durum wheat	1 738 824	2.2	3 903 100
Barley	468 412	3.5	1 642 700
Rice	187 187	6.4	1 203 000
Oats	184 237	2.1	388 740

2.1 Wheat

On wheat (both durum and bread), species of Fusarium are common. F. nivale (Fr.) Ces. F. moniliforme Sheld., F. culmorum (Smith) Sacc. and F. graminearum Shabe are frequently observed on seed and are isolated from root and foot rotted plants (2, 5, 16, 22, 24). Septoria nodorum (Berk.) Berk. is also widespread (6, 13, 14) and seed infection up to 17% was recently recorded on samples of durum wheat from northern Italy (Montorsi, personal information). Drechslera spp. were also recorded in the field (10) and observed on seed samples in the laboratory of our institute.

2.2 Barley

Drechslera graminea (Rabenh. ex Schlecht.) Shoemaker and Ustilago nuda (Jens.) Rostr. are present in several growing areas and they were observed in commercial seeds (27, 35) (Figure 2). The former caused severe outbreaks of stripe disease when organo-mercury fungicides were banned. The losses were particularly relevant in the Po Valley, where climatic conditions are very favourable to the transfer of the pathogen, both from infected seeds to seedlings, and from infected plants to the new generation seeds. In areas so congenial to the disease, if the seeds are not treated, a low initial level of seed infection can lead to a seed infection up to 60% within a few generations (23).

2.3 Maize and Sorghum

Stalk rot is widespread in most areas where maize is grown. In seasons when the conditions are favourable to the pathogens, plants may die before reaching full maturity or lodging may occur. The disease is attributed to a complex of fungi, among which Fusarium and Acremonium spp. play a major role (11). F. moniliforme, A. strictum Gams and A. sclerotigenum (F. et R. Moreau ex Val.) Gams were recorded on maize kernels (30).

F. moniliforme is also pathogenic to sorghum, causing root and stalk rot (4).

2.4 Rice

On rice, Pyricularia oryzae Cav. has been a cause of severe yield reduction for a long time. More recently, Drechslera oryzae (Breda de Hann) Subram. et Jain caused severe epidemics, particularly in the most important growing areas of the north west. Seeds produced in these areas in 1983 and 1984 were in some cases so heavily infected that they failed to reach the minimum percentage of germination required for certification. In fact, a high percentage of seedlings were killed because of fungal attacks during germination (3). Several other pathogens, including F. moniliforme, were detected on seeds (31).

2.5 Oats

Drechslera avenae (Eidam) Sharif was recorded on commercial seed samples, produced during the period 1977-1981, tested at our institute. Seed contamination was high and reached 100% in some samples (18). The fungus was mostly external, although in some cases even the embryo was infected (26).

3. WHAT HAS BEEN ACHIEVED?

Surveys of diseases, including seed transmitted diseases, are regularly conducted in the fields in several locations on the most widespread or promising cultivars and the results are published every year in a journal which has a large distribution among both technicians and farmers (8, 20). These, and other data concerning seed-transmitted diseases, are transmitted to the farmers through the national television service 'Videotel'.

Seed health testing has been conducted for many years at the Experimental Institute for Plant Pathology, in several universities and, more recently, by the Seed Certification Agency.

A special committee was appointed to propose up-to-date methods for seed testing, including seed health. The recommendations of the committee were submitted to the competent authorities and they are expected to be included in the official regulations within a short period. These methods should enable the two official seed-testing laboratories to perform routine seed-health testing.

Several investigations were carried out on chemical seed treatments, both by public institutes and pesticide producers (12, 15, 21, 33, 34).

Losses were assessed for leaf stripe disease of barley, in order to correlate the percentage of seed infection with yield reduction. In a cooperative research carried out by the Experimental Institute for Plant Pathology, the Experimental Institute for Cereal Crops and the University of Pisa for three years in three locations, it was shown that seed infection level over 14% causes remarkable yield losses (25). The effect of seed treatments on seeds infected to various degrees by D. graminea was also investigated. It was observed that standard chemical treatments fail to compensate fully for yield losses when the seed infection is over 30% (Delogu et al., unpublished).

As a result of such studies, through the regulatory action of the Ministry of Agriculture, treatment of infected seed lots has become widespread. Chemicals were applied on a large amount of commercial seed lots of barley (in north western Italy 74% of 8800 t of certified seeds were treated in 1985) and rice (13 400 t in 1984; 11 200 t in 1985).

Another contribution of seed pathologists to integrated control of cereal diseases could derive from recent studies on fungal antagonists to be applied as a seed dressing against seed-borne and/or soil-borne pathogens. The first promising results have already been published (32, 36).

The exchange of information among researchers dealing with seed pathology has increased. Foreign scientists were invited to visit Italian research institutes, such as, for instance, Dr. P. Neergaard, former Director of the Danish Government Institute of Seed Pathology for Developing Countries, Copenhagen, Denmark. Dr. Neergaard also contributed a report to a meeting held in Rome in 1981 (19). Italian researchers were sent to important institutions dealing with seed pathogens in Europe and in the United States of America and brought fresh impetus to the activities.

4. WHAT REMAINS TO BE DONE

Although research and application of seed pathology has greatly progressed in recent years, a lot remains to be done in order to assure a significant contribution of this branch of science and its related technologies to the application of integrated control of cereal diseases in Italy. Namely, there is a need for further studies on important seed-transmitted diseases aimed at defining thresholds for treatments of commercial seeds. The ideal approach, in order to limit chemical treatments to the actual needs, would be the establishment of a double threshold system: (i) a first level of seed infection or contamination below which no treatment is needed, (ii) a second level above which the seed should be rejected because the treatment is unable to restore the yield potential of the healthy seed. The seed lots ranged in between should be treated before commercialisation. Stricter requirements should be applied to pre-basic and basic seeds. Moreover, seed health testing should become a routine procedure under certification schemes, which still rely too much upon mere field inspections and laboratory tests for germination and vigour of seeds.

A greater effort is needed to shift seed production towards geographical areas less favourable to diseases. Most cereal seed production is concentrated in northern and central Italy (17), where infrastructures are the best and climatic conditions tend to favour seed infections.

5. CONCLUSIONS

Seed pathology has reached a good knowledge of many aspects of seed-transmitted diseases and, in recent years, has had some practical impact. Nevertheless, more knowledge is needed concerning the correlation between seed infection and yield losses in order to establish treatment or refusal thresholds for the most important pathogens.

Certification schemes should be integrated by routine seed health testing and the areas for seed multiplication should be chosen giving more consideration to the aspect of seed health. At the same time, greater care should be taken to prevent introduction of pathogens from abroad, through appropriate seed inspections.

From all these actions we can expect a significant contribution of seed pathology to integrated control of cereal diseases. Crops obtained from healthier seeds will need fewer treatments with positive effects both on the human diet and the environment.

REFERENCES

(1)   BALDACCI, E., 1956. Malattie del riso trasmesse per seme. Sementi elette, II (2), 56-60.
(2)   BOTTALICO, A. and PIGLIONICA, V., 1977. La sanità delle cariossidi di frumento e di mais con particolare riferimento alla presenza in esse di specie patogene di Fusarium. Rivista di Agronomia, 11, 146-152.
(3)   BUFFA, G. and MERISIO, G., 1985. Concia delle sementi di Riso in Italia. Sementi elette, 31 (1-2), 23-26.
(4)   CAPPELLI, C., TOSI, L. and ZAZZERINI, A., 1986. Fusarium moniliforme e Sclerotium bataticola agenti del marciume delle radici e del culmo del Sorgo. L'Informatore agrario, 42 (15), 61-64.
(5)   CARIDDI, C., 1982. Infezioni di Fusarium nivale (Fr.) Ces. in cariossidi di Frumento e moria dei seminati. Informatore fitopatologico, 32 (1), 59-60.
(6)   CASTELLANI, E. and GERMANO, G., 1976. Le Stagonospore graminicole. Annali Fac. Agraria Univ. Torino, 10, 1-135.
(7)   CHIAPPELLI, R., 1933. Malattie e nemici del riso. Quaderni Stazione sperim. Risicoltura, Vercelli. Serie I: Oryza sativa, 26 pp.
(8)   CORAZZA, L. and CHILOSI, G., 1986. Comportamento in campo dei varietà di Orzo nei confronti dei principali parassiti fungini nel 1985-86. L'Informatore agrario, 42 (34), 51-55.
(9)   CORBETTA, G., 1956. Un agente patogeno del Riso: Fusarium moniliforme Sh. Il Riso, 5 (9), 5-8.
(10)   DEL VESCOVO, M., 1962. Contributo alla conoscenza di alcune 'elmintosporiosi' di Graminacee spontanee e coltivate nella Regione appulo-lucana. Annali Fac. Agraria Univ. Bari, 16, 137-159.
(11)   D'ERCOLE, N., 1980. Fusariosi del Mais e del Sorgo. II. Incidenza della malattia sulla germinabilità dei semi. Consiglio Nazionale delle Ricerche. La Difesa dei Cereali, Atti Giornate fitopatologiche. Suppl. 3, 39-43.
(12)   FORMIGONI, A., et al., 1980. Concia del seme e trattamenti in vegetazione per la protezione dell'Orzo dalle malattie crittogamiche. Atti Giornate fitopatologiche, 2, 323-329.
(13)   FRISULLO, S., 1983. Parassiti fungini delle piante nell'Italia meridionale. III. Phaeosphaeria nodorum (Müll.) Hed. su Frumento duro. Phytopathologia mediterranea, 22, 194-198.
(14)   FRISULLO, S. and NALLI, R., 1982. Presenza di Septoria nodorum (Berk.) Berk. nell'Italia centrale e meridionale. Giornate Internazionali sul Grano Duro. Industria e Ricerca scientifica, Foggia, 3-4 May. Monografia di Genetica agraria, 305-307.
(15)   FRISULLO, S. and PIGLIONICA, V., 1975. Prove di lotta contro Helminthosporium gramineum Rabh. su Orzo. Atti Giornate fitopatologiche, 719-720.

(16) INNOCENTI, G., 1981. Ulteriori osservazioni sul 'mal del piede' del Frumento tenero. Atti Convegno 'La Difesa dei Cereali nell'ambito dei progetti finalizzati del C.N.R.'. Ancona, 10-11 December, 67-70.

(17) LOVATO, A. and MONTANARI, M., 1977. Zone di produzione di seme in Italia. Rivista di Agronomia, 11, 78-89.

(18) MONTORIS, F. and PORTA-PUGLIA, A., 1983. Importanza fitopatologica di Pyrenophora avenae Ito et Kuribay. su seme di Avena e problemi relativi, in Italia. Sementi elette, 29 (4), 13-16.

(19) NEERGAARD, P., 1981. Seed-borne diseases in European, especially Mediterranean crops. Problems and importance. Annali dell'Istituto Sperimentale per la Patologia Vegetale, Roma, 8, 5-26 (publ. 1982). Italian translation in: Informatore fitopatologico, 31 (12), 5-15.

(20) PASQUINI, M., et al., 1986. Prove epidemiologiche 1985-86: comportamento in campo rispetto alle malattie di Frumenti duri e teneri. L'Informatore agrario, 42 (39), 77-84.

(21) PEZZALI, M. and PORTA-PUGLIA, A., 1983. Trattamento delle smenti di Avena contro le infezioni primarie da Pyrenophora avenae Ito et Kuribay. Informatore fitopatologico, 33 (12), 47-49.

(22) PIGLIONICA, V., 1975. Il mal del piede nel Frumento. Italia agricola, 112 (1), 114-120.

(23) PIGLIONICA, V., CARIDDI, C. and FRISULLO, S., 1979. Helminthosporium gramineum e Ustilago nuda, due parassiti dell'Orzo (Hordium vulgare) oggi facilmente controllabili. La Difesa delle Piante, 2 (1), 5-14.

(24) PIGLIONICA, V., FRISULLO, S. and SNYDER, W.C., 1974. Osservazioni sul mal del piede del Grano duro. Italia agricola, 111 (6), 121-123.

(25) PORTA-PUGLIA, A., DELOGU, G. and VANNACI, G., 1986. Pyrenophora graminea on winter barley seed: effect on disease incidence and yield losses. Journal of Phytopathology, 117, 26-33.

(26) PORTA-PUGLIA, A. and MONTORSI, F., 1982. Localizzazione di Pyrenophora avenae Ito et Kuribay. nel 'seme' di Avena. Informatore fitopatologico, 32 (1), 35-38.

(27) PORTA-PUGLIA, A., SAPONARO, A. and MONTORSI, F., 1982. Alcune osservazioni sulla micoflora della semente di Orzo in Italia. Informatore fitopatologico, 32 (1), 39-45.

(28) RICHARDSON, M.J., 1979. An annotated list of seed-borne diseases. Commonwealth Mycological Institute, Kew, and International Seed Testing Association, Zurich, 320 pp.

(29) SAPONARO, A., 1981. Aspetti e problemi fitopatologici delle sementi dei Cereali e delle Foraggere. Informatore fitopatologico, 31 (12), 17-30.

(30) SAPONARO, A. and MONTORSI, F., 1979. Acremonium strictum Gams e Acremonium sclerotigenum (F. et R. Moreau ex Valenta) Gams su cariossidi di Mais. Phytopathologia mediterranea, 18, 221-224.

(31) SAPONARO, A., PORTA-PUGLIA, A. and MONTORSI, F., 1986. Alcuni importanti funghi patogeni del Riso trasmissibili per seme. Informatore fitopatologico, 36 (1), 40-43.

(32) SCARAMUZZI, G., VANNACCI, G. and PECCHIA, S., 1986. Valutazione dell'attività antagonista di microrganismi applicati al seme nei confronti di Rhizoctonia solani Kühn. Proceedings of the meeting on 'Advanced Biotechnologies and Agriculture', Bologna, 29th May/1st June, 1986.

(33) TANO, F. and BISIACH, M., 1982. Influenza della concia dell'Orzo (Hordeum vulgare L.) sul controllo di Helminthosporium gramineum Rabh. e di Ustilago spp. e su alcuni parametri agronomici. Sementi elette, 28 (6), 3-16.

(34) VANNACCI, G., 1981. Basi scientifiche e indicazioni pratiche sulla concia del seme nell'Orzo. Sementi elette, 27 (4), 3-5.

(35) VANNACCI, G., 1981. Indagini sperimentali per la migliore metodologia
      per il reperimento di due importanti parassiti, Drechslera graminea e
      Drechslera sorokiniana, su cariossidi di Orzo. Sementi elette, 27 (4),
      15-19.
(36) VANNACCI, G. and PECCHIA, S., 1986. Evaluation of biological seed
      treatment for controlling seed-borne inoculum of Drechslera sorkiniana
      on barley. 38th International Symposium on Crop Protection, Gent, 6th
      May.

Development of a quantitative method for assessment of *Polymyxa graminis* Led. inoculum potential in soils

H.Maraite, J.P.Goffart & V.Bastin
Laboratoire de Phytopathologie, Université Catholique de Louvain, Belgium

Summary

In the framework of a research programme on integrated control of *Polymyxa graminis* Led., vector of the barley yellow mosaic virus (BYMV) on winter barley in Western Europe, the plant-bait technique is optimised and standardised for identification of factors affecting inoculum potential and multiplication of this obligate parasite.

Barley seedlings are grown in glass tubes on a soil/sand mixture watered with half-strength Hoagland solution, at saturation during the second and third week and sparsely otherwise. The assay is performed in controlled environmental cabinets with a photoperiod of 12 h at 13 000 lux. Root infection is assessed, with the help of a microscope, in roots boiled in lactophenol with cotton blue, or in homogenates of roots treated with KOH.

The highest levels of thalli (zoosporangial or cystosoral) and of cystosori are observed after four and seven weeks, respectively, when the plants are grown at night/day temperatures of 15-20°C, compared to 10-15, 20-25 or 25-30°C.

Serial dilutions of infested soil with sterilised sand permit comparison of inoculum potential in soils.

1. INTRODUCTION

The plasmodiophoraceous fungus, *Polymyxa graminis* Led., is an intracellular obligate root parasite of Gramineae and a vector of several soil-borne viruses on cereals. The barley yellow mosaic virus (BYMV) has gained economical importance in the last ten years in several European countries: German Federal Republic (7), France (10), United Kingdom (5), German Democratic Republic (15), Netherlands (11). In Belgium, this virus was detected in 1977, causing up to 50% yield reduction on susceptible cultivars (13).

Observations in Japan and in Germany have shown that an eight to ten year crop rotation without barley is not enough to eliminate the virus from the soil (6). Control relies nearly exclusively on the use of cultivars, resistant or tolerant to the virus (14, 6). Different types of BYMV are reported (8, 3), but up to now no clear resistance breakdown has been noticed. Comparing the experience with other diseases, it appears, however, worthwhile to explore the possibility of integrated control, with other methods acting on the fungal vector and presumed field reservoir of the virus. Kusaba and Toyama (9) observed a decrease of disease incidence by a soil drench with the fungicide sodium p-dimethylaminobenzene diazosulfonate (Dexon). D'Ambra and Mutto (2) reported, on the other hand, degradation *in vitro* of cystosori of *Polymyxa betae* by *Trichoderma harzianum*.

In order to identify factors associated with differences of inoculum potential of P. graminis in Belgium, and in view of the analysis of the effect of farming practices and biological or chemical treatments on survival of cystosori and infection of barley roots, we are optimising and standardising the plant-bait bioassay (12, 1).

This paper presents the state of progress in developing a quantitative method for assessment of inoculum potential of P. graminis.

2. MATERIALS AND METHODS

2.1 Experimental Devices

The winter barley cultivar 'Gerbel', susceptible to BYMV in Belgium, (14) is used. Before planting, seeds are pre-germinated for one day on wet filter paper in Petri dishes.

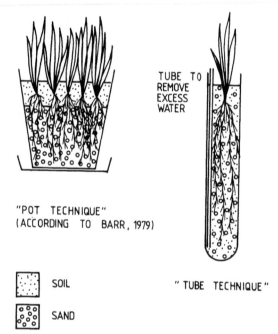

"POT TECHNIQUE"
(ACCORDING TO BARR, 1979)

TUBE TO
REMOVE
EXCESS
WATER

" TUBE TECHNIQUE"

SOIL

SAND

Fig. 1. Experimental devices for the assessment of inoculum potential of Polymyxa graminis in soils by a barley bioassay.

The 'pot technique' is an adaptation of the plant-bait method used by Barr (1). Eight cm Ø polyethylene pots are filled with autoclaved river sand (grain size between 0.5 and 2 mm) up to 5 cm deep and overlayered with 2 cm of infested dry soil, homogenised by sieving through a screen (2 mm aperture). A saucer prevents contacts between pots and allows, if necessary, soil humidity to be maintained at saturation (Figure 1). Five seeds are placed at 1 cm depth in the soil. Humidity of the pots is checked at least once a day and adjusted alternately with tap water and half-strength Hoagland solution.

For the 'tube-technique', glass tubes 2.5 mm Ø, 15 cm height) are filled with 50 ml of soil or soil/sand mixture. One seed is sown per tube.

Plants are watered as in the pot technique. During the period of high watering, the soil surface is kept covered with water. Excess of water is removed by vacuum through a 5 mm Ø tube plunging to the bottom of the culture tube.

Unless otherwise stated, the plants are grown in controlled environmental cabinets at 15°C during the night and 20°C during the day (12 h), with 13 000 lux of light.

2.2 Assessment of Root Infection

Soil and sand particles are washed from the roots by running water.

For the detection of root infection and determination of the percentage of infected roots, fungal structures are coloured by boiling the whole root system in lactophenol with 0.05% cotton blue for 2 min, followed by a storage in lactophenol for one day before observation. The roots are disentangled under water in a tray and surveyed with a stereomicroscope at a magnification of 60 x. In case of doubt, sections are analysed at higher magnifications under a microscope.

For counting the number of cystosori in the roots, the root systems of several plants are cut into 3 to 5 mm pieces, mixed with water and filtered on Whatman Nr 1 filter paper in a Buchner funnel. The fresh weight of the roots is determined and 500 mg samples macerated for 3 h at 25°C in 20 ml of 10% KOH in water. The macerate is homogenised for 1 min with a Virtis OmniMixer at 25 000 rpm, centrifuged for 20 min at 7500 rpm and the precipitate resuspended in 4 ml lactophenol. The number of cystosori in the suspension is determined with the help of a Fuchs-Rosenthal counting chamber and the number of cystosori/g fresh weight of roots calculated.

3. RESULTS AND DISCUSSION

3.1 Importance of the Watering Conditions for Completion of the Life-Cycle of P. Graminis

Free water appears esential for cystosori germination, infection of the roots by primary zoospores leading to the zoosporangial thallus, release of secondary zoospores and secondary infections leading eventually to formation of cystosoral plasmodia (12, 16, 1). However, neither the duration of the various steps of the life-cycle, the number of sporangial cycles, the factors initiating cystosoral formation, nor the multiplication factor are known. In order to provide first elements for optimising the plant-bait method, the effect of various watering conditions on evolution of root infection was analysed in a preliminary experiment with the pot technique (Table 1). Plants were regularly harvested up to 40 days after planting and the roots analysed for the presence of thalli and cystosori of P. graminis, as well as of other fungal infections.

Zoosporangial thalli corresponding to the description of P. graminis were observed by the seventh day for all watering conditions. The first cystosori were detected by the 23rd day in roots of plants with high watering during the first two weeks and the 33rd day for those with high watering between the eighth and the 21st day. On the 40th day, cystosori were detected neither in the roots of plants grown only for one week on soil saturated with water, nor in those watered just enough to permit plant growth. Thalli were, however, detected in roots for all watering conditions, except for plants grown on autoclaved sand. Frequently root infections by Olpidium brassicae (Wor.) Dang, Pythium spp., Phialophora radicicola Cain were also observed. Occasionally, Rhizophydium graminis Led. and Lagena radicicola Vanterpool and Led. were detected. The young thalli of O. brassicae were often difficult to distinguish under the stereomicroscope

TABLE 1. Effect of Watering Conditions on The Detection of Thalli and Cystosori of Polymyxa Graminis in Roots of Barley Seedlings Grown in Pots on Naturally Infested Soil (5 plants/condition and harvest).

Harvest	Watering condition (1) day						Observation of	
days after planting	1 to 7	8 to 14	15 to 21	22 to 28	29 to 35	36 to 40	thalli	cystosori
7 to 19	L	L	L	L	L	L	X	0
	H	H	L	L	L	L	X	0
	L	H	H	L	L	L	X	0
	L	H	L	L	L	L	X	0
23	L	L	L	L	L	L	X	0
	H	H	L	L	L	L	X	X
	L	H	H	L	L	L	X	0
	L	H	L	L	L	L	X	0
33 to 40	L	L	L	L	L	L	X	0
	H	H	L	L	L	L	X	X
	L	H	H	L	L	L	X	X
	L	H	L	L	L	L	X	0

(1) L = low, just enough to allow growth of the plant, without water remaining in the saucer.
    H = high, soil saturated with water during the whole period.
    X = presence
    0 = absence

from undifferentiated plasmodia of P. graminis. This reduces the value of assessment of P. graminis inoculum potential in the soil on the basis of the number of plasmodia present in the roots during the first weeks of growth. The minimum period of high watering for cystosore initiation is between one and two weeks. The optimum watering period allowing only one secondary zoospore cycle, will be determined after a study of factors affecting cystospore germination and cystosoral plasmodium initiation.

Young barley seedlings suffer in soil saturated with water and infections by other aquatic fungi are severe during this period. It appears preferable to start high watering only one week after planting.

3.2   Temperature Optimum

Ledingham (12) reported heaviest infections of wheat by P. graminis in the greenhouse at about 15 or 18°C. Rao (16) mentioned, on the other hand, that the highest soil-borne wheat mosaic virus transmission occurred when inoculum of rootlets with cystosori of P. graminis were incubated in water at 20°C for four days, compared to 17 and 28°C, before addint wheat seedling

262

for five days at 17°C and their transplantation in pots for growth at 17°C. Precise data are, however, lacking.

In order to determine optimum temperature conditions for rapid infection and cystosorus formation, barley seedlings were grown by the pot technique on naturally-infested soil (St. Amand) at four temperature conditions (Figure 2) under 8000 lux light. High watering was applied during the second and third week. Roots were harvested every week between the third and seventh week and the percentage of primary and secondary roots infected by plasmodia and/or cystosori were determined.

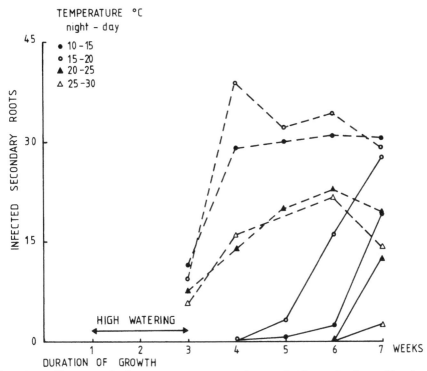

Fig. 2. **Influence of the temperature on the evolution of plasmodia (--) and cystosori (-) of P. graminis in secondary roots of barley seedlings grown on naturally infested soils** (Means of the analysis of two pots with five plants/temperature and duration).

At the 15-20°C night/day temperature condition, the first cystosori were detected in the primary roots one week after the end of the high watering period and on the secondary roots one week later (Figure 2), confirming the results of the watering experiment. Cystosorus formation was significantly delayed and lower at the other temperature conditions. Similar data were obtained in a repetition of this experiment running for a longer period; percentage of roots showing cystosori were similar after seven or ten weeks. Contrary to the data of the first experiment, presented in Figure 2, the percentages of roots with plasmodia were highest in the second experiment at the end of the high watering period, but decreased up to the fourth or fifth week and stayed at the same level afterwards.

With the pot technique, the analysis of the root system is hampered by interweaving of the roots. Percentage of infected roots varied, on the other

263

hand, by a factor of two among plants under the same growing conditions. For these reasons the pot technique was replaced in later studies by the more standardisable tube technique.

3.3 Storage Conditions

The assessment of inoculum potential in the soil can be affected by differences in degree of maturation of the fungal propagules. Ledingham (12) used soils that had been frozen or dried for a few months.

In order to evaluate the effect of storage conditions on subsequent infection of the bait plants, soil samples were taken on August 14, 1985, in two winter barley fields in areas with BYMV symptoms in March 1985 and in areas without symptoms. One part of the soils was humidified with tap water, the other air dried, before storage in clay pots in the dark at 4°C or 18°C for ten months. After one month of storage, barley seedlings were grown for seven weeks on soil samples by the pot technique and after ten months by the tube technique; the pot technique showed unacceptably large variations among replicates. The effect on the expression of the inoculum potential of a heat treatment (60°C for 20 h), just before planting was analysed for the same soils. High watering was applied between the second and third week.

The heat treatment of wet soil strongly decreased the inoculum potential; plants grown on five of the eight treated soil samples showed no infection. On dry soil, the heat treatment apparently did not affect inoculum potential (Table 2). Root infection was higher on soils dried before storage for ten months than on soil kept humid. Storage of dry soil at 4°C was not better than at room temperature.

TABLE 2. Effect of Storage Conditions and of a Heat Treatment (HT) of the Soil on the Amount of Cystosori in Roots of Barley Plants, After Seven Weeks of Growth in Tubes for Soil Samples Taken from an Area with (+) or Without (-) Symptoms of BYMV in Two Fields in 1985.

Storage conditions for ten months	Origin of the soil			
	Corroy		Hannut	
	+ BYMV	- BYMV	+ BYMV	- BYMV
	Number of cystosori/g fresh weight (1)			
Wet 4°C	4 000	4 700	6 000	7 300
Dry 4°C	24 000	5 300	27 000	30 000
4°C + HT (2)	39 000	1 300	23 000	29 000
18°C	38 000	3 300	27 000	13 000
18°C + HT	26 000	4 700	21 000	25 000

(1) Counted in root homogenates after KOH treatment, mean of three replicates of five plants.
(2) Heat treatment: incubation at 60°C for 20 h just before planting.

3.4 Assessment of Inoculum Potential by Dilution of the Soil

Because the multiplication rate of P. graminis, from cystospore germination to cystosorus formation, is not known and may depend on physical, chemical and biological soil factors, counting of cystosori in roots of the bait plant may not, by itself, be accurate enough for evaluation of inoculum potential in various soils. For this reason, the decrease of percentage of infected plants grown singly by the tube technique on serial dilutions of infested soil with sand was analysed. Soil samples taken in May 1986 in two winter barley fields with BYMV symptoms were stepwise diluted by a factor of 5 (volume : volume) by mixing with autoclaved sand in a Retch Universal Mixer. Tubes were filled with 50 ml of the infested soil, of autoclaved sand or of the various soil/sand dilutions. High watering was applied during the second and third weeks.

Analysis of the root system after seven weeks revealed for both soils a decrease of the percentage of infected plants only at concentrations of infested soil in the mixture lower than 4%, the dilution end point being around 0.01% (Figure 3). The most probable number of infection units (IU) calculated according to the method described by Gerard (4), was estimated at 1.2 and 1.4 IU/g dry soil for the soils of Hannut and Ambresin, respectively.

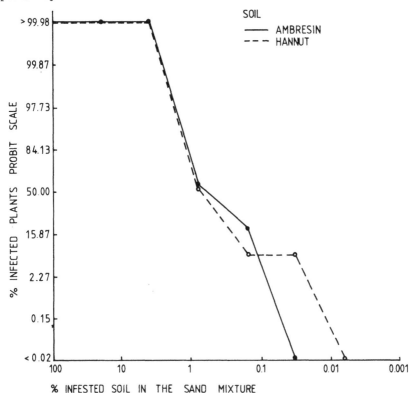

Fig. 3. Evaluation of the inoculum potential of Polymyxa graminis in soils from two fields showing BYMV symptoms, by growing barley seedlings singly in tubes containing serial dilutions of the soil with autoclaved sand. Analysis of the presence of cystosori in the roots after seven weeks of growth, 15 plants/dilution.

The influence of physical and chemical soil factors on expression of the inoculum potential is minimised in these conditions by the high dilution in sand and the watering with Hoagland solution. The decrease of the number of cystosori formed in the infected roots on the serial dilutions of the soil, is currently being analysed. We are, on the other hand, still working to improve the method by synchronising germination of cystospores, and subsequently optimising the period of high watering, in order to avoid overlapping zoosporangial cycles. The method will then be calibrated by soil infestation with known amounts of cystosori.

A systematic survey in various fields in Belgium is planned for 1987 in order to identify factors affecting inoculum potential and multiplication of P. graminis, as a basis for the elaboration of integrated control measures against BYMV.

ACKNOWLEDGEMENTS

This work is supported by the CEC Agricultural Research Programme on Energy in Agriculture, Contract VI/4741/84-F 1410.

REFERENCES

(1)   BARR, D.J.S., 1979. Morphology and host range of Polymyxa graminis, Polymyxa betae, and Ligniera pilorum from Ontario and some other areas. Canadian Journal of Plant Pathology 1, 85-94.

(2)   D'AMBRA, V. and MUTTO, S., 1986. Parassitismo di Trichoderma harzianum su cistosori di Polymyxa betae. J. Phytopathology 115, 61-72.

(3)   EHLERS, U. and PAUL, H-L., 1986. Characterization of the coat proteins of different types of barley yellow mosaic virus by polyacrylamide gel electrophoresis and electro-blot immunoassay. J. Phytopathology 115, 194-304.

(4)   GERARD, G., 1983. Eléments de statistique. Ciaco Edition, Louvain-la-Neuve, Belgium.

(5)   HILL, S.A. and EVANS, E.J., 1980. Barley yellow mosaic virus. Pl. Path. 29, 197-199.

(6)   HUTH, W., 1984. Die Gelbmosaikvirose der Gerste in der Bundesrepublik Deutschland -Beobachtungen seit 1978. Nachrichtenbl. Deutsch. Pflanzenschutzd. 31, 49-55.

(7)   HUTH, W. and LESEMANN, D.-E., 1978. Eine für die Bundesrepublik Deutschland neue Virose an Wintergerste. Nachtrichtenbl. Deutsch. Pflanzenschutzd. 30, 184-185.

(8)   HUTH, W., LESEMANN, D.-E. and PAUL, H.-L., 1984. Barley yellow mosaic virus: purification, electron microscopy, serology, and other properties of two types of the virus. Phytopath. Z. 111, 37-54.

(9)   KUSABA, T. and TOYAMA, A., 1970. Transmission of soil-borne barley yellow mosaic virus. 1. Infectivity of diseased root washings. Ann. Phytopath. Soc. Japan 36: 214-222.

(10)   LAPIERRE, H., 1980. Les virus des céréalès à paille. Phytoma-Défense des cultures 321, 34-38.

(11)   LANGENBERG, W.G. and VAN DER WAL, D., 1986. Identification of barley yellow mosaic virus by immunoelectron microscopy in barley but not in Polymyxa graminis or Lagena radicicola. Neth. J. Pl. Path. 92, 133-136.

(12)   LEDINGHAM, G.A., 1939. Studies on Polymyxa graminis, n. gen. n. sp., a plasmodiophoraceous root parasite of wheat. Canadian Journal of Research 17, Sec. C., 38-51.

(13)   MAROQUIN, C., CAVELIER, M. and RASSEL, A., 1982. Premières observations sur le virus de la mosaïque jaune de l'orge en Belgique, Bull. Rech. Agron. Gembloux 17, 157-176.

(14)   MAROQUIN, C., CAVELIER, M. and CROHAIN, A., 1983. Sensibilité relative de onze variétés d'orge d'hiver au virus de la mosaïque jaune de l'orge. Med. Fac. Landbouw. Rijksuniv. Gent 48/3, 801-806.

(15)   PROESELER, G., STANARIUS, A. and KÜHNE, T., 1984. Vorkommen des Gerstengelbmosaik-Virus in der DDR. Nachrichtenblatt für den Pflanzenschutz in der DDR: 38, 89-91.

(16)   RAO, A.S., 1968. Biology of Polymyxa graminis in relation to soil-borne wheat mosaic virus. Phytopathology 58: 1516-1521.

Biological control of weeds: Present status in Italy

P.Del Serrone & A.Quacquarelli
Istituto Sperimentale per la Patologia Vegetale, Rome, Italy

Summary

A new strategy for weed control has recently been considered in Italy. The Experimental Plant Pathology Institute is carrying out experiments dealing with the use of fungal pathogens as possible agents for biological control of weeds. Special attention is being paid to two troublesome cereal weeds: velvetleaf (Abutilon theophrasti Medicus) and poppy (Papaver rhoeas Linnaeus). Two fungi are under study: Alternaria tenuissima (Kunze ex Nees et Nees) Wiltshire on velvetleaf and the anamorph of Pleospora papaveracea (De Notaris) Saccardo on poppy. Preliminary results on biomass reduction caused by a conidial spray of A. tenuissima on velvetleaf are reported in comparison with a herbicide treatment.

1. INTRODUCTION

Arable land surfaces in Italy are about 9 200 000 hectares and represent nearly 30% of the total area of the country. Fifty two per cent of this arable land is cultivated with cereals of which wheat and maize represent 30% and 10%, respectively. Weed management in Italy is mainly performed by chemical control. In our country 110 active ingredients are registered as herbicides, among which 37 can control weeds in wheat, 25 in barley, 22 in maize, 20 in oats and 15 in rice.

The total amount of herbicides used in Italy is very impressive. For instance, 250 000 tons of them were utilised in 1982, and molinate, atrazine and paraquat reached totals of 63, 25 and 11 thousand tons respectively (1).

As for many kinds of pesticides, the side effects of herbicides are also well known. Among them, it is worthwhile to mention the possible toxic effect on producers, farmers and consumers. The environment is also badly affected by the continual use of herbicides (19) which can persist and accumulate in the soil with subsequent contamination of drinkable water supplies. Last summer, in a maize growing area of northern Italy, some water supplies were contaminated by atrazine residues. This circumstance forced the Italian authorities to cautiously suspend the use of atrazine-containing herbicides. Selection pressure towards resistant weed populations is another hindrance which continually calls for adjustment of weed management (29). Negative effects are also known on soil microflora and microfauna (9).

All the above considerations push many scientists to find other strategies for weed control in order to reduce and, hopefully, eliminate the use of chemicals in agriculture.

The biological control of weeds is a new approach towards a modern concept of integrated control. Biocontrol foresees the use of agent organisms, like viruses, bacteria, fungi, nematodes and insects, against other target organisms, for instance, weeds.

Initially, this tactic was used to reduce weed populations occasionally introduced in new areas by exotic natural enemies. This is known as 'classical biological control'.

More recently the use of biological control of weeds by fungal pathogens has also been aimed at fighting indigenous weeds by exotic and indigenous fungi. Weed pathogenic fungi, occurring naturally in the environment, are used for massive spore production. These spores are released as herbicides, generally once a year, so that they can inundate each plant with the optimum inoculum concentration at the best time to obtain the maximum epidemic pressure. In this way the natural endemic conditions are changed into epidemic outbreaks. This is known as 'inundative biological control'.

Fungi, especially those belonging to <u>Deuteromycetes</u>, are studied in international programmes because of their common occurrence on higher plants and their ability to grow and sporulate in culture (28). Nowadays, mycoherbicides are on the market. For instance, in the USA two mycoherbicides were registered in 1981 and 1982, respectively. Universities, USDA and different firms are interested to develop biological control of weeds by fungal pathogens because of the big economic impact of weed problems. The cost for chemical control of velvetleaf alone in maize, soyabean, sugarbeet and cotton, was estimated at 500 million dollars in 1982 (24).

2. BIOLOGICAL CONTROL IN ITALY

Several biological control aspects have been considered in Italy against crop pests like insects, fungi, mites and nematodes, obtaining promising results that are encouraging the implementation of several researches and practical applications (3, 5, 10, 11, 12, 18, 26, 31).

The biocontrol of weeds, however, is a very new field of research in our country even if some pathogens of weeds have been occasionally recorded (14, 15, 16, 20).

Our institute is carrying out a research programme on biological control of weeds in cereal systems. It is supported by the Commission of the European Communities and by the Italian Ministry of Agriculture and Forestry, and pays special attention to fungi as agent organisms against weeds. The programme foresees two target weeds: velvetleaf (<u>Abutilon theophrasti</u> Medicus) and poppy (<u>Papaver rhoeas</u> Linnaeus).

2.1 Velvetleaf

<u>A. theophrasti</u> (<u>Malvaceae</u>) is one of the most widespread weeds in maize growing areas, and some populations are atrazine-resistant. This herbicide, in fact, is applied as the only active ingredient on 50% of the treated area. Eighty per cent of the remaining area is treated with mixtures containing atrazine and other active ingredients (pendimathalin, alachlor, metachlor and cyanazine) (21, 22).

Velvetleaf is an annual herbaceous weed which overwinters on grains. It germinates in spring with scalar emergence, and blooms from June to September. It also infests sugarbeet, soyabean, tomato and tobacco and it has been recorded in different regions, especially in northern Italy (17).

This weed is reported as belonging to the crop-specific gregarious flora in maize. Velvetleaf seeds, produced in very large amounts, are able to inhibit germination of seeds of other plant species by producing free amino acids (8, 13). This property, recorded already in other weeds, can explain the tendency of velvetleaf to grow in high density (carpet-like effect). Also single plants can be very competitive with the crop because of their impressive canopy, reaching 3 m in height.

Alternaria tenuissima (Kunze ex Nees et Nees) Wiltshire (Dematiaceae) (7) was recorded in Latium. This species has a cosmopolitan diffusion, being isolated from a large number of hosts, on which it is considered to be a secondary parasite. In field, greenhouse and laboratory observations, however, this isolate behaved as a primary pathogen.

The fungus causes interveinal yellow spots with irregular margins progressively enlarging and necrotising. In a more advanced stage, leaves become curled and die (6).

Specificity tests were made on different varieties of several crop and ornamental plants phylogenetically close to velvetleaf. All plants were resistant, even if some of them showed a few very tiny burnt spots with a yellow halo, possibly caused by phytotoxins (2, 25). Anyway, the fungus was never reisolated from these spots.

This fungus has been used in preliminary field tests to 1) establish the most suitable growth stage of plants to be inoculated for obtaining the maximum biomass reduction, in comparison with a herbicide; 2) study the possible additive effect of herbicide and mycoherbicide applied pre-emergence and in post-emergence, respectively.

One year old velvetleaf seeds were broadcast-sown in plots of 20 m^2. Velvetleaf plants were sprayed at the beginning of May, early in the morning, with a water suspension (1.2 litres/plot) of conidia (1 x 10^6 conidia/ml) with a surfactant added (Table 1). Conidia were obtained from the fungus grown on modified potato dextrose agar at 25°C, in dark conditions, inside one litre Roux bottles.

TABLE 1. Biomass Production of Velvetleaf Treated at Different Ages with A. tenuissima, as Compared with a Herbicide

	Treatment	(*) Inoc. concen.	Time of applic.	(**) Biomass (kg)
Control 1	water	-	post-emerg.	29.3 a A (***)
Control 2	teepol	1%	post-emerg.	26.9 a B
Herbicide	atrazine + alachlor	0.9 + 2.2 kg/ha	pre-emerg.	6.3 b C
Myco- herbicide	A. ten	10^{6}con/ml	post-emerg. (3-6 leaves)	6.7 b C
Myco- herbicide	A. ten	10^{6}con/ml	post-emerg. (6-10 leaves)	14.9 c D
Myco- herbicide	A. ten	10^{6}con/ml	post-emerg. (10-15 leaves)	21.7 d B

(*) the volume of sprayed liquid was 1.2 litres for each plot
(**) average of five replications
(***) treatments followed by the same letters are not significantly different (lower case, P=0.05; upper case, P=0.01)

Results showed that the mycoherbicide affected significantly the biomass of velvetleaf at the 3-6 and 6-10 leaf stages; it was more efficient than herbicide.

The second experiment confirmed the efficacy of the mycoherbicide either when applied alone or associated with a herbicide (Table 2).

TABLE 2. Biomass Production of Velvetleaf Treated at the 6-10 Leaf Stage with A. tenuissima as Compared with Herbicide

	Treatment	(*) Inoc. concen.	Time of applic.	(**) Biomass (kg)
Control 1	water	-	post-emerg.	27.2 a A (***)
Control 2	surfac.	1%	post-emerg.	26.5 a A
Herbicide	atrazine + alachlor	0.9 + 2.2 kg/ha	pre-emerg.	8.2 b B
Myco-herbicide	A. ten.	10^6con/ml	post-emerg.	6.7 b B
Herbicide + Myco-herbicide	atrazine + alachlor A. ten.	0.9 + 2.2 kg/ha 10^6con/ml	pre-emerg. post-emerg.	5.7 b B

(*) the volume of sprayed liquid was 1.2 litres for each plot
(**) average of five replications
(***) treatments followed by the same letters are not significantly different (lower case, P=0.05; upper case, P=0.01)

2.2 Poppy

Genus Papaver L. includes ten species and P. rhoeas is one of the most widespread in Italy. It is a herbaceous annual plant with germination occurring during the entire year and with scalar emergence. It infests several crops. At present no particular resistance problem has been recorded. It produces the highest biomass among wheat weeds (23).

During the last two years several fungi have been collected from poppy. One of them, isolated from leaves, was identified as Dendryphion papaveris (Sawada) Sawada (Dematiaceae) anamorph of Pleospora papaveracea (De Notaris) Saccardo (Pleosporaceae). This fungus is considered strictly specific for the Papaveraceae family (7).

The results of preliminary pathogenicity trials carried out in greenhouse conditions suggest a possible use of this fungus as an agent for biological control of poppy. In fact, when poppy seedlings were inoculated with a conidial suspension of about 1 x 10^6 conidia/ml, 70% of plants were killed, and the others were severely damaged.

3. CONCLUSIONS

Weed biocontrol is a necessity for more efficient integrated weed management. Several examples are already available indicating the validity of this strategy (4, 27).

Fungi can play a major role in this strategy, particularly when 'inundative biological control' can be applied against indigenous weeds. The first step requires the finding of pathogens on weeds.

In Italy, two pathogenic fungi have been found recently on velvetleaf and poppy, respectively. Even though more work is still to be done before releasing mycoherbicides, there is no doubt that these results are encouraging for further scientific and practical programmes.

The search for solutions for the following points are in progress:

- better knowledge of host/pathogen relationships;
- optimisation of production, harvesting and storage of conidia;
- timing of field applications.

The development of this project also needs intervention by the Government. An open question to be solved, in fact, is the issue of legislation, which is not yet available either in Italy or in some other European countries.

What has been presented in this paper is very little when compared to the many outstanding questions of biological control. We are endeavouring to find pathogenic fungi on other troublesome weeds, paying special attention either to gramineaceous weeds, like Avena spp. (in wheat), Sorghum halepenses (in maize), Echinocloa crus-galli (in rice), or dicotyledons, like Amaranthus spp., Solanum spp. and Chenopodium spp. All these weeds are of great concern in our country, since their control is becoming more difficult, due to their resistance to chemicals (21, 30).

REFERENCES

(1)   ANONYMOUS, 1984. Annuario di Statistica Agraria, 30. Edition 1983. Sagraf Napoli, 341 pp.
(2)   BALLIO, A., 1981. Structure - activity relationships. In: Toxins in plant diseases. (R.D. Durbin, Editor). Academic Press, New York, 451 pp.
(3)   BISIACH, M., MINERVINI, G. and VERCESI, A., 1985. Indagini sulla attività di Trichoderma spp. verso Pythium ultimum Trow. La Difesa delle Piante, 2, 115-126.
(4)   CHARUDATTAN, R. and WALKER, H.L., 1982. Biological Control of Weeds by Plant Pathogens, J. Wiley and Sons, New York, 293 pp.
(5)   CICCARESE, F., FRISULLO, S. and CIRULLI, M., 1985. Impiego di Trichoderma spp. nella lotta biologica delle malattie radicali da funghi della fragola. La Difesa delle Piante, 2, 1147-1156.
(6)   DEL SERRONE, P. and IALONGO, M.T., 1984. Abutilon theophrasti Medicus (Malvaceae) nuovo ospite di Alternaria tenuissima (Kunze ex Nees et Nees) Wiltshire. Phytopathologia Mediterranea, 23, 91-93.
(7)   ELLIS, M.B., 1971. Dematiaceous Hyphomycetes. Commonwealth Mycological Institute, Kew, England. 608 pp.
(8)   ELMORE, C.D., 1980. Free- aminoacids of Abutilo theophrasti Medicus seed. Weed Res., 20, 63-64.
(9)   FAVILLI, P. and FLORENZANO, A., 1982. Atrazina ed ecologia microbica del suolo. Convegno sul tema: Inquinamento del terreno. Portici, 15 May 1981, 89-94.
(10) FOSCH, S., KOVACKS, A. and DESEO, K.V., 1985. Sviluppo della lotta microbiologica contro ifitofagi delle piante agrarie. La Difesa delle Piante, 2, 233-254.
(11) GAMBARO, P.I., 1984. Possibilità e limiti della lotta biologica. L'Informatore Agrario, 18, 61-63.
(12) GARIBALDI, A., BRUNATTI, F. and ALLOCCHIO, A., 1985. Terreni repressivi verso Fusarium oxysporum f. sp. dianthi: isolamento di

microorganismi e loro attività in vaso. La Difesa delle Piante, 2, 101-106.

(13) GRESSEL, J.R. and HOLM, L.G., 1964. Chemical inhibition of crop germination by weed seeds of Abutilon theophrasti. Weed Res., 4, 44-53.

(14) IALONGO, M.T., 1979. Una popolazione di Erysiphe cichoracearum (DC.) Mérat specifica per il Sonchus. Informatore Fitopatologico 29 (10), 9-11.

(15) IALONGO, M.T., 1983. Una popolazione di Erysiphe cichoracearum specifica per il papavero. Annali dell' Istituto Sperimentale per la Patologia Vegetale, 8, 27-30.

(16) IALONGO, M.T., TEDESCHI, S. and PECORA, P., 1983. Una popolazione di Puccinia suaveolens (Pers.) Rostr. specifica per Cirsium arvense (L.) Scop. Annali dell' Istituto Sperimentale per la Patologia Vegetale, 8, 81-87.

(17) IALONGO, M.T. and DEL SERRONE, P., 1984. Alcuni aspetti biologici del Cencio molle, una infestante che si sta diffondendo in Italia. L'Informatore Agrario, 42, 105-107.

(18) MAGNOLER, A., 1985. Valutazioni in campo di un baculovirus contro le larve di malacosoma neustria L. in Sardegna. La Difesa delle Piante, 2, 329-338.

(19) MAINI, P., COLLINA, A., CHIANELLA, M. and SGATTONI, P., 1986. Persistenza di alcuni erbicidi nel terreno: relazione tra analisi chimiche, biologiche e parametri fisico-chimici in funzione della successione colturale. Giornate Fitopatologiche 1985, 3, 427-438.

(20) QUACQUARELLI, A., RANA, G.L. and MARTELLI, G.P., 1976. Weed hosts of plant pathogenic viruses in Apulia. Poljoprivredna Znastvena Smotra, 39, 561-563.

(21) RAPPARINI, G., 1986. Il diserbo chimico del mais: i problemi sollevati dall' impiego dell' atrazina. L'Informatore Agrario, 7, 127-137.

(22) RAPPARINI, G., 1986. Anche il diserbo dei cereali diventa sempre più impegnativo. L'Informatore Agrario, 5, 71-77.

(23) SGATTONI, P., MALLEGNI, C., ORSI, E. and KOVACS, A., 1984. Weeds and herbicides in wheat in Italy. Proc. EWRS 3rd Symp. on Weed Problems in the Mediterranean area, 419-427.

(24) SPENCER, N.R., 1983. Velvetleaf, history and economic impact. Proc. South. Weed Soc., 36, 352.

(25) STOESSEL, A., 1981. Structure and biogenetic relations: fungal non-host specific. In: Toxins in Plant Disease (R.D. Durbin, Editor). Academic Press, New York, 451 pp.

(26) SURACI, D., 1985. E' possibile la lotta biologica alle piante infestanti. L'Informatore Agrario, 50, 73-75.

(27) TEBEEST, D.O. and TEMPLETON, G.E., 1985. Mycoherbicides: Progress in Biocontrol of Weeds. Plant Disease, 69, 6-10.

(28) TEMPLETON, G.E., 1982. Status of weed control with plant pathogens. In: Biological Control of Weeds by Plant Pathogens. (R. Charudattan and L.H. Walker, Editors). J. Wiley and Sons, New York. 29-44.

(29) ZANIN, G. and LUCCHIN, C., 1980. Resistenza delle infestanti agli erbicidi con particolare riferimento alle atrazine: situazione attuale e prospettive future. Rivista di Agronomia, 3, 330-345.

(30) ZANIN, G., CANTELE, A., DELLA PIETA', S., LORENZONI, G.G., TEI, F. and VASSANA, C., 1985. Le erbe infestanti graminacee nella moderna agricoltura: dinamica, problemi e possibili soluzioni. Atti, Convegno Verona 14 November 1985. 14-19.

(31) ZAZZERINI, A. and TOSI, L., 1985. Osservazioni sull' efficacia antagonistica di alcuni funghi e batteri nei confronti della Sclerotinia sclerotiorum (Lib.) de Bary. La Difesa delle Piante, 2, 163-168.

Session 3
Farming systems

Chairman: A.El Titi

Economic and ecological implications of different agricultural systems in northern Germany

Th.Basedow

Institute of Phytopathology and Applied Zoology, University of Giessen, FR Germany

Summary

Different ways, by which farmers in northern Germany try to minimise their costs and maximise their yields were analysed from the viewpoints of economy and of ecology.

Growing wheat repeatedly can enhance the saddle gall midge, but apparently not the wheat blossom midges. The main limiting factors preventing the permanent growing of wheat in northern Germany are cereal diseases rather than insect pests.

It is shown by comparative studies, that growing wheat at a high level of intensity depresses numbers of beneficial arthropods, favours the cereal aphids, but does not give the best economic profit (though the yield may be best, the costs are too high). Routine insurance sprayings against cereal aphids are shown to be uneconomic. Cereal aphid control should rather be performed according to critical thresholds. Another study revealed that the repeated routine application of insecticides, which is not economic on the one hand, may enhance pest populations in sugar beet on the other. The reduced number of polyphagous predators seems to be responsible for this.

In consequence of the findings presented it is stressed that a lower level of intensity in agriculture is advisable to farmers from the economic view. Furthermore, the antagonists of pests will then be able to act more efficiently in favour of the farmers.

1. INTRODUCTION

Growing cereals tends to be more and more a financial problem for farmers in the European Community: the costs of production are rising, especially those of pesticides (1), while the prices paid to the farmers for cereals are lowered. In this situation, farmers have to find a way to minimise their production costs or to maximise their yield, optimally to do both. Different ways by which farmers try to reach these goals in northern Germany, were examined. The results are presented here.

2. RESULTS

A report is given here on the results of different studies, each with its own experimental design. A short note on the methods applied during each of the investigations will therefore be given in advance of the presentation of the respective results obtained.

2.1 Repeated Growing of Wheat

2.1.1 Entomological aspects

2.1.1.1 Wheat blossom midges

The infestation of winter wheat by wheat blossom midge larvae (Dipt., Cecidomyiidae) was measured (2) on neighbouring fields at Dettum near Brunswick (FRG), 1970-75 (Table 1).

TABLE 1. Infestation of Winter Wheat by Wheat Blossom Midge Larvae on Fields, With and Without Crop Rotation. Dettum/Brunswick, FRG.

| Year | Number of fields | | Wheat blossom midge larvae per 100 ears | | | |
| | | | Contarinia tritici | | Sitodiplosis mosellana | |
	Wheat after wheat	Other previous crop	Wheat after wheat	Other previous crop	Wheat after wheat	Other previous crop
1970	1	2	490	1085	110	105
1972	2	4	2232	2133	379	459
1973	1	5	1505	2703	533	590
1975	3	3	1813	2067	679	2721
average	n=7	n=14	1510	1997	425	967

No significant differences were obtained concerning the attack by Contarinia tritici (Kirby) or by Sitodiplosis mosellana (Géhin), whether wheat was grown after wheat or after another crop, e.g. sugar beet (3).

2.1.1.2 Saddle gall midge

This cereal pest, Haplodiplosis marginata (v. Roser) is known to be favoured by repeated cereal cropping (4). But it was shown (5) that growing wheat after what is not always necessarily followed by an outbreak of the saddle gall midge. This observation was confirmed by a two year study of larval populations of Haplodiplosis marginata with the method quoted above (2), on neighbouring fields of winter wheat at Dettum near Brunswick, grown either after winter wheat or after sugar beet, respectively (Table 2). In the first year of the study wheat as previous crop favoured the occurrence of the saddle gall midge larvae, but in the second year it did not.

TABLE 2. Infestation of Winter Wheat by Haplodiplosis marginata on Fields With and Without Crop Rotation. Dettum (FRG)

| Year | Number of fields | | Larvae of saddle gall midge per 1000 cm^2 | |
	Wheat after wheat	Wheat after sugar beet	Wheat after wheat	Wheat after sugar beet
1979	4	4	38.0	13.5
1980	3	5	3.7	3.8

When spring wheat was grown on a small plot permanently for more than 20 years, larval populations of H. marginata did not remain on a high level of abundance but fluctuated like insect populations under normal conditions (6).

So although gall midges are often limiting factors when wheat is grown repeatedly, this is not necessarily the case.

2.1.2 Phytopathological aspects

No special investigations were performed by the author on this subject, but a farmer at Brodersdorf/Kiel (FRG) was interviewed every year. On two small fields of 1.5 ha each, with loamy soil, he grew wheat continuously for 11 successive years, waiting for the 'decline-effect'. But the yield of the two plots - which were treated intensively with pesticides every year - decreased continuously during this period, apparently mainly due to the take-all-disease (Gaeumannomyces graminis), so that the farmer stopped this 'experiment', and after 11 years turned back to crop rotation.

2.2 Growing Wheat at Different Intensities - Comparative Studies

2.2.1 Effects on parts of the wheat field fauna

To investigate this topic, two extremes were chosen. In one area of investigation, Brodersdorf/Kiel (FRG), chemical plant protection in fields of winter wheat was performed intensively by farmers. All treatments have been supervised there by the author since 1974 (Table 3). From 1974 to 1983 the number of active ingredients applied per season increased from six to 11 per wheat field, and insecticides were sprayed on every wheat field every year. Wheat was grown on one third of the study area of 140 ha. The other two thirds were winter rape (also sprayed intensively), and winter barley.

TABLE 3. Treatments of Winter Wheat Fields at Brodersdorf/Kiel (FRG) with Plant Protection Ingredients Between 1974 and 1983

	1974	1979	1983
Number of treatments* with:			
herbicides	2	2	3
growth regulators	1	1	2
fungicides	2	5	5
insecticides	1	1	1
Sum of plant protection ingredients	6	9	11

* partly combined

In another village, Passade, 5 km distant from Brodersdorf, one farmer had kept an area of 10 ha free of agrochemicals since 1974 ('organic-biological farming'). Winter wheat, field beans, spring oats and winter rye were grown on this area.

In 1984, 20 pitfall traps were placed in a wheat field in both areas, from 4th May to 8th August, at least 50 m apart from the field border, with 10 m distance from each other. Traps were half-filled with 0.5% formaldehyde plus detergent and had a diameter of 10 cm; they were emptied weekly. Additionally, cereal aphids were counted weekly on 200 tillers per field. Since the wheat stands had a different density (781 ears per square metre at Brodersdorf, and 483 at Passade), aphid numbers were calculated per square metre, for comparison.

From Table 4 it can be seen that the beneficial arthropods were nearly twice as numerous on the field free of pesticides, as compared with the field with a high pesticide level.

TABLE 4. Beneficial Arthropods and Cereal Aphids on Fields of Winter Wheat near Kiel (FRG), 5 km Distant from Each Other, Under Significantly Different Regimes of Agrochemicals Applied

Specimes in 20 pitfall traps 4/5-8/8	Passade 10 ha, 10 years free of pesticides, low level of N	Brodersdorf 140 ha, intensely treated with pesticides, high level of N
Carabidae	11 672	2 802
Staphylinidae	3 819	4 466
Araneae	7 943	5 181
Coccinellidae	464	30
Sum of 'beneficials'	23 898	12 479 (-48%)
Cereal aphids per square metre (maximum, July)	10 452	50 968 (+390%) (if not sprayed with insecticide)

In contrast, the cereal aphids reached nearly five-fold on the field with pesticides, as compared with the pesticide-free field (this was actually confirmed by leaving, on the pesticide field, four small plots of 150 m² without insecticides; the rest of the field received an insecticidal spray) (parathion).

But it is not thought, in this case, that the lack of beneficials is the main reason for the very high number of the cereal aphids observed in the wheat field treated with pesticides. Since Macrosiphum avenae was the most common species, it is more likely that the high level of nitrogen has enhanced the increase of the aphid numbers (7, 8).

2.2.2 Yields and net monetary output

The beneficial role of the predators, though reliably stated (9, 10) is not fully appreciated by farmers and by advisors. Therefore, routine insurance applications of insecticides are still usual, mostly for cereal aphid control. But it is known that the cheapest insecticides, which are often preferred, are harmful to predator populations (11, 12, 13).

So the widespread use of broad spectrum insecticides might favour the occurrence of pests. Therefore, it seems desirable to have other arguments to convince advisors and farmers that spraying pesticides as a routine is not the best method of plant protection. The most effective argument should be that routine sprayings are not economic. To test this hypothesis, two series of experiments were initiated.

2.2.2.1 Economics of cereal aphid control

On fields of winter wheat near Kiel (FRG), it was tested on plots of 25 m² with four replicates, whether or not a routine insecticidal spray for cereal aphid control would result in yield increases every year.

Table 5 shows that between 1979 and 1984 an insecticidal spray gave a positive result in three out of six years only. So a routine spray cannot be recommended. Cereal aphids in winter wheat should rather be controlled - under the growing conditions in the FRG - only when, at the end of flowering, the number of one aphid per ear and flag leaf is exceeded (14), as can be derived from Table 5, also.

TABLE 5. Effects of Insecticidal Sprays (Pirimicarb, 150 g a.i./ha at the end of flowering) on Cereal Aphid Numbers and on Yield of Winter Wheat Grown near Kiel (FRG) on Sandy Loam

Year	Variety	Cereal aphids per ear and flag leaf (unsprayed)		Yield (dt/ha) (unsprayed) 84% dry matter	Increase of yield by spraying Pirimicarb dt/ha
		at end of flowering	at milky ripeness		
1979	Topfit	1.5	39.0	68.0	+14.3
1980	Vuka	0.1	5.3	63.1	0
1981	Vuka	0.7	1.0	85.0	0
1982	Tabor	0.1	0.4	93.9	0
1983	Vuka	10.5	21.6	73.8	+10.2
1984*	Kanzler	1.5	21.1	90.2	+8.1

* Results of an experiment by Dr. Chr. Bauers, Kiel

So the first set of experiments gave an argument against routine applications of insecticides in cereals. The second experimental set intended to include more variables than insecticides alone.

2.2.2.2 Economy of growing wheat intensively

On fields of winter wheat at Passade/Kiel (FRG), the amount of fertiliser, growth regulator and of pesticides given was varied on plots of 0.2 ha with three replicates. The infestation of the plots by cereal diseases and by aphids was measured on 50 or 100 tillers per plot weekly, and all plots were harvested separately.

Table 6 gives the result of the first experiment, performed in 1982, when cereal aphids were few (see Table 5). Also the degree of infestation by cereal diseases did not vary significantly between treatments. This was partly due to nearly zero infection (mildew) and to the fact that the fungicides available have no effect, e.g. on the take-all disease (Gaeumannomyces graminis).

TABLE 6. Yield and Net Monetary Output of Winter Wheat ('Disponent') Grown at Different Intensities at Passade/Kiel (FRG), 1982. Previous Crop (1981): Winter Rape.

Treatments	I	II	III
N (kg/ha)	152	199	240
Herbicide	+	+	+
Chlormequat	-	+	+
Fungicides	-	3	4
Insecticide	-	-	+
Yield (dt/ha, 84% dry matter)	86.6*	95.7	98.0
Gross output (DM/ha)	4 654.75	5 143.88	5 267.50
Costs (DM/ha) for seed, fertiliser, growth regulator, pesticides and for application of the chemicals	995.69	1 333.53	1 515.82
Net output (DM/ha)	3 659.06	3 810.35	3 741.68
Profit (DM/ha), as compared with treatment I		+151.29	+82.62

* significant difference from the other treatments at $P=0.05$

From Table 6 it can be seen that, under the conditions described, the highest input (III) resulted in the highest yield (dt/ha). However, when all costs were taken into account, it was the moderate treatment (II) that gave the highest profit (DM/ha). The evaluation of further experiments with more variables is in progress. The same tendency appears as above: the highest input does not give the highest profit. This is an important argument towards integrated plant protection.

2.3 Growing Sugar Beet Intensively: Effects on Noxious and Beneficial Arthropods and on Yield

Investigations took place on a field of 60 ha at Grünholz/Kappeln, FRG, 1981-86. The field was divided into three plots of 20 ha each. Farming was performed intensively on all plots, and only the insecticidal treatments were varied: one plot did not receive any insecticide, the second received selective insecticides once per season, the third received broad spectrum insecticides twice per season. The crops grown were winter barley (1981 and 1986, without insecticides), winter rape in 1982, winter wheat (1983 and 1985) and sugar beet in 1984.

Predator populations were sampled by 20 pitfall traps and by removal trapping on 20 m^2 per plot (June and July). Phytophagous insects were counted on 200 plants per plot, weekly. To measure the yield of sugar beet,

40 m^2 per plot were harvested by hand and, after cleaning the beets, by weighing them in the field without leaves.

Table 7 shows that populations of polyphagous predators had declined after three years' use of insecticides in 1984, as compared with the plot free of insecticides. In contrast to this, pest populations, in spite of the application of insecticides, were higher on the insecticide plot than on the plot free of insecticides. When yields were measured, it was interesting that no significant differences existed between plots (Table 7). In consequence it must be stated here that the repeated use of insecticides was not economic in this case.

TABLE 7. Effects of Insecticides, Applied in Normal Dosage for Three Years on 20 ha Plots of a 60 ha Field, on Beneficial and Noxious Arthropods in Sugar Beet. Grünholz/Kappeln, 1984

Specimens per 10 m^2	Without insecticides	Pirimicarb once per season (1982: Phosalon, twice)	Parathion twice per season (1982: + Methoxychlor)
Carabidae	554	220	234
Staphylinidae	340	400	308
Araneae	173	156	110
All predators (June/July)	1 067	776	652
Aphis fabae			
31 July	2 584	1 835	5 510
18 September	3 226	5 865	4 272
Pegomyia betae			
3 July	7.7	8.4	7.0
11 September	4.9	36.4	39.9
Yield (dt/ha)	601.8	540.0	548.6

3. DISCUSSION AND CONCLUSIONS

The data presented here have shown that it is uneconomic to:

- grow wheat continuously on sandy loam in northern Germany;
- apply insecticides to winter wheat as a routine;
- grow wheat and sugar beet at very high intensities.

The number of beneficial arthropods present in the fields is greatly diminished by intensive agriculture. This fact may create new pest problems. On the other hand, growing wheat on a level of low input allows a lot of beneficials to survive, but it is not the most profitable way of growing wheat, as has been shown (Table 6). However, it proved most economic to use insecticides according to critical thresholds and to grow wheat on a medium level of intensity. It has been shown elsewhere (15) that 'integrated management' is the most profitable system for other agricultural crops as well. So, in the present difficult situation for farmers (which is, to a great extent, due to unexpected problems of a political origin) the advice to farmers must be to perform 'integrated cereal growing' with the aid of advisers and scientists who are required to develop more economic thresholds. In this way, not only will the farmers benefit, but also the predatory arthropods. It is most likely that these will then, as antagonists of the pests, be even more successful in suppressing the pests, initiating a beneficial cycle.

ACKNOWLEDGEMENTS

The author is grateful to the Federal Ministry for Research and Technology at Bonn (FRG), for financial support of a part of the studies presented here; also to Harold Rzehak and Wolfgang Liedtke for their help and Karen Kähler and Karen Diedrichsen for excellent technical assistance during these studies. Dr. Chr. Bauers, Plant Protection Office of Schleswig-Holstein, Kiel (FRG), kindly allowed me to quote his result of 1984 in Table 5.

REFERENCES

(1)   METCALF, R.L., 1980. Changing role of insecticides in crop protection. Ann. Rev. Entomol. 25, 219-256.
(2)   BASEDOW, Th. and SCHÜTTE, F., 1973. Neue Untersuchungen über Ei-ablage, wirtschaftliche Schadensschwelle und Bekämpfung der Weizengallmücken (Dipt., Cecidomyidae). Z. angew. Entomol. 73, 238-251.
(3)   BASEDOW, Th. and SCHÜTTE, F., 1982. Die Populationsdynamik der Weizengallmücken Contarinia tritici (Kirby) und Sitodiplosis mosellana (Géhin) in zwei norddeutschen Weizenanbaugebieten von 1969 bis 1976. Zool. Jahrb. Syst. 109, 33-82.
(4)   GOLIGHTLY, W.H. and WOODVILLE, H.C., 1974. Studies of recent outbreaks of saddle gall midge. Ann. appl. Biol. 77, 97-101.
(5)   SPITTLER, H., 1969. Beiträge zur Morphologie, Biologie und Ökologie des Sattelmückenparasiten Platygaster equestris nov. spec. (Hymenopt., Proctotrupoidea, Scelionidae) unter besonderer Berücksichtigung seines abundanzdynamischen Einflusses auf Haplodiplosis equestris Wagner (Diptera, Cecidomyidae). II. Z. angew. Entomol. 64, 1-34.
(6)   BASEDOW, Th., 1986. Die Abundanzdynamik der Sattelmücke, Haplodiplosis marginata (von Roser) (Dipt., Cecidomyiidae), bei Fruchtwechsel, bei wiederholtem und bei permanentem Anbau von Weizen. Z. angew. Entomol. 102, 11-19.
(7)   HINZ, B. and DAEBELER, F., 1976. Der Einfluss der Stickstoffdüngung auf die Vermehrung der Grossen Getreideblattlaus, Macrosiphum (Sitobion) avenae (F.), auf Winterweizen. Wiss. Z. Wilhelm-Pieck-Univ. Rostock, Math.-Nat. R. 25, 653-655.
(8)   HANISCH, H.-Ch., 1980. Untersuchungen zum Einfluss unterschiedlich hoher Stickstoffdüngung zu Weizen auf die Populationsentwicklung von Getreideblattläusen. Z. Pflanzenkrankh. Pflanzenschutz 87, 546-556.
(9)   EDWARDS, C.A., SUNDERLAND, K.D. and GEORGE, K.S., 1979. Studies on polyphagous predators of cereal aphids. J. appl. Ecol. 16, 811-823.
(10)   DECLERCQ, R. and PIETRASZKO, R., 1983. Epigeal arthropods in relation to predation of cereal aphids. Aphid Antagonists. Proceedings of a

Meeting of the EC Experts' Group, Portici, Italy, 23-24 November, 1982. Editor, CAVALLORO, R. - A.A. Balkema, Rotterdam, 88-92.

(11) BASEDOW, Th., BORG, Å. and SCHERNEY, F., 1976. Auswirkungen von Insektizidbehandlungen auf die epigäischen Raubarthropoden in Getreidefeldern, insbesondere die Laufkäfer (Coleoptera, Carabidae). Entomol. exp. appl. 19, 37-51.

(12) VICKERMANN, G.P. and SUNDERLAND, K.D., 1977. Some effects of Dimethoate on arthropods in winter wheat. J. appl. Ecol. 14, 767-777.

(13) BASEDOW, Th., BORG, Å. and SCHERNEY, F., 1981. Auswirkungen von Insektizidbehandlungen auf die epigäischen Raubarthropoden in Getreidefeldern, insbesondere die Laufkäfer (Col., Carabidae). II. Acta Agriculturae Scandinavica 31, 153-164.

(14) BASEDOW, Th., BAUERS, Chr. and LAUENSTEIN, G., 1985. The preliminary control threshold for cereal aphids in winter wheat in Western Germany. IOBC/WPRS Bull. VIII, 3, 36-39.

(15) ZEDDIES, J., JUNG, G. and EL TITI, A., 1986. Integrierter Pflanzenschutz im Ackerbau: Das Lautenbach-Projekt. II. Ökonomische Auswirkungen. Z. Pflanzenkrankh. Pflanzenschutz 93, 449-461.

Influence of straw incorporation, and associated herbicide and molluscicide treatments on fauna in cereal crops

B.D.Smith & D.A.Kendall

Department of Agricultural Sciences, Long Ashton Research Station, University of Bristol, UK

Summary

Straw burning destroys some wildlife which would benefit if burning was restricted by law. Burning also reduces weed and cereal regrowth populations and in its absence their control may require additional herbicide treatment, for example with paraquat, which kills plants quickly and is toxic to aphids, thus preventing the spread of Barley Yellow Dwarf Virus. Slugs are favoured by straw either on the surface or incorporated into the soil and recently there has been a large increase in the use of molluscicides, especially methiocarb. The influence of three methods of straw disposal were studied in relation to two drilling techniques and to the use of methiocarb. More aphids and more virus was found in direct drilled winter wheat following methiocarb treatment. In winter barley more virus was found on ploughed than on direct drilled plots. It is suggested that these effects result from the impact of treatments on the natural enemies of aphids. The implications of using broad-spectrum pesticides in relation to straw disposal are discussed.

1.1   Introduction

The adoption of direct drilling and minimum cultivation in the 1970s facilitated increased production of winter cereals in the UK, and the disposal of surplus straw has become a problem as the area of cereals has increased. Field burning is the cheapest and most convenient way of disposing of the current straw surplus but this has caused widespread concern to the general public. The only alternative to burning, in the short term, is chopping and soil incorporation. Pressures to reduce, and even to ban, straw burning have stimulated AFRC/ADAS studies on straw incorporation and its implications for farmers and for farmland ecology. Straw burning is declining: in 1984, 37% of the total straw in England and Wales was burnt and this decreased to 28% in 1985. The proportion of straw incorporated into the soil increased from 7% in 1984 to 10.5% in 1985 (1).

Burning straw and stubble decreases the populations of many animals which live on the soil surface or in the soil, e.g. mites, spiders, Collembola and some earthworms. This may be due to direct killing by burning or to the amount of organic matter which they require being reduced. Such effects are useful when pests are killed but damaging if the species concerned are natural enemies of pests or contribute to soil fertility.

One beneficial effect of straw burning is the destruction of weed seeds, the proportion killed depending on the temperature achieved and the position of the seeds in the soil. On farm sites with natural weed infestations burning was found to kill 40-80% of blackgrass and 40-97% of barren brome seeds (2). Restrictions on straw burning encourage weed and cereal regrowth in the stubble and these plants form a 'green bridge' that enables some pests and diseases to persist in the inter-crop period and attack the next crop. Other pests are favoured by the presence of additional straw on the soil surface or incorporated into the soil. Extra pesticide treatments may be required to control pests favoured by the absence of burning.

The effects on invertebrate fauna of some agrochemicals used to control slugs and aphids are described in relation to straw disposal and cultivation methods.

1.2 Effects on slugs

When straw is incorporated farmers become more concerned about slugs than any other pests (3). Hence the use of molluscicides has increased over the last decade, mainly through the use of methiocarb on cereals (4).

Field plots in Oxfordshire, where shallow cultivations had been used to incorporate straw for 4 years, Glen, Wiltshire and Milsom (5) found that slug numbers were much higher ($180/m^2$) than in plots where straw was burnt ($14/m^2$). Damage to seeds and seedlings was correspondingly worse where straw was chopped and spread compared with burnt areas.

It is not known how fast slugs increase in the presence of straw residues because most straw incorporation experiments have only recently started; monitoring in the first two years of incorporation in field experiments at Long Ashton has not indicated significant increases in numbers.

1.3 Effects on aphids and virus

Aphids cause damage to cereal crops in autumn mainly through transmitting Barley Yellow Dwarf Virus (BYDV). The aphids which transmit this virus to cereal crops also live on the cereal regrowth and grass weeds found between harvest and drilling, the so called 'green bridge'. Many species of grass are good hosts for cereal aphids (6). Winged aphids rapidly colonise plants in stubbles after harvest and, if the vegetation is not killed by herbicides or cultivations, the aphids can survive and infect the new crop with BYDV; they can also survive on plant tissue under the soil for several weeks and move upwards to the sown crop (7).

The 'green bridge' is usually larger where straw is not burnt because burning kills seeds. The weed problem, especially that of annual grass weeds, is worse where straw is incorporated by shallow cultivation rather than by ploughing.

Cereal regrowths and grass weeds were counted in 11 fields in August and early September in S.W. England; regrowths accounted for more than 90% of the total plants in nine of the fields. BYDV was found in either regrowths or grasses in almost half of the fields sampled, and more than 15% of grass plants were infected in some stubbles (7).

In experiments done jointly with ADAS it was found that herbicides containing paraquat or glyphosate applied before ploughing killed plants, and reduced aphid numbers and BYDV spread. Paraquat is the faster acting herbicide and it is now known that, unlike glyphosate, it is directly toxic to aphids (8). Thus the control of BYDV spread by paraquat is probably due to a combination of direct toxicity and rapid destruction of the aphid food source. These effects could be particularly beneficial when the interval between treatment and drilling is very short. Early drilling of cereals increases the risk that emerging crops will coincide with flights of viruliferous aphids and in virus prone sites late October spraying with pyrethroid insecticides has become routine practice. However, insecticide used at this time may be too late to prevent virus spread by aphids which colonise the crops from cereal regrowth or grasses. An insecticide could be used on the 'green-bridge' but the risk of aphids surviving to infect the new crop depends on how well grasses and regrowth are killed and buried before drilling, on how well the seed bed was consolidated, and on the length of time between cultivation and drilling.

1.4 Environmental effects

In a comparison of methods of straw disposal, Edwards found that whilst removal by baling decreased numbers of Collembola by approximately 30%, burning had a bigger impact reducing numbers to 15% of the pre burn population. Similar decreases in spiders, thrips and other insects occurred. Earthworm populations were not affected much by straw removal in the first year but after 3 years of either baling or burning there were pronounced effects; the deep burrowing species Lumbricus terrestris was eliminated by burning. Barnes and Ellis (10) found that this species increased when chopped straw was left on the surface.

Thus some wildlife will clearly benefit from restrictions on straw burning, but such benefits may be offset if the pesticides used to control the pests which are favoured by the absence of burning, and/or the incorporation of straw, also kill non-target species.

Both of the most commonly used molluscicides, methiocarb and metaldehyde, have broad-spectrum activity but the consequences of their widespread use are unknown. Adverse effects of one or the other have been reported on earthworms, beetles, leatherjackets, woodlice, birds, small mammals and hedgehogs.

Paraquat can be toxic to a range of animal species in laboratory tests, and the work at Long Ashton on its field effectiveness against aphids, has been widened to include assessments of its activity against other fauna. Paraquat treated stubbles are being compared with untreated ones, and others treated with glyphosate which also removes cover for fauna but without risk of direct toxicity.

The effects of methiocarb on pests and other fauna are currently being investigated in relation to cultivation treatments associated with straw disposal on wheat and barley. The treatments were:
(a) Straw baled and removed followed by ploughing and drilling.
(b) Straw chopped and spread   "   "   "   "   "
(c) Straw baled and removed   "   "   direct drilling.
(d) Straw burnt   "   "   "   "
Methiocarb was applied in early October to plots of each treatment on winter wheat following drilling in the last week of September. Effects were measured on slugs (Dr Glen will report these elsewhere), aphids, BYDV, aphid predators and other invertebrates. Aphid predators were counted as they emerged from field collected soil samples which were subjected to gradual flooding (5).

Method of cultivation and straw disposal did not have significant effects on aphids but there were some interactions between methiocarb treatment and husbandry practices; significantly more BYDV was found on the two direct drilled treatments where methiocarb had been applied in one wheat experiment. In a second wheat experiment there were significantly more aphids on direct drilled plots treated with methiocarb, although no effect was seen on BYDV because virus infection was very low (0.01%). In an experiment on barley neither the method of cultivation nor straw disposal had a significant effect on aphids but there was significantly

more BYDV on plots ploughed after baling compared with plots where straw
was chopped and incorporated. Also there was more BYDV on the plots which
had been baled before direct drilling than on plots burnt before direct
drilling. Overall, and in contrast to the situation on wheat, there was
more virus in those straw disposal treatments that were followed by
ploughing than in those followed by direct drilling.

In wheat significant effects of individual treatments were not found
on Carabid or Staphylinid adults or larvae but fewer larvae were found on
direct drilled plots in one of the experiments. In barley these insects
were not affected significantly but centipedes appeared to be more
abundant on direct drilled plots (11). Centipedes are known to predate
aphids. The method used for predator assessment in these experiments is
probably not as efficient as some other methods and this is being investigated. Nevertheless, clear negative correlations were found between
aphids, BYDV, Carabids and Staphylinids in the wheat experiments and it is
suggested that the larger aphid populations and greater virus spread found
in the direct drilled plots where methiocarb was used was mainly due to
toxic effects of this molluscicide on aphid predators. The larger effects
seen on the direct drilled plots compared to ploughed plots could have
been due to the methiocarb pellets being more accessible to ground surface
predators on the relatively undisturbed soil surface. In the barley
experiment populations of Carabids and Staphylinids were half those in the
wheat but centipedes were six times more abundant. It is possible that
the lower numbers of centipedes on plots which had been ploughed may have
allowed aphids to spread more virus there.

In these experiments attention has been focused on aphid predators
extracted from soil samples but other invertebrates not sampled, such as
spiders, could have contributed to aphid control and also been affected by
methiocarb applications.

However, the inferred removal of at least some of the natural enemies
of aphids in wheat by methiocarb suggests that appreciable natural control
can occur during the autumn and early winter period. This effect was
sufficiently pronounced in one wheat experiment to allow BYDV spread to
almost reach its economic damage threshold of 5% plant infection, a level
which would warrant a specific aphicide application.

When Glen, Wiltshire and Milsom attempted to control slugs using
methiocarb in different ways, they found that whilst drilling pellets with
the seed did not result in significantly less damage there was
significant benefit from using the chemical as a seed dressing. However,
it is unlikely that this method of application will become acceptable farm
practice because of the risk to birds and other fauna from spillage of
treated grain or from its exposure during drilling.

The effects of broad-spectrum pesticides on non-target fauna in early
autumn may be greater than if the same chemicals were applied later when
populations have declined, and species are less active in the lower soil
and air temperatures. If the same species are affected by pesticides
applied post-harvest and by later applications, e.g. late October
pyrethroid treatments to control BYDV spread, then the recovery of numbers
could be further delayed with possible consequences for the effectiveness
of natural pest control in the following spring.

Clearly there is scope for further studies to measure interactions
between pesticide applications and other husbandry practices. Such
studies could lead to better ways of controlling pests with combinations
of chemical and cultural methods which have less impact on natural
control.

REFERENCES

1. MAFF Statistics (1986). Straw Survey 1985 England & Wales.
2. Moss. S. (1985). Straw burning and its effect on weeds. Straw, Soils and Science. Agricultural and Food Research Council Report, 18-19.
3. National Farmers Union (1986). Proceedings of the Second Straw Incorporation Register Conference, Jan. 1986.
4. MAFF (1986). Review of usage of pesticides. In: Agriculture, Horticulture and Animal Husbandry in England and Wales, 1980-3. Reference Book 541.
5. GLEN, D.M., WILTSHIRE, C.W. and MILSOM, N.F. (1984). Slugs and straw disposal in winter wheat. Proceedings of 1984 British Crop Protection Conference - Pests and Diseases, 139-143.
6. SMITH, B.D., KENDALL, D.A. and WRIGHT, M.A. (1984). Weed grasses as hosts of cereal aphids and effects of herbicides on aphid survival. Proceedings of the 1984 British Crop Protection Conference - Pests and Diseases, 19-24.
7. KENDALL, D.A. (1986). Volunteer cereals, grass weeds and swards as sources of cereal aphids and Barley Yellow Dwarf Virus. Proc. European Weed Research Society Symposium 1986, Economic Weed Control, 201-208.
8. WRIGHT, M.A., KENDALL. D.A. and SMITH, B.D. (1985). Toxicity of paraquat, paraquat + diquat and glyphosate to the cereal aphid *Rhopalosiphum padi*. Tests of Agrochemicals and Cultivars no. 6, 8-10. (Ann. appl. Biol. 106, supplement).
9. EDWARDS, C.A. (1977). Investigations into the influence of agricultural practice on soil invertebrates. Ann. appl. Biol. 87, 515-520.
10. BARNES, B.T. and ELLIS, F.B. (1979). Effects of different methods of cultivation and direct drilling, and disposal of straw residues on populations of earthworms. Journal of Soil Science, 30, 669-679.
11. KENDALL, D.A., SMITH, B.D., CHINN, N.E. and WILTSHIRE, C.W. (1986). Cultivation, straw disposal and Barley Yellow Dwarf Virus infection in winter cereals. Proceedings of 1986 British Crop Protection Conference - Pests and Diseases, 981-987.

Long-term influences of an integrated farming system on some cereal diseases

A.El Titi
Landesanstalt für Pflanzenschutz, Stuttgart, FR Germany

Summary

The incidence of fungus diseases on cereal crops is highly dependent upon the growing conditions and consequently on the farming system practised. Following centuries of fundamental research many interactions between single husbandry techniques and different cereal pathogens became clear. A major goal of integrated management approaches is to make use of such interactions in order to decrease the susceptibility of the crop plants to the infesting pathogens on one hand and to improve the antagonistic regulation mechanisms on the other hand, when planning overall farm strategies.

In long-term studies at Lautenbach, FRG, an integrated farming system based on a minimised soil tillage, reduced nitrogen fertilisation, band-sowing and an intensive green-manure regime, is compared with a current farming system. Stem base diseases, powdery mildew, brown and yellow rust and Septoria leaf spot are monitored in these studies.

On the integrated cropped winter wheat the incidence of infestation by stem base diseases, mainly caused by eye-spot disease, was found to be lower than on the conventionally cropped wheat of the same variety. However, during the first four experimental years a heavier mildew infestation was assessed on the integrated plots, thereafter significantly lower, on the average of 15%. Serious infestations by brown rust occurred only during 1983. The initial infestation level was higher on the integrated fields followed by a decrease of the leaf damage area, on average by 7%.

1. INTRODUCTION

The list of pathogens attacking cereal crops in south-west Germany includes mainly parasitic fungi. Roots, stems, leaves and glumes may be attacked, often causing considerable yield losses. The fact that a pathogen is present in a field, does not mean that the crop will necessarily suffer seriously. Disease severity will depend on various factors such as weather conditions, cultivar, soil conditions, etc. Outbreaks are known to arise when most suitable conditions coincide with the reproduction demands of the pathogen sp. involved. These include the inhibition of antagonistic agents. Different soil organisms are known to produce substances with fungicidal effects, feed on mycelia of fungi or to compete with the parasitic fungi. The complex nature of such relationships can hardly allow an evaluation of single components involved. In integrated farming systems there is a high emphasis on making the best possible use of antagonistic agents. This means, very simply, that husbandry technique has to be manipulated:

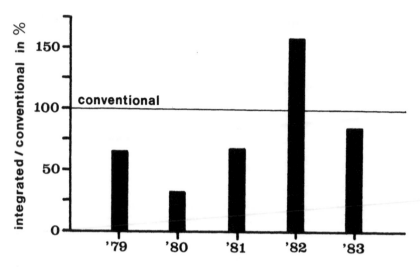

Fig. 1. The relative infestation incidence of stem base disease in integrated and conventionally cropped winter wheat at Lautenbach.

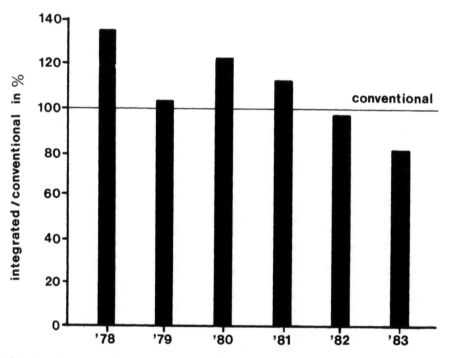

Fig. 2. The percentage of powdery mildew (<u>Erysiphe graminis</u>) damaged leaf area in integrated related to conventionally cropped winter wheat.

- to increase the antagonistic agents;
- to decrease the build-up of pathogen inoculum;
- to decrease the susceptibility of the host crop to attack.

These elements are considered to build up the 'framework' of any integrated approach before a decision is made on whether or not chemical control measures are to be recommended. After evaluating the effects of common farming techniques, a package of methods and measures have been put together to form an integrated farming system.

On a commercial farm of 245 ha (Lautenbach), six plots of eight hectares each, located within the large fields of Lautenbach, were used to study the economical and ecological effects of this integrated farming system. Additionally, six adjacent plots of four hectares, with conventional farming, were set out for comparison. The major characteristics of the integrated system are reduced soil tillage, double row drilling, reduced N-fertilisers, trefoil undersowings and high threshold values for the application of pesticides.

2. RESULTS

The results considered in this paper deal mainly with cereal diseases, assessed on fields of both farming systems over a six year period. Routine monitoring was carried out at occasions coinciding with the timing of fungicide applications. One hundred plants or tillers/plot were sampled on each occasion. The infestation incidence was calculated as the mean of 100 single assessments/plot.

2.1 Stem Base Disease

The monitored systems -pale-buff-brown lesions at Feekes: 32- are known to be caused by a complex of pathogens, but mainly by Pseudocercosporella herpotrichoides, Rhizoctonia spp., Fusarium spp. The incidence of attack in integrated and conventional fields is shown in Figure 1.

2.2 Powdery Mildew

Higher infestations by powdery mildew were assessed on the integrated plots in the first four years (Figure 2). Lower infestation incidence was, however, estimated thereafter. The difference was significant for all sampled leaves in 1983 (Figure 3).

3. DISCUSSION AND CONCLUSIONS

The integrated farming system - as it is conceived and implemented at Lautenbach - seems to suppress some major diseases of cereal crops. This suppression can be related to the improved soil structure, and consequently growing conditions, of cereals. It can also be due to increased activity of the soil biota in the integrated fields. This can be the case for soil-borne pathogens, such as Ps. herpotrichoides, Fusarium, Ophiobulus or Septoria sp. In the case of E. graminis this theory does not apply, since the inoculum is 'air-borne'. The reduced nitrogen supply might be an explanation for the lower infestation levels. For the practical implementation of integrated farming systems it is not of major importance to evaluate the effects of the single farming techniques, but the long-term effects of the whole system.

293

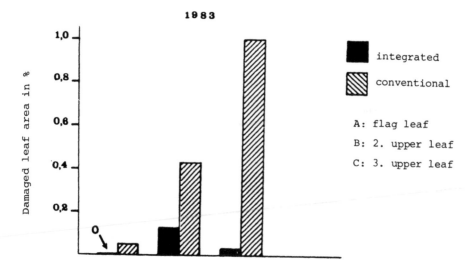

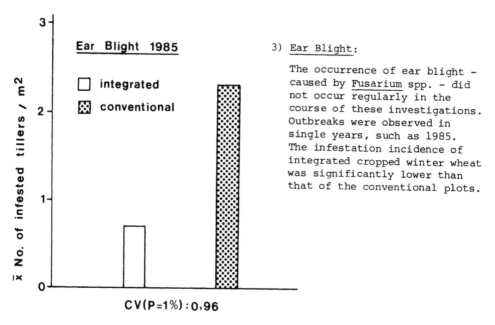

3) Ear Blight:

The occurrence of ear blight -
caused by Fusarium spp. - did
not occur regularly in the
course of these investigations.
Outbreaks were observed in
single years, such as 1985.
The infestation incidence of
integrated cropped winter wheat
was significantly lower than
that of the conventional plots.

Fig. 3. Damaged leaf area in integrated and conventionally cropped winter
wheat.

REFERENCES

(1)   EL TITI, A. and RICHTER, J. Integrated management of arable farming
systems. The Lautenbach Project. (In press).
(2)   EL TITI, A. Environmental manipulation detrimental to pests.

Integrated cropping of winter wheat and its perspectives

F.G.Wijnands & P.Vereijken
Experimental Station for Arable Farming and Field Production of Vegetables, Lelystad, Netherlands

Summary

Since 1979 on the national experimental farm for development and comparison of alternative farming systems (DFS) at Nagele, The Netherlands, an integrated farming system has been developed. As an example, the integrated cropping of wheat is discussed and placed in the context of the whole farm. The preliminary results show that the conventional wheat cropping has the highest physical yields and gross margins. Lower yields in the integrated system could not be sufficiently compensated for by cost reduction based on a considerably lower input of fertilisers and pesticides. However, the yield level can further be improved, mainly by a better timing of the N-fertilisation. Detailed cropping models are being developed to be used in microcomputer or view-data systems, to provide for the higher level of knowledge that is needed to practise integrated farming.

1. INTRODUCTION

The Dutch experimental farm for development and comparison of alternative farming systems was started in 1979 under the name 'Development of Farming Systems (DFS)'. It is situated near the village of Nagele in the north east polder, 3-4 m below sea level on heavy sandy marine clay (lutum fraction of 24%). The size of the farm is 72 ha.

Three farming systems - biodynamic, integrated and conventional - are being studied. They are run independently of each other on a commercial basis by only one manager and four co-workers. The biodynamic farm (22 ha) began as a mixed farm with 20 dairy cows and a ten year rotation. Its main objective is to be self-supporting in fertilisers and fodder. No pesticides are allowed. The conventional and the integrated farms are 17 ha each, with the same four year rotation of arable crops. The plot size is 2 ha. The conventional farm serves as a reference. Its main aim is a maximum financial return. The integrated farm should produce a corresponding financial output, but is also aimed at minimum input of fertilisers, pesticides and machinery to avoid pollution of the environment and save non-renewable resources.

The main objective of research in the first ten years will be the development of the two alternative systems. Comparative research between the systems will become more important when they are nearing a more or less optimum state. For more details concerning design, research and results of the project see Vereijken (2).

1.1 The Concept of Integrated Farming

The concept of integrated farming is extensively described in a recent IOBC-bulletin (1). It is stated that, "An integrated farming system relies

295

as much as possible on cultural and biological inputs, with chemicals as integrated supplements. The main aims are to minimise inputs of non-renewable resources and to provide a better balance between adequate production of yields and farm income on one hand and ecological, environmental and sociological aims on the other. All these considerations must be compatible with cost effectiveness." Integrated crop protection should be considered as part of an integrated crop production system. Considering the problem of overproduction and production-limiting measures, cost reduction and improvement of efficiency should be emphasised more than raising yields.

2. INTEGRATED CROPPING OF WINTER WHEAT

The main elements of the integrated cropping of wheat will be highlighted in comparison with the conventional approach.

2.1 Rotation

The crop rotation is a major element in an integrated approach. It is a highly effective preventive measure against diseases, pests and weeds. Moreover it is of great significance for the maintenance or improvement of the structure and fertility of the soil. Unfortunately the use of a healthy rotation is limited by the demands of the economy.

Therefore, and for reasons of comparative research, the integrated farm on DFS has the same four year rotation of high yielding crops as the conventional farm (Table 1). Potatoes are the most profitable, followed by sugar beet and the vegetables, carrots and onions. Peas and wheat are financially less attractive but especially the latter is needed as a break crop. Potatoes need a good soil structure and react very well to fresh organic matter in the soil. From this point of view wheat is an excellent preceding crop. Grown after sugar beet it restores the structure of the soil. Secondly it offers good opportunities for the undersowing of a green manure crop. This provides, together with the straw of the wheat, a lot of fresh organic matter and additional N, if it concerns a legume.

TABLE 1. Crop Rotations*

Conventional + Integrated
1. ½ ware potato, ½ seed potato
2. ½ pea, ¼ onion, ¼ winter carrot
3. sugar beet
4. winter wheat

* Rotations from 1985 onwards. Before this vegetables were not grown. On the conventional and integrated farms winter barley was grown instead of pea and winter carrot.

2.2 Choice of Cultivar

Winter hardiness and baking quality are a prerequisite for both systems. From Table 2 it appears that in the integrated system more attention is given to disease resistance and weed suppression than to yield potential, contrary to the conventional approach. Generally, tall cultivars with a greater distance between leaves and between flag leaf and ear are more 'resistant' to the vertical spread of diseases, in particlar Septoria nodorum blotch (Leptosphaeria nodorum). A growth regulator enhances the chances of such diseases. So, to limit the risk of lodging without using growth regulators, the cultivar to be chosen should be firm enough. Leaf abundance and early soil covering are important for effective weed suppression. Yield capacity is less important, but nevertheless it should be not less than 95% of that of top cultivars.

TABLE 2. Differences in Criteria for the Choice of Cultivar

Conventional	Integrated
Yield capacity	Resistance - lodging - major pathogens - pre-sprouting
Resistance - pre-sprouting - major pathogens	Weed suppression - leaf abundance - early soil covering
	Yield capacity

2.3   Fertilisation

High input, especially of N-fertilisers, may lead to a heavy and dense crop which is more susceptible to lodging and attack by pests and diseases. Moreover, it may encourage abundant weed growth. Therefore, we decided to restrict the first N-application in the integrated system by some 40 kg compared to the conventional system (Table 3). The N-fertilisation in both systems is based on the N-min method. The P- and K-fertilisation for the whole rotation is applied on the potatoes.

TABLE 3. Fertilisation

	Conventional	Integrated
Mineral N dressing in kg/ha		
first	140 - N-min	100 - N-min
second		60 - (30)*
third		40 dependent on crop and weather condition
Mineral P and K		0
Green manure		Sugar beet leaves and tops

* Second application can be reduced by 30 kg N/ha because sugar beet leaves and tops are incorporated

2.4   Undersowing of Green Manure Crops

From Table 4 it appears that in the integrated system a mixture of grass and clover is preferred for undersowing instead of only grass. The green manure crop should be sown as soon as possible, so it can establish before the light interception by the wheat impairs further growth. Then it

297

only has to maintain itself during the crop phase to start off in the stubble with two to three weeks lead compared to sowing after harvest. Red clover is preferred because it combines a good production (biological N-fixation) and weed suppression in the stubble with a tolerance for certain herbicides which may have to be used.

TABLE 4. Criteria for a Green Manure Crop as Undersow

Conventional	Integrated
Criteria	
Chance for success and productivity	The same
No limitations for chemical weed control in crop and stubble	Good weed suppression in the stubble
	Preferably legumes
Choice	
Perennial or Italian ryegrass	Mixture of red clover with Perennial or Italian ryegrass

2.5   Disease and Pest Control

As shown in Table 5, the conventional system mainly relies on chemicals for the control of pests and diseases. The integrated crop protection starts with the choice of a highly resistant cultivar in both the physiological and phenological sense. Postponement of the sowing until around the first of November reduces the risk of autumn infections by pests and diseases, for example aphids, slugs and foot rot diseases. According to our experience, attack by Fusarium spp. can be reduced by abandoning seed treatment. A possible lower emergence can be compensated for during tillering, ear development and grain filling. The abandoning of growth regulation and the somewhat lower N-fertilisation makes the crop less susceptible to attack by many pests and diseases.

TABLE 5. Pest and Disease Control

	Conventional	Integrated
Choice of cultivar	Yield capacity more important than resistance	The reverse
Sowing date	October	± 1 November
Seed treatment	Yes	No
N-fertilisation	Economically optimal	Restricted first application
Growth regulator	If necessary	No
Chemical control	Supervised control, EPIPRE	Adapted EPIPRE

For the additional chemical control of aphids and diseases the so-called EPIPRE-system is used (3). It is based on field observations, epidemiological simulation models, yield expectations and economic damage thresholds. We have adapted EPIPRE somewhat for the integrated system by relating the growth and multiplication rate of obligate diseases and aphids to the N-fertilisation. A second adaptation was the introduction of an extra 'levy' on the use of biocides, thus raising the damage thresholds. And finally, the yield expectations in the integrated system are somewhat lower.

2.6   Weed Control

Also in integrated weed control many elements are combined to restrict the use of herbicides as much as possible (Table 6). Delaying ploughing and sowing until around the first of November has generally no negative impact on yield and is such an effective prevention of weeds in our region that autumn applications of herbicides are not necessary. The lower N-level together with the strong weed-suppression of the chosen cultivar usually makes spring applications also superfluous.

TABLE 6. Weed Control

	Conventional	Integrated
Choice of cultivar	Yield capacity more important than weed suppression	The reverse
Sowing date	October	± 1 November
N-fertilisation	Economically optimal	Restricted first application
Mechanical control	Harrowing at undersowing	
Chemical control in crop and stubble	Preventive Broad spectrum Full-field	Curative Selective Spot treatments

Perennial weeds are controlled during the whole rotation by spot treatments of chemicals, especially in the stubbles of wheat and peas. In the choice of herbicides to be used besides their spectrum of activity their toxicological properties are considered.

2.7   Stubble Management

Contrary to the conventional system, straw in the integrated system is usually chopped and left on the field to serve as a carbon source for the soil micro-organisms after ploughing. Thus they can incorporate mineral nitrogen and prevent it from leaching. However, when straw prices exceed DFL 70 per ton, the straw is sold (Table 7).
On the conventional wheat stubble, liquid chicken manure is applied to give the potatoes, the following crop, fresh organic matter, P and K. Most of the N is lost by volatilisation of ammonia and leaching of nitrate. Only a part of the N is taken up by the green manure crop. To avoid such

TABLE 7. Management of the Stubble and Green Manure Crop

	Conventional	Integrated
Straw	Selling	Chopping, as Conv. if prices exceed DFL 70/t
Stubble	Short	Long, unless straw is sold
Organic manure after harvest	Liquid chicken manure	No
Herbicides	Preventive Broad spectrum Full-field	Curative Selective Spot treatments
Cultivation of stubbles and green manure	Ploughing in autumn under good soil/ weather conditions	Rotary cultivation before ploughing

undesirable effects in the integrated system, organic manure is only applied in spring shortly before seed-bed preparations.

In the conventional system the grass green manure is ploughed directly under, together with the wheat stubbles and straw. In the integrated system however, the mixture of straw, stubble and grass-clover is first rotary cultivated to enhance the aerobic and fast decomposition, and to avoid disturbance of the air and water status, and root development of the next crop.

3. PRELIMINARY RESULTS

Table 8 shows that over the years 1982-1986 on average only 2.0 chemical treatments per field were applied in the integrated system against 5.7 in the conventional system. Expressed in active ingredients the pesticide input was reduced from 6.23 to 1.69 kg/ha.

TABLE 8. Chemical Control over the Years 1982-1986

	Average number of treatments per field		Active ingredients (kg/ha)	
	Conventional	Integrated	Conventional	Integrated
Herbicides	1.5	0.4	3.76	0.59
Growth regulator	0.8	0.0	0.43	0.00
Fungicides	2.4	0.8	1.85	0.97
Insecticides	1.0	0.8	0.19	0.13
Total	5.7	2.0	6.23	1.69

Figure 1 shows the physical yields in both systems over the years 1982-1986. The yield stability in the integrated system is somewhat lower than in the conventional system (variation coefficient of quantitative production is 14.7% and 10.5% respectively).

300

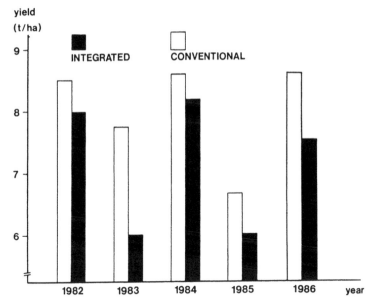

Fig. 1. The physical yields of winter wheat in the integrated and conventional system during the years 1982-1986.

From Table 9 it appears that over the same period the wheat in the integrated system yielded on average 0.9 ton/ha less, eventually leading to a DFL 340/ha lower gross margin. Table 9 also shows that at world market prices for the grain (DFL 0.32/kg) the difference between the two systems was reduced to DFL 150/ha. So, even then the lower physical yields were insufficiently compensated for by cost reduction, based on the lower input of pesticides and fertilisers.

TABLE 9. Yields over the Years 1982-1986

		Conventional	Integrated
Physical yield	1)	8.03	7.13
Total returns	2)	4.38	3.90
Allocated costs	3)	1.22	1.08
Gross margin	4)	3.16	2.82
World market prices			
Total returns	2)	2.57	2.28
Allocated costs	3)	1.22	1.08
Gross margin	4)	1.35	1.20

1) t/ha
2) price/t x t/ha in 1000 guilders
3) costs of pesticides, fertilisers, hired labour, sowing seed, insurance, interest
4) total returns minus allocated costs

Nevertheless, it should be mentioned that in the same period over the whole rotation the integrated farm had slightly better economical results, thanks to large reduction in costs (1).

4. PERSPECTIVES

Although in recent years, the yields of winter wheat in The Netherlands have increased considerably, the inputs, especially of pesticides, have remained at a rather low level. The average number of chemical treatments per field is only some 6-7 (including herbicides, seed treatment and growth regulators).

In a situation like this with very efficient use of inputs there is little opportunity left for substantial decrease of allocated costs. However, we tried in our experimental integrated approach to abandon most of the chemical control treatments and to reduce the N-fertilisation by some 40 kg/ha.

However, economically, this only meant an average input reduction of 140 guilders/ha, equalling 250 kg wheat at EC prices, or 435 kg at world market prices. This was not enough to compensate for the average yield loss of about 900 kg/ha over the years 1982/1984. So, it has to be concluded that economically integrated wheat cropping cannot yet compete with a conventional approach, unless the average yield loss can be brought back to a lower level.

How can the yield level in the integrated system be improved? In our opinion mainly by better dosing and timing of the second N-application. A result of the reduction of the first application by some 40 kg/ha with the normally timed second application around GS 31, is that the development of side-tillers is reduced. To avoid this, the second application should be given earlier.

Progress on this major point may also enhance the yield stability. This can be further improved by helping the farmer with his decision making by a computerised guidance system, which we are developing.

These systems enable the farmers to act on a higher level of knowledge by providing detailed cropping programmes, based on a systematic survey of all possible decisions, with a logic key to choose in every situation the right one, depending on the specific farm, soil and weather conditions.

More specifically, these guidance systems contain the following elements:

- Warning procedures; depending on crop attributes, growth stage and previous cropping measures, attention is called to certain decisions that have to be considered.
- Decisions rules and advice models; for control of pests, weeds and diseases but also for fertilisation, timing of operations, etc.
- General information; for example on disease observations, use of pesticides, properties of cultivars, susceptibility of diseases, pests and weeds for biocides, etc.
- Registration; input/output, feedback to system.

Automated cropping guidance systems offer probably the most efficient way to introduce new and highly sophisticated systems, such as integrated farming, into practice and to keep them updated.

Therefore, we have decided to participate in a regional view-data project. At first a conventional wheat cropping system will be implemented, to get acquainted with the specific problems of this new type of extension. Next, an integrated version will be tried out within the framework of a farmers' study club. If, in cooperation with such a small group of highly motivated and skillful farmers, a sufficiently successful and safe integrated wheat cropping system is elaborated, it will be put at the disposal of the farmers, as a final step.

REFERENCES

(1)   VEREIJKEN, P., 1986. From conventional to integrated agriculture.
       Netherlands Journal of Agricultural Science, 34, 387-393.
(2)   VEREIJKEN, P., EDWARDS, C., EL TITI, A., FOUGEROUX, A. and WAY, M.,
       1986. Report of the study group 'Management of farming systems for
       integrated control'. IOBC-WPRS Bulletin 1986/IX/2. ISBN 92-9067-001-0.
(3)   ZADOKS, J.C., 1981. EPIPRE: A disease and pest management system for
       winter wheat developed in The Netherlands. EPPO Bulletin 11, 365-369.

Investigation of farming systems on integrated crop protection in cereals: The work of the Game Conservancy, UK

N.W.Sotherton
The Cereals and Gamebirds Research Project, The Game Conservancy, Fordingbridge, Hampshire, UK

G.R.Potts
The Game Conservancy, Fordingbridge, Hampshire, UK

Summary

The entomological studies of the Game Conservancy began in the 1960's when work to assess the effects of modern agriculture on wild gamebird populations began. Interest in gamebird ecology, stimulated many studies to be carried out. As a result, long-term monitoring of the status of insects in cereal fields began in 1969 and has been undertaken annually since that time.

Autecological studies of non-target arthropod species have been carried out on both cereal aphid predators eg. Carabidae and Staphylinidae and on important species of prey for gamebird chicks eg. Chrysomelidae. Work has begun to screen the spectrum of the insecticidal activity of pesticides used in cereal fields, especially foliar fungicides against non-target, beneficial arthropods.

Management of crop and non-crop habitats to encourage cereal aphid natural enemies or chick food species, or both has been carried out. This includes the provision of overwintering refuges in field boundaries and favourable conditions on the crop floor by not using certain herbicides or by undersowing. Restricting the use of pesticides on cereal field headlands has produced weedy, insecticide free, areas at the crop margin. As a result, an increase has been observed in numbers of beneficial insects especially phytophagous species (and partridges).

1. Introduction

The Game Conservancy is an independent research organisation whose work is funded from membership subscription, and by grants from charitable trust funds and Government/Research Council grants. The aim of the research is to improve the status of wild populations of many (quarry and non-quarry) species found in both agricultural and non-agricultural land. It may therefore appear paradoxical for a research remit of this nature to encompass entomology, farming systems and integrated pest management as well as the vertebrates themselves. However, it was as a result of work on a quarry species, the grey, English or Hungarian partridge (Perdix perdix L.) that studies of the cereal ecosystem began.

305

In Britain, the grey partridge has suffered a severe decline in numbers over the past thirty years. Spring pair densities have been surveyed by the Game Conservancy since 1933 and the national decline since 1952 has been about 80%. Similar declines have been observed in North America and Eastern Europe (1). The reason for this decline in Britain has been shown to be the increased rates of chick mortality (1).

For the first two to three weeks of life, grey partridge chicks rely almost entirely on insects (some of them predators of cereal aphids) in the diet, before switching to a plant diet at ten days to three weeks of age (2). Poor chick survival rates are caused by a lack of sufficient insect food in the diet of chicks during these crucial early weeks (1) (Figure 1).

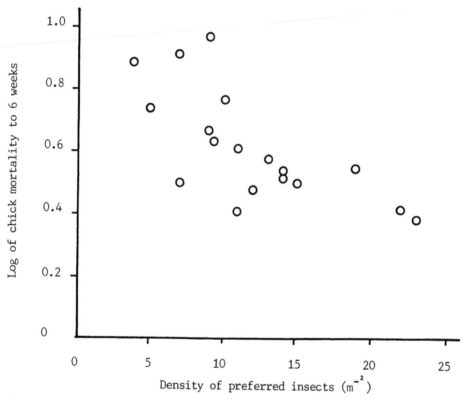

Figure 1. Annual chick mortality in relation to the density of preferred insects from 1969-1986, third week of June, Sussex, United Kingdom.

Gamebird chicks forage predominantly in cereal crops, which have become more and more intensively farmed since the 1950's. Intensification over this period has been associated with an increased use of pesticides especially (3) which are responsible for this lack of sufficient insect food.

This then was the background for much of the Game Conservancy's research efforts from 1969 to the present day on insect ecology. In fact the broadly based studies carried out by Potts & Vickerman in the early 1970's were some of the first to be undertaken in the cereal ecosystem (4). Their remit was to discover how intensive agriculture was affecting the chick food insects vital in the survival of young gamebirds and to devise cereal management systems more sympathetic to non-target insects.

2. Approaches

The Game Conservancy's entomological studies have approached the problem from several directions. Firstly, there was a commitment to long-term insect monitoring in order to quantify changes in numbers on intensively farmed cereal growing areas coupled with the detailed recording of farming practices and pesticide use. Secondly, groups of insects that were considered most important as either cereal aphid natural enemies or preferred food items of chicks were identified. Especially important species have then been subject to detailed autecological studies to quantify the impact of intensive farming methods especially pesticide use. Other species were used to screen compounds to discover the spectrum of their insecticidal activity against beneficial arthropods. Finally, large-scale, on farm, research has been carried out to manage both crop and non-crop habitats to create optimum conditions for the important beneficial insects.

2.1 Long-term monitoring

Every year since 1969, annual estimates of insect density have been made on a 62 km² mixed and arable study area in Sussex in southern England. Details of this area are described elsewhere (4). Samples are taken with a Dietrick vacuum insect sampler (D-vac) from between 100-150 cereal fields scattered through the 17 farms on the study area. Details of the sampling methods are also given elsewhere (5, 6). Monitoring has enabled us to describe between crop and between year variations in many species of both pest (7, 8) and beneficial arthropods (9) and even in one case to propose forecasting schemes based on spring temperatures (7).

As a result of monitoring, based on the first 11 years data (1970-1980) from 104 taxa, 50% appeared to have declined whereas only 4% had increased (5). Some of the conclusions reached over the full 17 year period have been that annual variations in the abundance of insects have been dominated by cereal aphids but in the longer-term, the densities of most species were much lower than before the widespread use of herbicides and in many cases, numbers of these were still declining. This was especially the case for leaf-feeding Coleoptera (mainly Chrysomelidae and Curculionidae) and the Tenthredinidae and Lepidoptera (1) (Figure 2).

Monitoring work is funded for a further two years, during which time the computerised data bank will be analysed in greater detail in order to correlate trends in insect numbers with changes in agricultural practices especially in crop rotation type and crop edge effects (6), to create hypotheses to test in replicated farm scale experiments and to give direction to our future entomological studies.

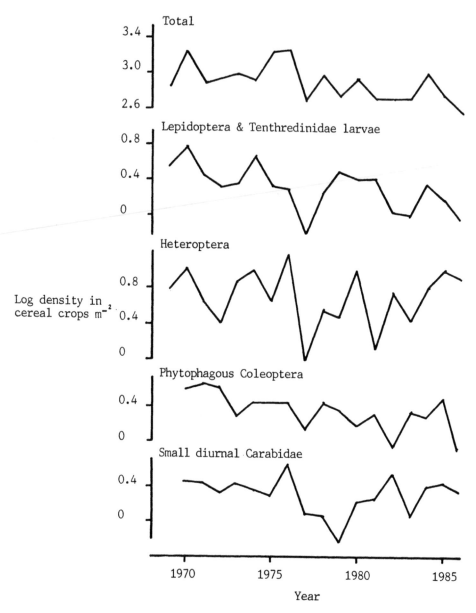

Figure 2. Trends in densities of preferred insects in cereal crops from 1969-1986, third week of June, Sussex, United Kingdom. (Total non-preferred insects for comparison).

2.2 Autecological and group studies

Whilst insect monitoring continued, research interests began to focus on attempting to identify and rank the most important species of beneficial insects. Over 300 species of polyphagous cereal aphid predators have been identified (K.D. Sunderland Pers. comm.), but which were the candidate species that merited detailed investigation for their ability to reduce populations levels of pest species?

Work at The Game Conservancy tended to concentrate on the feeding habits of generalist arthropod predators, including studies on their diet, temporal and spatial distribution within crops and their importance as aphid predators (10, 11, 12).

In 1980, Sunderland & Vickerman made an attempt to rank species of generalist predator for their importance as cereal aphid natural enemies by using a predation index. The ranking was based on two criteria; namely predator density and the proportion of individuals captured during the increase phase of the aphid population found (by gut dissection) to contain aphid remains (13). Six of the 16 species investigated were shown to have a relatively high predation index. These were three species of Carabidae (Demetrias atricapillus (L.), Agonum dorsale (Part) and Bembidion lampros (Herbst.) two species of Staphylinidae of the genus Tachyporus (T. hypnorum (F.) and T. chrysomelinus (L.)) and the common earwig (Forficula auricularia L.). As a result, autecological studies have been carried out at Southampton University in association with The Game Conservancy on A. dorsale (14) and D. atricapillus (15) whilst currently a study has began on the genus Tachyporus. A subsequent study has now also been completed at Southampton on F. auricularia with informal assistance from The Game Conservancy (16).

Recently, other criteria for the ranking of polyphagous predators have been added to aid our appraisal of species as cereal aphid predators. For example, the ability of the predators to overwinter on farmland (17, 18) and subsequently the extent and timing of their penetration into cereal crops in the spring from their overwintering refuges (6). Related work has also included the strategies adopted by the cereal aphids themselves to overwinter and the consequences for the impact of natural enemies (19, 20).

Studies on other groups of the natural enemies of cereal aphids have focussed on the quantification of the importance of aphid-specific predators by gut dissection (11) and predator exclusion using cages (21).

Other non-pest species of arthropods that have been identified as preferred food items of wild gamebirds chicks (2, 22) have also been studied in more detail, for example, the ecology of the chrysomelid beetle, Gastrophysa polygoni (L.) has been investigated (23). Because of its poor powers of dispersal and dependence on species of weeds as host plants, this beetle has been used as an indicator species to detect the effects of agricultural practices on the non-target arthropod fauna (23, 24, 25) as well as the investigation of the role of the predatory fauna of cereal fields on a non-pest species (26, 27).

2.3 Studies on intensive farming practices

Throughout the period of insect monitoring, farming practices and pesticide use   data have been gathered for every field sampled on the study area in Sussex (1).   Studies on the effects of modern farming practices were carried out and began by quantifying effects of the use of herbicides and especially their impact on partridges via the removal of the host plants and thus preferred insect food items (28).   Work later progressed to investigate the effects of specific compounds on predatory and non-target arthropods (24, 29) and the effects of general and specific-herbicide use on particular species (24).

The major change in pesticide use over the period of our monitoring has been the use of foliar fungicides (1).   Research into the effects of these compounds has been undertaken on both pests (30) and beneficial species (25, 31, 32, 33). G. polygoni appeared to be susceptible species to fungicide use (25).   The impact of insecticides of chick survival has been studied (34,35) as well as the impact of individual compounds on the predatory fauna (36, 37).   Dimethoate was shown to have a wide spectrum of activity against cereal aphid natural enemies (36).

The effects of other agricultural practices on beneficial arthropods have also been studied.   These include straw burning which was shown to decrease numbers of beneficial insects (38, 39, 40) and the establishment of grassland by undersowing cereal crops which was shown to favour · both natural enemy and chick food insect groups (4, 41).

Currently, pesticides are being screened, both in the laboratory and in the field in order to be able to advise farmers on the spectrum of activity of the compounds used in cereals.   In the laboratory, this has involved the assessment of pesticides, especially foliar fungicides for their insecticidal activity against both predatory species and chick food insects.   Mortality is measured following pesticide contact or ingestion of contaminated material.   Sub-lethal effects on longevity, fecundity, fertility and predatory potential of surviving insects are also measured.

Field work has involved the estimation of the extent and consequences of the insecticidal properties of the foliar fungicide, pyrazophos (32,33). Under field conditions, pyrazophos has been shown to reduce numbers of beneficial arthropods by up to an average of 96% relative to untreated areas (Table I).

More recently, work has involved the assessment of the spectrum of activity of four aphicides approved to control cereal aphids in the summer.   Laboratory   screening   using   the   carabid beetle   Trechus quadristriatus (Shrank), an aphid predator and also a preferred chick food item has shown some of these compounds to be harmless under laboratory conditions (Table II).

Studies on the impact of pyrethroid insecticides on non-target, beneficial species undertaken at Southampton University are currently being undertaken in association with The Game Conservancy.

2.4 Habitat management

Having recognised the importance of groups of beneficial arthropods

Table I. The mean percentage reduction of beneficial arthropod groups in cereal field twenty days after treatment with pyrazophos relative to adjacent unsprayed plots of cereals. Data from daytime ground-zone searches, Hampshire, southern England, June 1984, May 1985.

Percentage reduction

Crop	Winter wheat	Spring barley	Winter barley	Winter wheat
Year	1984	1984	1985	1985
Growth stage of application	50	60	37	37
Total predatory arthropods	65	74	76	74
Carabidae	69	65	74	65
Staphylinidae	80	75	83	84
Tachyporus sp.	80	79	88	86
Highly ranked polyphagous predators	87	70	86	83
Araneae	37	62	49	57
Total aphid specific predators	89	94	88	96

Table II. Laboratory screening of four aphicides against Trechus quadristriatus. Pesticides applied at field dose rate and dilution via Potter tower (n=70).

	% mortality after 48 hours
Control (distilled water)	0.0
pirimicarb	1.5
demeton-s-methyl	16.9
phosalone	0.0
dimethoate	100.0

to both pest suppression and survival of wild gamebird chicks, research attention has recently become focussed on the management of the arthropod habitats in both the crop and adjacent non-cropped areas. Our work has demonstrated the importance of field boundaries for the overwintering sites of the six most highly ranked polyphagous predators (17) and has revealed the preferred type of field boundary structure for five of them. For example, significantly higher densities of the premier ranked predator, D. atricapillus were found in hedge banks than in grass banks, grass strips or shelterbelts (18) and the mid-field, mid-summer density of this species was significantly (positively) correlated to its overwintering densities in adjacent field boundaries. We also can relate overwintering densities to various structural vegetational and micro-climatic parameters describing the refuge sites (15, 42). Studies in collaboration with Southampton University have now begun to experimentally manipulate field boundaries to improve their potential as overwintering sites and to create new boundary/refuge sites in fields to increase overwintering success and thus augment predator pressure effectiveness.

Selective use of pesticides at the crop margin has produced insecticide free, relatively weedy areas in which chick food insect numbers have been significantly increased including many smaller species of Carabidae and Staphylinidae (43) but particularly Heteroptera and Chrysomelidae (44). This has been of very great short- and long-term benefit to partridges and pheasants (44, 45, 46).

3. Conclusions and Future

The entomological work of The Game Conservancy have become an integral part of the studies of wild gamebird populations, farming systems and the cereal ecosystem ecology. Our present studies are being carried out under the auspices of The Cereals and Gamebirds Research Project. One of the most important aspects of this Project is that it is funded primarily by farmers. Future work will focus on providing them directly with information to help them farm efficiently and profitably in ways that are sympathetic to the needs of farmland wildlife, not only gamebirds but also non-target, non beneficial species of farmland wildlife such as Lepidoptera (47, 48).

The late 1980's will also see a greater collaboration with Europe. This involves our continued participation in I.O.B.C. groups but also with the anticipated funding of a joint study with the National Agency of Environmental Protection in Denmark.

4. Acknowledgments

The authors would like to thank those who have been involved with the entomological work of The Game Conservancy over the last 20 years. Most of these people are listed in the reference section following. We would also like to thank the following people who will add to this list over the next few years; N. Carrick, A. Cherry, P. Cuthbertson, P. Dennis, M. Langley, S. Moreby, G. Owen, A. Pullen and S. Southway.

5. References

1. POTTS, G.R. (1986). The Partridge: Pesticides, Predation and Conservation. Collins, London, U.K.
2. GREEN, R.E. (1984). The feeding ecology of and survival of partridge chicks (Alectoris rufa and Perdix perdix) on arable farmland in East Anglia, UK. Journal of Applied Ecology, 21, 817-830.
3. RANDS, M.R.W. and SOTHERTON, N.W. (in press). Gamebirds, Agriculture and Conservation. In: Population, Biology and Management of British Gamebirds, Eds. P.J. Hudson and M.R.W. Rands, Collins, London.
4. POTTS, G.R. and VICKERMAN, G.P. (1974). Studies on the cereal ecosystem. Advances in Ecological Research 8, 107-197.
5. POTTS, G.R. (1984). Monitoring changes in the cereal ecosystem. In: Proceedings of the N.E.R.C. I.T.E. Symposium No.13 'Agriculture and the Environment'. Ed. D. Jenkins. Monks Wood Experimental Station, Huntingdon, 128-134.
6. COOMBES, D.S. and SOTHERTON N.W. (1986). The dispersal and distribution of polyphagous predatory Coleoptera in cereals. Annals

of Applied Biology 108, 461-474.

7. VICKERMAN, G.P. (1977). Monitoring and forecasting insect pests of cereals. Proceedings 1977 British Crop Protection Conference - Pests and Diseases, 227-234.

8. VICKERMAN, G.P. (1982). Distribution and abundance of adult Opomyza florum (Diptera: Opomyzidae) in cereal crops and grassland. Annals of Applied Biology, 101, 441-447.

9. VICKERMAN, G.P. (1982). Distribution and abundance of cereal aphid parasitoids (Aphidius spp.) on grassland and winter wheat. Annals of Applied Biology, 101, 185-190.

10. SUNDERLAND, K.D. (1975). The diet of some predatory arthropods in cereal crops. Journal of Applied Ecology 12, 507-515.

11. VICKERMAN, G.P. and SUNDERLAND, K.D. (1975). Arthropods in cereal crops: nocturnal activity, vertical distribution and aphid predation. Journal of Applied Ecology 12, 755-766.

12. CARTER, N. and SOTHERTON, N.W. (1983). The role of polyphagous predators in the control of cereal aphids. Xth International Congress of Plant Protection, 1983, p. 778.

13. SUNDERLAND, K.D. and VICKERMAN, G.P. (1980). Aphid feeding by some polyphagous predators in relation to aphid density in cereal fields. Journal of Applied Ecology 17, 389-396.

14. GRIFFITHS, E., WRATTEN, S.D. and VICKERMAN, G.P. (1985). Foraging by the carabid beetle Agonum dorsale in the field. Ecological Entomology, 10, 181-189.

15. COOMBES, D.S. (1986). Factors limiting the effectiveness of Demetrias atricapillus (L.) (Coleoptera: Carabidae) as a predator of cereal aphids. Unpublished Ph.D. Thesis, University of Southampton.

16. CARILLO, R. (1985). The ecology of, and aphid predation by the European earwig Forficula auricularia L. in grassland and barley. Unpublished Ph.D. thesis, University of Southampton.

17. SOTHERTON, N.W. (1984). The distribution and abundance of predatory arthropods overwintering on farmland. Annals of Applied Biology, 105, 423-429.

18. SOTHERTON, N.W. (1985). The distribution and abundance of predatory arthropods overwintering in field boundaries. Annals of Applied Biology, 106, 17-21.

19. HAND, S.C. (1982). Overwintering and dispersal of cereal aphids. Unpublished Ph.D. thesis, University of Southampton.

20. WILLIAMS, C.T. (1982). The overwintering and low temperature biology of cereal aphids. Unpublished Ph.D. thesis, Southampton University.

21. CHAMBERS, R.J., SUNDERLAND, K.D., WYATT, I.J. and VICKERMAN, G.P. (1983). The effects of predator exclusion and caging on cereal aphids in winter wheat. Journal of Applied Ecology, 20, 209-225.

22. HILL, D.A. (1985). The feeding ecology and survival of pheasant chicks on arable farmland. Journal of Applied Ecology, 22, 645-654.

23. SOTHERTON, N.W. (1982). Observations on the biology and ecology of the chrysomelid beetle Gastrophysa polygoni in cereal fields. Ecological Entomology 7, 197-206.

24. SOTHERTON, N.W. (1982). Effects of herbicides on the chrysomelid beetle Gastrophysa polygoni (L.) in laboratory and field. Zeitschrift für angewandte Entomologie, 94, 446-451.

25. VICKERMAN, G.P. and SOTHERTON, N.W. (1983). Effects of some foliar fungicides on the chrysomelid beetle Gastrophysa polygoni (L.). Pesticide Science, 14, 405-411.

26. SOTHERTON, N.W. (1982). Predation of a chrysomelid beetle (Gastrophysa polygoni) in cereals by polyphagous predators. Annals of Applied Biology 101, 196-199.

27. SOTHERTON, N.W., WRATTEN, S.D. and VICKERMAN, G.P. (1984). The role of egg predation in the population dynamics of Gastrophysa polygoni (Coleoptera) in cereal fields. Oikos 43, 301-308.

28. POTTS, G.R. (1970). The effects of the use of herbicides in cereals on aphids and on the feeding ecology of partridges. Proceedings of the 10th British Weed Control Conference, 1970, 299-302.

29. VICKERMAN, G.P. (1974). Some effects of grass weed control on the arthropod fauna of cereals. Proceedings of the 12th British Weed Control Conference 1974, 929-939.

30. VICKERMAN G.P. (1977). The effects of foliar fungicides on some insect pests of cereals. Proceedings of the 1977 British Crop Protection Conference - Pests & Diseases 1, 121-128. Brighton, Sussex.

31. SOTHERTON, N.W. and MOREBY, S.J. (1984). Contact toxicity of some foliar fungicides sprays to three species of polyphagous predators found in cereal fields. Tests of Agrochemicals and Cultivars 5, Annals of Applied Biology Supplement, 16-17.

32. SOTHERTON, N.W., MOREBY, S.J. and LANGLEY, M.G. (in press). The effects of the foliar fungicide pyrazophos on beneficial arthropods in barley fields. Annals of Applied Biology.

33. SOTHERTON, N.W. and MOREBY, S.J. (in press). The effects of foliar fungicides on beneficial arthropods in wheat fields. Entomophaga.

34. POTTS, G.R. (1977). Population dynamics of the grey partridge: overall effects of herbicides and insecticides on chick survival rates. In: Proceedings of the XIIIth International Congress of Game Biologists, 1977, 203-211. Atlanta, Georgia.

35. POTTS, G.R. (1981). Insecticides sprays and the survival of partridge chicks. Game Conservancy Annual Review 12, 39-48.

36. VICKERMAN, G.P. and SUNDERLAND, K.D. (1977). Some effects of dimethoate on arthropods in winter wheat. Journal of Applied Ecology, 14, 767-777.

37. MOREBY, S.J. and POTTS, G.R. (1985). Insecticides and the survival of gamebird chicks in 1984. Game Conservancy Annual Review 16, 47-49.

38. VICKERMAN, G.P. (1974). Some effects of grass weed control on the arthropod fauna of cereals. Proceedings of the 12th British Weed Control Conference 3, 929-39.

39. SOTHERTON, N.W. (1980). The ecology of Gastrophysa polygoni (L.) (Coleoptera: Chrysomelidae) in cereals. Unpublished Ph.D. Thesis, University of Southampton.

40. POTTS, G.R. (1974). The grey partridge: problems of quantifying the ecological effects of pesticides. In: Proceedings of the XIth International Congress of Game Biologists, 1973, 405-413, Stockholm, Sweden.

41. VICKERMAN, G.P. (1978). The arthropod fauna of undersown grass and cereal fields. Scientific Proceedings of the Royal Dublin Society 6, 156-165.

42. SOTHERTON, N.W. (1984). Distribution and abundance of predatory insects on farmland in relation to habitat. Report of The Game Conservancy, pp. 62.

43. SOTHERTON, N.W., RANDS, M.R.W. and MOREBY, S.J. (1985). Comparisons of herbicide treated and untreated headlands on the survival of game and wildlife. Proceedings of the 1985 British Crop Protection Conference - Weeds 3, 991-998. Brighton, Sussex.

44. RANDS, M.R.W. (1985). Pesticide use on cereals and the survival of partridge chicks: a field experiment. _Journal of Applied Ecology_, 22, 49-54.
45. RANDS, M.R.W. (1985). The survival of gamebird (Galliformes) chicks in relation to pesticide use on cereals. _Ibis_, 128, 57-64.
46. RANDS, M.R.W. (1986). The effect of hedgerow characteristics on partridge breeding density. _Journal of Applied Ecology_, 23, 479-487.
47. RANDS, M.R.W. & SOTHERTON, N.W. (1985). Pesticide use on cereal crops and changes in the abundance of butterflies on arable farmland in England. _Biological Conservation_, 36, 71-82.
48. DOVER, J.W. (1986). An investigation into the effects of pesticide-free headlands in cereal fields on butterfly populations. _Annual Review of the Game Conservancy_, 17, 65-69.

Closing session

Chairman: K.D.Sunderland

Conclusions and recommendations on general discussion and sessions' report

R.Cavalloro
CEC, Joint Research Centre, Ispra, Italy

A.El Titi
Landesanstalt für Pflanzenschutz, Stuttgart, FR Germany

P.Lucas
Station de Pathologie Végétale, INRA, Le Rheu, France

S.D.Wratten
Department of Biology, University of Southampton, UK

Introduction

The experts's group meeting on "Integrated crop protection in cereals" was organized by the Commission of the European Communities, Directorate-General Agriculture, in collaboration with the Glasshouse Crops Research Institute, Department of Entomology & Insect Pathology.

The principal aim was to discuss the use of integrated methods with a view to the reduction of pesticide inputs and increases in the efficiency of pest control, in the framework of the European agricultural policy.

The experts presented their results divided into three principal lines: "Pests", "Diseases and Weeds", and "Farming Systems". At the closing session each chairman presented, on the basis of an analysis of the communications, some suggestions for consideration by the participants.

The meeting was concluded with a one day technical-scientific excursion to the agricultural evaluation unit at Southampton university (hosted by Dr. S.D. Wratten) and to the game conservancy cereals and gamebirds research project (hosted by Dr. G.R. Poots and Dr. N.W. Sotherton at Fordingbridge).

At Southampton university the participants were given two excellent presentation concerning:
- the use of computer technology in information transfer of research findings into practical agricultural usage;
- novel computer methodology for studying the behaviour of predatory insects.

At Fordingbridge the participants enjoyed an impressive wide-ranging exposition of the research activities of the game conservancy, including effect of pesticides on beneficial arthropods and the results of long-term studies monitoring the cereal ecosystem.

Report

After a wide-ranging discussion, the participants agreed on the following:

"Pests"
-----

a) Conclusions

(i) Changes in pest status

In some EC-Member States there has recently been a considerable increase in the acreage of cereal production, with concomitant increases in fertilizer and pesticide usage; this has resulted in changes in the status of some pests and the creation of new pests. Effective control of some major pests in some countries (e.g. Sunn pests (<u>Aelia</u> spp., <u>Eurygaster</u> spp.: Heteroptera) in Southern Europe) still relies exclusively on insecticides.

(ii) Forecasting

Progress has been made in the development of short-term forecasts of peak abundance of some aphid species (e.g. <u>Sitobion</u> <u>avenae</u> (F.)). Captures of adults in light-traps, together with examination of plants for oviposition, allows some foreward estimation of damage to maize by corn stalk borer (<u>Sesamia</u> <u>nonagrioides</u> (Lef.)).

(iii) Virus vectors

There is now a better understanding of how the continuity of some aphid-transmitted virus diseases is achieved in some areas. For example, the PAV strain of barley yellow dwarf virus (BYDV or VJNO) in Western (oceanic) France is maintained sequentially on barley and maize with cultivated grasses forming a long-term reservoir.

(iv) Natural enemies

Evidence is now available for the value of parasites and stenophagous and polyphagous predators in pest control. Emphasis in previous studies was given to Carabidae, but it is now realised that Staphylinidae and Araneae are also important groups of predators. The importance of the total assemblage of polyphagous predators, rather than component species, is stressed.

(v) Partially-resistant varieties

Varieties of maize exhibiting some degree of resistance to pests have proved useful in restricting attack by the corn stalk borer and in reducing aphid multiplication and virus spread. Although many different mechanisms can underly partial resistance in cereals, there is a clear negative correlation between the level of hydroxamic acids in seedling wheat and the intrinsic rate of increase of the aphid <u>Sitobion</u> <u>avenae</u>. It is now appreciated that a wide range of interactions between the effects of partial resistance and natural enemies are possible and that many could be of benefit for pest control.

(vi) Pesticides and natural enemies

Autumn applications of pyrethroids have short-term effects on populations of some predators, but on others the reduction in numbers persists well into summer. Broad-spectrum insecticides applied during the summer are detrimental to most natural enemies. More selective compounds (such as pirimicarb) spare some of the natural enemies but result in starvation of stenophagous predators. One solution to this problem (currently studied for late-season aphid attack) is to use reduced dosages of selective pesticides; many stenophages survive, yield loss is prevented and there is a good economic return with minimal environmental damage. Another means of ameliorating the undesirable effects of pesticides is to omit spraying the headlands; this permits the survival and overwintering of at least a proportion of the natural enemy fauna and also benefits other wildlife. Because of the current wide acceptance of the importance of natural enemies, much thought has been given to devising rigorous, repeatable protocols for assessing the many potential effects of agrochemicals on them; blueprints are now available.

b) Recommendations

(i) Changes in pest status

In EC-Member States where there has been a recent large-scale increase in cereal production, the basic data on many aspects of cereal pest control are sparse or lacking. There is a need to establish economic thresholds, develop forecasting schemes, identify natural enemies and resistant varieties and experiment with integrated control. The prospects for controlling corn stalk borer using pheromone lures and mass-trapping appear to be good, but studies on flight activity and dispersal behaviour of the adult moths are needed.

(ii) Forecasting and virus vectors

It is anticipated that the accuracy of forecasting both aphid attack in summer and virus (BYDV) incidence in the autumn would be improved if long-term continuous monitoring of these pests in fields (in addition to "Euraphid" suction traps) were carried out at a range of national and European sites and the results made available annually for analysis by interested parties. There is a need for work on the MAV and RPV strains of BYDV, as they could have a different epidemiology from the better-known PAV strain.

(iii) Natural enemies

There is a need to model the effects if natural enemies on pest populations and also to find techniques for enhancing populations of species considered to have especially high potential. Studies aimed at understanding the role of alternative prey in affecting the efficiency of pest control by polyphagous predators are required.

(iv) Partially-resistant varieties

It is desirable to initiate a system of collaborative interchange of genetic material and methods to facilitate the rapid accumulation of a collection of varieties having useful resistance characteristics (especially multiply-based ones) against pests and virus diseases. There is a need to understand and quantify those interactions between partially-resistant varieties and natural enemies which promote pest control.

(v) Pesticides and natural enemies

Laboratory toxicology studies, used in isolation, are considered to be inadequate. Field studies (preferably on a wide range of temporal and spatial scales and taking account of differences in the mobility of various natural enemies) should be implemented to determine the levels and routes of exposure of natural enemies to pesticides. The use of reduced-dose selective pesticides should be studied in relation to early-season aphid attack.

(vi) Integration of new methods of pest control

It is now considered timely to evaluate the effectiveness of packages of multiple pest control measures; e.g. a combination of natural enemies, partially-resistant cereal varieties, reduced-dose selective pesticides and unsprayed headlands.

"Diseases and Weeds"
----

a) Conclusions

(i) Disease forecasting

A forecasting system for leaf spot (<u>Septoria tritici</u> Rob. <u>ex</u> Desm.), which incorporates monitoring of inoculum levels and amount of rain splash, is being tested; initial results are encouraging. The movements, across Europe, of strains of barley mildew (<u>Erisyphe graminis</u> DC. f. sp. <u>hordei</u>), varying in resistance to fungicides and virulence against barley resistance, have been mapped and are reasonably predictable. Thus, a Member State can have advance warning of the characteristics of an imminent strain of barley mildew and be ready to deploy the appropriate varieties and fungicides.

(ii) Techniques of disease limitation

Several diseases can be transmitted through the seed and the use of fungicide-treated seeds can reduce the need for fungicide sprays in the field. Some seed treatments reduce the incidence of take-all (<u>Gaeumannomyces graminis</u> (Sacc.) von Arx et Olivier var. <u>tritici</u> Walker) and eyespot (<u>Pseudocercosporella herpotrichoides</u> (Fron) Deighton). Some winter wheat varieties are partially-resistant to these diseases and the incidence of take-all (but not eyespot) is reduced if ammonium sulphate fertiliser is used instead of ammonium nitrate. Diversification of varieties and fungicides within fields can reduce the rate of spread of barley mildew and wheat yellow rust (<u>Puccinia striiformis</u>) Westend). Yield of barley under attack by barley mildew was the same whether the seeds of just one component of a three variety mixture were treated of whether the mixture was conventionally treated. For wheat under attack by yellow rust, some mixtures of susceptible and resistant varieties gave the same yield, when sprayed once with fungicide, as did susceptible monocultures sprayed three times. Although foliar diseases can adapt very rapidly to single control measures, they have difficulty in adapting to simultaneous multiple control measures; this weakness is exploited in integrated disease control by ensuring that adjacent plants are different in terms of either variety or fungicide treatment.

(iii) Weeds

The extensive use of herbicides in some EC-Member States is selecting for resistant weeds and can also result in contamination of bodies of water; cases are on record of herbicides, containing certain active ingredients, being withdrawn because of contamination of water supplies. Inundative biological control of weeds using pathogens may eventually reduce the reliance on chemical pesticides. As an example, the troublesome weeds, velvetleaf (Abutilon theophrasti Medicus), is controlled as reliably by a pathogenic fungus (Alternaria tenuissima (Kunze ex Nees et Nees) Wiltshire) as by several types of herbicide.

b) Recommendations

(i) Changed status of diseases

Associated with the recent increase in cereal production in some Member States there has been a reduction in the number of varieties grown and an increasing reliance on foreign varieties. There is therefore a need to monitor disease levels, which could increase because of monoculture, and to breed varieties suitable for local conditions.

(ii) Methodology and taxonomic studies of diseases

There is a need to develop effective methods for the evaluation of the inoculum potential of Polymyxa in soil and for determining levels of contamination of seeds by diseases. More quantitative information is needed on the relationship between seed infection and yield loss to establish treatment and refusal thresholds. Progress with taxonomic studies of diseases would be greatly facilitated by the establishment of a European library of cereal diseases and by greater exchange of material between European scientists. More information is needed concerning the virulence and susceptibility to fungicides of existing strains.

(iii) Disease forecasting

More information is required concerning the epidemiology of glume blotch (Septoria nodorum) to enable the development of a combined forecasting scheme for Septoria spp. which is also integrated with control options for other diseases.

(iv) Techniques of disease limitation

More studies are needed on the integration of cultural and chemical methods for controlling foot and root diseases of wheat; the screening of varieties for resistance to take-all is also an important requirement in this context. More research is needed to establish whether integrated disease control (by variety and fungicide diversification) does reduce selection for fungicide insensitivity by foliar diseases. There is an opportunity for cooperation in this venture at the European level; it is suggested that variety mixture and fungicide combinations be tested, in a range of EC-Member States, in relation to European survey data (to be published by CEC) for virulence and insensitivity in barley and wheat mildew.

(v) Weeds

It is important to actively search for and document naturally-occurring pathogens, e.g. fungal, bacterial and viral antagonists.

"Farming systems"

a) Conclusions

(i) In England

Long-term declines in the density of polyphagous predators have been recorded in areas receiving high inputs of agrochemicals as compared with areas receiving low inputs. Reductions in the degree of natural control of aphids have recently been recorded in the high-input areas. It has been shown that some fungicide active ingredients drastically reduce the numbers of natural enemies leading to pest increase; other fungicide active ingredients are relatively benign. Increases in the acreage of cereals subject to straw incorporation rather than strawburning has led to an increased incidence of slugs and greater use of methiocarb. An increased incidence of aphids and virus damage (BYDV) have been recorded in direct drilled wheat after the use of methiocarb; this has been tentatively attributed to a reduction of natural enemies caused by methiocarb. Monitoring of the invertebrate fauna in an extensive area of farmland in southern England over a period eighteen years has demonstrated that long-term declines in the numbers of some natural enemies have begun to reverse in recent years; the reasons for this are unknown.

(ii) In Germany

The incidence of eyespot was less in areas of low-input "integrated" farming than in conventionally-farmed areas. In the first four years after the establishment of the "integrated" areas, mildew proliferated more in integrated than in the conventional areas; thereafter this was reserved. In northern Germany, intensive systems were found to have reduced numbers of natural enemies and more pests than systems making use of economic thresholds. Although the intensive systems had better yields, they were not as profitable as systems using economic thresholds (because of greater variable costs).

(iii) In the Netherlands

Conventional wheat systems gave better yields and gross margins than integrated systems. However, when the full rotation was considered integrated systems were found to be economically superior to conventional systems.

b) Recommendations

(i) Specific

There is a need for studies of autumn control of slugs and aphids, especially in the context of straw incorporation, to find combinations of chemical and cultural methods with minimum adverse impact on natural control. Studies of improved timing of nitrogen fertilization are needed to improve yields in integrated farming systems.

(ii) General

Because of the economic viability and ecological importance of the integrated approach to farming, it is desirable to establish model systems in a greater range of Member States. Studies in all EC-Member States should be expanded to consider the long-term effects of the whole farming system rather than just single farming techniques. The development of computer-based systems for speeding the flow of information from researcher to farmer should be encouraged.

The rate of implementation of the above recommendations would be increased by the following aspects of collaboration between EC-Member States:

- exchange of scientists;

- organisation of meetings in specialist subject areas as round-table discussions and workshops (e.g. epidemiology and forecasting systems of Septoria diseases, epidemiology of rust diseases, soil-borne diseases, weeds and integrated control, role of generalist arthropod predators in pest control);

- coordination of field experiments.

List of participants

Belgium:

DE CLERCQ R.
State Nematology and Entomology Research Station
Burg Van Gansberghelaan, 96
9220 Merelbeke

MARAITRE H.
Université Catholique de Louvain
Laboratoire de Phytopathologie
Place Croix du Sud, 3
1348 Louvain-La-Neuve

Bundesrepublik Deutschland:

BASEDOW T.
Justus-Liebig-Universitaet
Institut fuer Phytopathologie und Angewandte Zoologie
Ludwigstrasse 23
6300 Giessen

EL TITI A.
Landesanstalt fuer Pflanzenschutz
Reinsburgstrasse 107
7000 Stuttgart

LIMPERT E.
Technische Universitaet Muenchen
Lehrstuhl fuer Pflanzenbau und Pflanzenzuechtung
8050 Freising-Weihenstephan

POEHLING H.M.
Universitaet Hannover
Institut fuer Pflanzenkrankheiten
Herrenhaeuserstrasse 2
3000 Hannover

Denmark:

SAMSOE-PETERSEN L.
Research Centre for Plant Protection
Zoological Department
Lottenborgvej, 2
2900 Lyngby

France:

DEDRYVER C.A.
Institut National de la Recherche Agronomique
Laboratoire de Zoologie
Domaine de la Motte-au-Vicomte
35650 Le Rheu

FOUGEROUX A.
Association de Coordination Technique Agricole
149, rue de Bercy
75595 Paris

LUCAS P.
Institut National de la Recherche Agronomique
Laboratoire de Pathologie Végétale
Domaine de la Motte-au-Vicomte
35650 Le Rheu

MOREAU J.P.
Institut National de la Recherche Agronomique
Station de Zoologie
Route de Saint-Cyr
78000 Versailles

VALLAVIEILLE-POPE C.
Institut National de la Recherche Agronomique
Laboratoire de Pathologie Végétale
Centre de Grignon
78850 Thiverval

Great Britain:

BURN A.
University of Cambridge
Department of Applied Biology
Pembroke Street
Cambridge CB7 SSJ

CARTER N.
Rothamsted Experimental Station
Harpenden, AL5 2JQ

CHAMBERS R.J.
Glasshouse Crops Research Institute
Worthing Road
Littlehampton BN17 6LP

CHERRY A.
University of Southampton
Department of Biology
Southampton SO9 5NH

DENNIS P.
University of Southampton
Department of Biology
Highfield
Southampton SO9 5NH

DIXON T.
University of East Anglia
School of Biological Sciences
Norwich

FRAMPTON G.K.
University of Southampton
Department of Biology
Southampton SO9 5NH

HALSALL N.
University of Southampton
Highfield
Southampton SO9 5NH

JEPSON P.
University of Southampton
Department of Biology and Chemistry
Southampton SO9 5NH

KENNEDY P.
University of Bristol
Department of Agricultural Sciences
Long Ashton Research Station
Bristol BS18 9AF

KING, N.
University of Southampton
Highfield
Southampton SO9 5NH

LINTON W.
University of Southampton
Department of Biology
Southampton S09 5NH

MANN B.P.
University of Southampton
Bitterne Manor
Southampton S09 5NH

MAUREMOOTOO J.
University of Southampton
Department of Biology
Southampton S09 5NH

POTTS G.R.
The Game Conservancy
Cereals and Gamebirds Research Project
Burgate Manor
Fordingbridge SP6 1EF

POWELL W.
Rothamsted Experimental Station
Department of Entomology
Harpenden AL5 2JQ

ROYLE D.J.
University of Bristol
Department of Agricultural Sciences
Long Ashton Research Station
Bristol BS18 9AF

SMITH B.
University of Bristol
Department of Agricultural Sciences
Long Ashton Research Station
Bristol BS18 9AF

SOPP P.I.
Institute of Horticultural Research
Worthing Road
Littlehampton BN17 6LP

SOTHERTON N.W.
The Game Conservancy
Cereals and Gamebirds Research Project
Burgate Manor
Fordingbridge SP6 1EF

SUNDERLAND K.D.
Glasshouse Crops Research Institute
Department of Entomology & Insect Pathology
Worthing Road
Littlehampton BN17 6LP

THACKRAY D.J.
University of Southampton
Highfield
Southampton S09 5NH

VICKERMAN P.
University of Southampton
Department of Biology
Highfield
Southampton S09 5NH

WILDING N.
Rothamsted Experimental Station
Harpenden AL5 2JQ

WOLFE M.S.
Plant Breeding Institute
Maris Lane, Trumpington
Cambridge CB2 2LQ

WRATTEN S.
University of Southampton
Department of Biology
Southampton S09 5NH

Greece:
MAZOMENOS B.
National Research Centre "Democritos"
Biology Department
15310 Aghia Paraskevi

TSITSIPIS J.A.
National Research Centre "Democritos"
Biology Department
15310 Aghia Paraskevi

Ireland:
FEENEY A.
An Foras Taluntais
Oak Park Research Station
Plant Pathology & Entomology Department
Carlow

Italy:

BARBAGALLO S.
Università degli Studi
Istituto di Entomologia Agraria
Via Valdisavoia, 5
95123 Catania

DEL SERRONE P.
Istituto Sperimentale per la Patologia Vegetale
Via C.G. Bertero, 22
00156 Roma

PORTA-PUGLIA A.
Istituto Sperimentale per la Patologia Vegetale
Via C.G. Bertero, 22
00156 Roma

QUACQUARELLI A.
Istituto Sperimentale per la Patologia Vegetale
Via C.G. Bertero, 22
00156 Roma

Netherlands:

WIJNANDS F.G.
Research Station for Arable Farming and
Field Production of Vegetables
Edelhertweg, 1
8200 Lelystad

Spain:

CASTANERA DOMINGUEZ P.
Instituto Nacional de Investigaciones Agrarias
Jose Abascal, 56
28003 Madrid

C.E.C.:

CAVALLORO R.
Commission of the European Communities
"Integrated Plant Protection" Programme
Joint Research Centre
21020 Ispra (Italy)

Index of authors